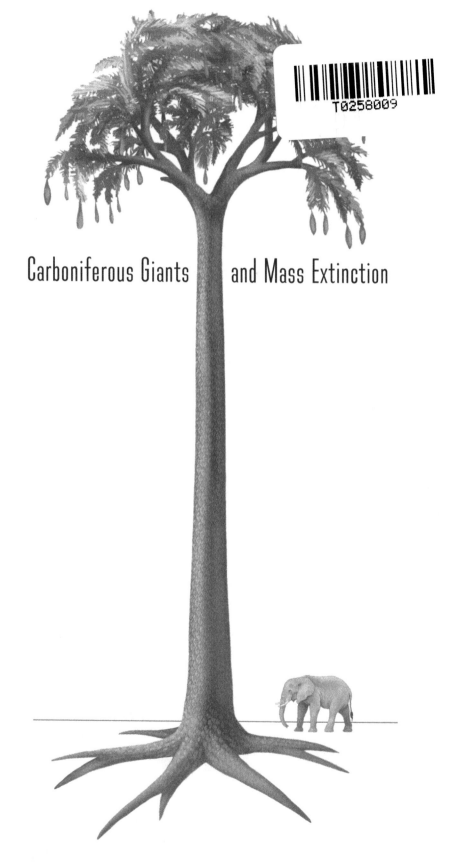

Carboniferous Giants   and Mass Extinction

Carboniferous Giants

George R. McGhee Jr.

Columbia University Press   New York

and Mass Extinction

THE LATE PALEOZOIC ICE AGE WORLD

Columbia University Press
Publishers Since 1893
New York   Chichester, West Sussex
cup.columbia.edu

Library of Congress Cataloging-in-Publication Data
Names: McGhee, George R., author.
Title: Carboniferous giants and mass extinction : the late Paleozoic Ice Age world /
    George R. McGhee Jr.
Description: New York : Columbia University Press, [2018] | Includes bibliographical
    references and index.
Identifiers: LCCN 2017056021 (print) | LCCN 2017060808 (e-book) | ISBN 9780231543385 |
            ISBN 9780231180962 (cloth) | ISBN 9780231180979 (pbk.) |
    ISBN 9780231543385 (e-book)
Subjects: LCSH: Mass extinctions. | Extinction (Biology) | Climatic changes. |
    Glacial epoch. | Paleobotany—Carboniferous. | Paleontology—Carboniferous. |
    Paleoclimatology—Paleozoic.
Classification: LCC QE721.2.E97 (e-book) | LCC QE721.2.E97 M388 2018 (print) |
    DDC 560/.175—dc23
LC record available at https://lccn.loc.gov/2017056021

Cover design: Lisa Hamm
Cover image: Richard Bizley / Science Source
Carboniferous insects. Artwork of a millipede (Arthropleura) and a dragonfly (Meganeura)
in the forests of the Carboniferous Period (354–290 million years ago).
Illustrations: (pages i, ii, and xiii) Mary P. Williams

For Marae

Tha gràdh agam ortsa.

Contents

Preface

I magine living in a world in which a dragonfly-like insect with a wing-span similar to that of a seagull zooms past your head; multilegged millipedes as long as a small car and as wide as a truck tire are crawl-ing along on the ground; venomous scorpions, stingers held curved over their backs, are as long as mid-sized dogs; and salamander-like animals are neither cute nor tiny—they are as long as alligators and have mouths full of sharp-pointed, deadly teeth.

Imagine living in a world with huge expanses of land covered by tropical rainforests, a green band that stretches around the middle of the planet across all of the continents. Rainforests in which you dis-cover, when you go hiking through them, very strange-looking trees: from a distance they look as though they have no leaves, even though the upper part of the trunk and the drooping branches of the tree are green. On closer examination, through a pair of field glasses, you can see that the tops of these trees do have leaves, but they are peculiar small bladelike leaves that partially overlap each other and wrap spirally around the branches and upper trunks. The tops of the trees thus have a somewhat fuzzy look, as though they were wearing green sweatshirts of leaves with multiple sleeves-of-leaves extending out along all of their branches. These peculiar sleeves-of-leaves look somewhat like those we see in our club mosses, but our little club mosses stand only 12 to 13 centimeters (five inches) high and these strange trees are gigantic, towering 50 meters (164 feet) into the sky.

Imagine living in a world in which a giant continent sits over the South Pole, a continent over five times larger than Antarctica—over twice the size of Africa, our largest single free-standing landmass surrounded by water—that is covered with vast rivers of ice sheets stretching away from a huge, white, glacial polar cap.

That world is not imaginary; it is the Earth—not our modern Earth, obviously, but rather the Earth as it existed in the Carboniferous Period of the Paleozoic world, over 300 million years ago. All of the things described here actually existed on that Earth—the giant animals, the giant trees, the giant continent, the giant glacial ice cap. There were giants in the Earth in those days.

This book is about that ancient Earth: how it came to be, and how it ended. We will examine the various hypotheses as to how animals, insects in particular, could have become so gigantic. We will examine the evolution of the great tropical rainforests that at first glance appear similar to those of our modern Earth, but that turn out to have been very different. The legacy of those great rainforests can be found around the world today in the massive deposits of coal that are found in Carboniferous strata—the huge volume of organic carbon that give the geologic period *Carboniferous* its name. Those coal deposits have been one of our major sources of energy since the beginning of the industrial era, and the burning of those coal deposits is releasing 300-million-year-old carbon dioxide back into our atmosphere and heating up the planet. The ancient Carboniferous Earth is transforming the modern Earth.

The Paleozoic world was totally destroyed in the end-Permian mass extinction, the greatest biodiversity crisis to occur since the evolution of animal life on Earth. The great ice caps of the Late Paleozoic Ice Age melted, the Earth began to warm, and the most catastrophic volcanic eruptions in Earth history began to inject trillions of tons of greenhouse gases into the Earth's atmosphere. The resultant "Hot Earth" lasted for five million years, an apocalyptic world in which the tropics were lethally hot—a world in which vertebrate land animals could only survive in cooler latitudes higher than 30° in the Northern Hemisphere and higher than 40° in the Southern Hemisphere. In stark contrast to our modern world—where the equatorial tropics harbor the highest diversity of life on the planet—the equatorial region of that Hot Earth was almost totally barren.

Even in the oceans, the equatorial temperatures were so high that fish and other vertebrate animals could only exist in the cooler waters of the high latitudes of the Earth. How could this have happened to the Earth? In this book we will examine in detail the triggers of the end-Permian mass extinction, and the nightmarish world that resulted.

This book is a summary of over four decades of my research into the ecological consequences of the Late Paleozoic extinctions, and I thank Columbia University Press, publishers of my previous books *The Late Devonian Mass Extinction* (1996) and *When the Invasion of Land Failed: The Legacy of the Devonian Extinctions* (2013), for making this third volume possible. I thank three anonymous reviewers of the manuscript for comments and suggestions that have helped me to improve the final book. And last, I thank my wife, Marae, for her patient love.

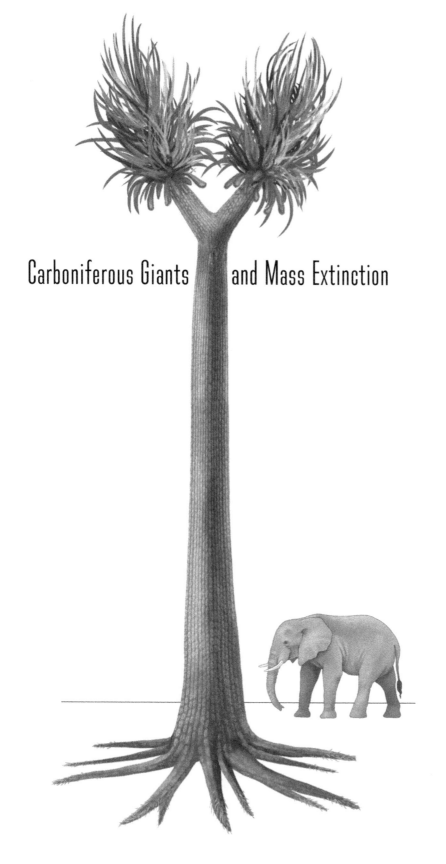

Carboniferous Giants and Mass Extinction

1 | Harbingers of the Late Paleozoic Ice Age

*The late Paleozoic ice age lasted for ~67 m.y. [million years] in eastern
Australia, and as such, it was the longest-lived icehouse interval in the
Phanerozoic.*

—Fielding, Frank, Birgenheier, et al. (2008, 55)

ICE AGES IN EARTH HISTORY

We know that at least seven ice ages have occurred in the past
4,560 million years, time periods during which the Earth partially—
or almost entirely—froze over. The history of the Earth is divided
into four eons: the Hadean, Archaean, Proterozoic, and Phanerozoic, from oldest to youngest (for the geologic timescale, see
table 1.1). These eons are themselves divided into smaller time units,
the eras, and they are divided into still smaller time units, the periods (table 1.1), much as our calendar years are divided into months,
weeks, and days. We know from fossil evidence that ancient bacterial life was present on the Earth 3,450 million years ago, during the
Paleoarchaean Era, and we have geochemical evidence that indicates that life was present even earlier during the Eoarchaean Era,
some 3,830 million years ago (McGhee 2013; Nutman et al. 2016).
The first four of the known ice ages occurred during the Proterozoic Eon, much later in time than the appearance of life on Earth,
and three more ice ages occurred even later during the Phanerozoic
Eon; thus life on Earth has successfully survived them all—but at

TABLE 1.1 The geologic timescale and ice ages.

Eon	Era	Period	Time of Onset (Ma)	Ice Ages (ICE)
Phanerozoic	Cenozoic	Quaternary	2.59	ICE
		Neogene	23.03	ICE
		Paleogene	66.0	ICE
	Mesozoic	Cretaceous	145.0	
		Jurassic	201.3	
		Triassic	252.2	
	Paleozoic	Permian	298.9	ICE
		Carboniferous	358.9	ICE
		Devonian	419.2	ICE
		Silurian	443.8	
		Ordovician	485.4	ICE
		Cambrian	541	
Proterozoic	Neoproterozoic	Ediacaran	635	ICE
		Cryogenian	850	ICE
		Tonian	1,000	
	Mesoproterozoic	Stenian	1,200	
		Ectasian	1,400	
		Calymmian	1,600	
	Paleoproterozoic	Statherian	1,800	
		Orosirian	2,050	
		Rhyacian	2,300	ICE
		Siderian	2,500	
Archaean	Neoarchaean		2,800	
	Mesoarchaean		3,200	
	Paleoarchaean		3,600	
	Eoarchaean		4,000	
Hadean			4,560	

Source: Timescale modified from Gradstein et al. (2012).

Note: Ma = millions of years before the present for the start of each time unit listed.

a cost. Many of the ice ages are associated with periods of extinction and large losses of biological diversity, a topic that will be explored in more detail in the next section of this chapter.

A major difference between the four Proterozoic and three Phanerozoic glacial episodes is reflected in their formal names, given in table 1.2. The four

TABLE 1.2 Ice ages in Earth history.

Ice Age	Position in Geologic Time (table 1.1)
A. Phanerozoic ice ages:	
1. Cenozoic Ice Age	Late Paleogene to Recent
2. Late Paleozoic Ice Age	Late Devonian to Late Permian
3. End-Ordovician Ice Age	Late Ordovician
B. Proterozoic ice ages:	
1. Gaskiers Snowball Earth	Ediacaran
2. Marinoan Snowball Earth	Cryogenian
3. Sturtian Snowball Earth	Cryogenian
4. Huronian Snowball Earth	Rhyacian

Proterozoic episodes are called snowball Earths, whereas the three Phanerozoic episodes simply are called ice ages. The snowball-Earth glaciations were much more geographically extensive than any seen in the Phanerozoic in that continental-covering, sea-level-reaching ice sheets extended from the poles of the planet all the way down to the equator. From space, the entire planet may have looked like one giant snowball, hence the name "snowball Earth." The first known snowball Earth, the Huronian, occurred some 2,300 million years ago during the Rhyacian Period (tables 1.1 and 1.2). Surprisingly, it is now thought that life may have triggered not only this massive freezing of the Earth during the Rhyacian but also the first mass extinction in Earth history. How could this be?

About 200 million years earlier than the Huronian freezing, at the beginning of the Siderian Period of the Paleoproterozoic Era (table 1.1), an event of major importance in the evolution of life on Earth occurred—the Great Oxygenation Event, or GOE for short. The very atmosphere of the Earth has been radically transformed by the presence of life. The original atmosphere of the Earth was probably very similar to that of its sister rocky volcanic planets Mars and Venus—that is, composed mostly of carbon dioxide. The atmosphere of Mars today is 95 percent carbon dioxide and the atmosphere of Venus is 97 percent, whereas the atmosphere of the pre-industrial-age Earth was only 0.03 percent carbon dioxide.[1] Both anaerobic and aerobic photosynthesizing bacteria[2] actively remove carbon dioxide from the atmosphere and use the carbon to form complex hydrocarbons for food. Thus, on Earth, life has been removing carbon dioxide from the atmosphere for the last 3,830 million years, contributing to the transformation of the Earth's atmosphere to its present carbon-dioxide-depleted state.

The aerobic photosynthesizing cyanobacteria not only remove carbon dioxide from the atmosphere but also add oxygen to the atmosphere.[3] About 2,500 million years ago, the oxygen-producing activity of the ancient cyanobacteria was finally to have its first major impact on the atmosphere of the Earth: it triggered the GOE. For the future evolution of complex—and large—life-forms with aerobic metabolism, the GOE was good news indeed as these organisms need free oxygen. For the ancient anaerobic life-forms—the original inhabitants of Earth—the GOE was a disaster because oxygen is a poison to them. The first mass extinction in the history of life on Earth probably occurred 2,500 million years ago, when vast unknown numbers of species of anaerobic bacteria and archaea perished by oxygen poisoning (McGhee 2013).

Diffusion of oxygen into the atmosphere was most probably the trigger of the Huronian Snowball Earth in the Rhyacian Period of the Paleoproterozoic Era (table 1.1), which occurred about 200 million years after the GOE (Lane 2002; Kopp et al. 2005). The steady drawdown of carbon dioxide, caused by the photosynthetic activity of life, was already reducing the greenhouse effect of this gas in the atmosphere. The presence of free oxygen in the atmosphere may then have begun to remove an even more powerful greenhouse gas—methane—via oxidation.[4] The resulting sharp drop in the greenhouse capacity of the atmosphere of the Earth may have triggered this first great snowball in Earth history.

Some 300 million years later, in the Orosirian Period of the Paleoprotoerozoic Era (table 1.1), oxygen concentrations in the atmosphere had reached high enough levels to begin oxidizing iron on land, and the first redbed strata appear in the terrestrial rock record. Redbeds are just that: layered beds of red sandstones and shales in which the iron has been oxidized, giving the strata their characteristic rusty-red color. By this time, the aerobic photosynthetic activity of cyanobacteria had raised the amount of oxygen in the atmosphere to something between 1 and 4 percent.[5]

Much later, the Earth froze in three snowball phases in quick succession—quick on a geologic timescale (tables 1.1 and 1.2). Two of these phases, the Sturtian and Marinoan Snowball Earths, occurred at 717 and 640 million years ago, respectively, during the Cryogenian Period of the Neoproterozoic—an aptly named period as "cryogen" comes from the Greek for "beginning of freezing cold."[6] The Sturtian Snowball Earth lasted some 57 million years; the Marinoan, about five million years; and the last snowball Earth, the Gaskiers, began 580 million years ago but lasted only 340,000 years during the Ediacaran Age (table 1.1).[7]

The earliest soft-tissued marine animals evolved about 780 million years ago (Erwin et al. 2011) during the Cryogenian, but it was only 541 million years ago that the first marine animals with mineralized tissues—skeletons—evolved. This evolutionary event marks the beginning of the Phanerozoic Eon (table 1.1), the geologic eon of the "visible animals."[8] The event is also often called the Cambrian Explosion, in recognition of the evolutionary pulse in the diversification of animal life that occurred at the beginning of the Cambrian Period of the Paleozoic Era (fig. 1.1). Large macroscopic animals with skeletons made of calcium, silica, and phosphorous compounds suddenly are present in rocks that previously contained only the impressions made by a few types of soft-bodied animals. All in all, almost 100 families of marine animals originated in the pulse of diversification that occurred at the beginning of the Paleozoic Era.

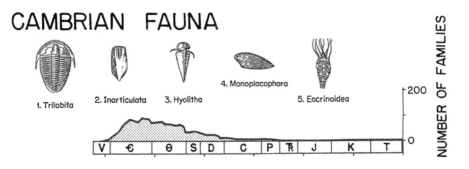

FIGURE 1.1 An evolutionary pulse in the diversification of large-bodied animals with skeletons occurred in the early Cambrian (graph); illustrated are some of the characteristic skeletonized marine animals of the Cambrian fauna (see text for discussion). Geologic timescale abbreviations: V, Ediacaran; barred-C, Cambrian; Θ, Ordovician; S, Silurian; D, Devonian; C, Carboniferous; P, Permian; T_R, Triassic; J, Jurassic; K, Cretaceous; T, Tertiary (Paleogene and Neogene).

Source: From *Paleobiology*, volume 10, pp. 246–267, by J. J. Sepkoski Jr., "A Kinematic Model of Phanerozoic Taxonomic Diversity: III. Post-Paleozoic Families and Mass Extinctions," copyright © 1984 The Paleontological Society. Reprinted with permission of Cambridge University Press.

Figure 1.1 illustrates some of the early skeletonized members of the major evolutionary clades of animals that were to become characteristic of the Paleozoic marine world: the Arthropoda (trilobites), Lophophorata (inarticulate brachiopods, hyoliths[9]), Mollusca (monoplacophorans), and Echinodermata (eocrinoids). We will examine the ecology and further evolution of all of these animal groups in greater detail in the chapters that follow. For example, the lophophorates are of particular interest in that they were later to evolve gigantic brachiopod shellfish in the Carboniferous.

Why did so many types of large animals with different skeletal chemistries appear so suddenly in the fossil record? A clue may be seen in some modern-day animals, such as bivalve molluscs that have the capacity to switch from aerobic metabolism to anaerobic metabolism when the oxygen content in water becomes depleted. While in the anaerobic-respiration mode, these bivalves also produce acid metabolites as a byproduct, and these acids actually begin to etch away and dissolve the calcium-carbonate shell of the animal (Lutz and Rhoads 1977; Babarro and De Zwaan 2008). Thus it has been proposed that the ability to grow and maintain mineralized skeletal tissues is a trait that is found in the aerobic-respiring organisms, meaning there has to be enough oxygen present in the environment to allow organisms to respire aerobically. Atmospheric modeling of the evolution of the Earth's atmosphere indicates that around 540 million years ago, just before the dawn of the Cambrian, oxygen levels in the atmosphere had finally risen to around 16 to 18 percent (Berner 2006), as compared to present-day levels of 21 percent. (We will consider these models in detail in chapters 3 and 4.) Atmospheric oxygen levels of 16 to 18 percent may have been the final trigger for animals belonging to numerous disparate phylogenetic lineages to simultaneously achieve sustained aerobic respiration, increase in size, and secrete mineralized skeletal tissues of several different chemical compositions in the different lineages. After the GOE and the Huronian Snowball Earth, the fact that the planet was hit by at least three more snowball-Earth glaciations in the latest Neoproterozoic (tables 1.1 and 1.2), immediately before the Cambrian animal diversification, is evidence for the continued effects of the drawdown of the greenhouse-gas carbon dioxide from the atmosphere and the injection of oxygen into the atmosphere by aerobic photosynthetic life.

In contrast to the snowball Earths, only the poles and higher-latitude regions of the Earth were covered by ice in the three Phanerozoic ice ages (tables 1.1 and 1.2). Our modern world is the product of the Cenozoic Ice Age, and polar ice and high-latitude continental glaciation still exist on the Earth—but perhaps not for long. The Cenozoic Ice Age began about 34 million years ago in the late Paleogene,[10] and its last glacial interval ended about 10,000 years ago (Lewis et al. 2008). It remains to be seen whether the Earth is still in the Cenozoic Ice Age or our warmer modern world is merely a interglacial interlude and whether the great ice sheets will return in the near future (on a geologic timescale). An alternative view is that the Earth is now in a major and unusual warming phase, triggered by the injection of carbon dioxide into the atmosphere by the human burning of fossil hydrocarbons for fuel. Humans began to mine and burn coal strata—the overwhelming majority of which are Carboniferous in age—in a big way during the Industrial Revolution and later added oil and natural-gas extraction and combustion through subsurface drilling. The subsequent increase in the carbon dioxide content of the atmosphere parallels an increase in the atmospheric temperature of the Earth, resulting in the accelerated melting of the residual ice that still exists from the last glacial phase of the Cenozoic Ice Age. Continued warming may result in an ice-free hot Earth in the future, and the Cenozoic Ice Age will be history.

The first Phanerozoic glacial phase, the end-Ordovician Ice Age, was peculiar in that it occurred during a period in geologic time when the Earth was in a long-term greenhouse phase and was quite warm, with an atmosphere still rich in carbon dioxide. The ice age was also quite short, consisting of an intense glaciation phase that lasted only 1.9 million years within a longer-term period of glacial advances and retreats in the latter part of the Ordovician and early part of the Silurian (McGhee et al. 2012, 2013). The evolution of the first land plants from water-dwelling plants had occurred by the middle of the Ordovician, but by the end of the Ordovician the continents of the Earth were still populated only by very small liverworts and related simple plants. The Ordovician is sometimes called the "Lilliputian plant world" because these tiny land plants—averaging under three centimeters (a little over an inch) in height—constituted the only plant cover of the planet Earth at that time. A few multilegged marine animals occasionally ventured out of the oceans onto the tidal flats and dry land near the end of the Ordovician—we have their fossil footprints preserved in stone—but they quickly returned to water and did not live on the land. (For an extensive discussion of the invasion of land by plants and animals, see McGhee 2013.) And last, in the south polar region of the planet, a vast ice sheet covered at least

30 million square kilometers (almost 12 million square miles) of land (Sheehan 2001); otherwise the land areas of the Earth were empty.

This book is about the second of the Phanerozoic glacial episodes (tables 1.1 and 1.2), the Late Paleozoic Ice Age—the longest ice age on Earth for the past 541 million years since the pulse of diversification of complex animal life in the Cambrian Explosion. If the world of the Ordovician Ice Age seems strange, the world of the Late Paleozoic Ice Age was far stranger, as we will see in this book.

ICE AGES AND EXTINCTIONS

All of the Phanerozoic ice ages (and probably all of the snowball Earths) are associated with the extinction of large numbers of species and major losses of biodiversity—but not all biodiversity crises in Earth history are associated with ice ages. Of the eight largest Phanerozoic biodiversity crises, five are argued to have been related to ice ages—one with the short end-Ordovician Ice Age and the other four with the longest glaciation in the Phanerozoic, the Late Paleozoic Ice Age (table 1.3). Curiously, these are also the first five biodiversity crises in Phanerozoic history (table 1.4) since the Great Ordovician Biodiversification

TABLE 1.3 Ecological-severity ranking from most severe (#1) to least severe (#7) of the eight largest Phanerozoic biodiversity crises since the beginning of the Ordovician.

| #1. End-Permian (Changhsingian Age) |
| #2. End-Cretaceous (Maastrichtian Age) |
| #3. End-Triassic (Rhaetian Age) |
| #4. **Late Devonian (Frasnian Age)** |
| #5. **End-Middle Permian (Capitanian Age)** |
| #6. **Early Carboniferous (Serpukhovian Age)** |
| #7. **End-Devonian (Famennian Age)**, End-Ordovician (Hirnantian Age) |

Source: From McGhee et al. (2004, 2013).

Note: The four biodiversity crises that are thought to be related to the Late Paleozoic Ice Age are listed in bold type; see text for discussion.

TABLE 1.4 Epoch and age divisions of the geologic timescale in the critical time interval leading up to and immediately following the Late Paleozoic Ice Age.

Period	Epoch	Age	Time of Onset (Ma)	Time of Crises
Jurassic (pars)		Hettangian	201.3	
Triassic	Late	Rhaetian	209.5	← Biodiversity Crisis
		Norian	228.4	
		Carnian	237	
	Middle	Ladinian	241.5	
		Anisian	247.1	
	Early	Olenekian	250.0	
		Induan	252.2	
Permian	Late (Lopingian)	Changhsingian	254.2	← Biodiversity Crisis
		Wuchiapingian	259.8	
	Middle (Guadalupian)	Capitanian	265.1	← Biodiversity Crisis
		Wordian	268.8	
		Roadian	272.3	
	Early (Cisuralian)	Kungarian	279.3	
		Artinskian	290.1	
		Sakmarian	295.5	
		Asselian	298.9	
Carboniferous	Late (Pennsylvanian)	Gzhelian	303.7	
		Kasimovian	307.0	
		Moscovian	315.2	
		Bashkirian	323.2	
	Early (Mississippian)	Serpukhovian	330.9	← Biodiversity Crisis
		Visean	346.7	
		Tournaisian	358.9	
Devonian	Late	Famennian	372.2	← Biodiversity Crisis
		Frasnian	382.7	← Biodiversity Crisis
	Middle	Givetian	387.7	
		Eifelian	393.3	
	Early	Emsian	407.6	
		Pragian	410.8	
		Lochkovian	419.2	
Silurian	Late	Pridolian	423.0	
		Ludfordian	425.6	
		Gorstian	427.4	
	Middle	Homerian	430.5	
		Sheinwoodian	433.4	

(continued)

Period	Epoch	Age	Time of Onset (Ma)	Time of Crises
	Early	Telychian	438.5	
		Aeronian	440.8	
		Rhuddanian	443.8	
Ordovician	Late	Hirnantian	445.2	← Biodiversity Crisis
		Katian	453.0	
		Sandbian	458.4	
	Middle	Darriwilian	467.3	
		Dapingian	470.0	
	Early	Floian	477.7	
		Tremadocian	485.4	

Source: Timescale modified from Gradstein et al. (2012).

Note: The temporal position of the seven major biodiversity crises (table 1.3) that occurred in this time interval are indicated with arrows. Ma = millions of years before the present for the start of each time unit listed.

Event (GOBE) (Webby et al. 2004; Servais et al. 2010), the second massive diversification of animal life in the oceans following the Cambrian Explosion (we will consider the evolutionary significance of the GOBE in detail in chapter 6). Just as curious is the fact that the end-Permian, end-Triassic (table 1.4), and end-Cretaceous biodiversity crises were clearly not associated with ice ages. We will consider the causes of the end-Permian mass extinction—the largest in Earth history—in detail in chapter 6, along with the end-Triassic extinction, as these two catastrophes appear to have had a similar causal mechanism. The end-Cretaceous mass extinction—the event that destroyed the dinosaur ecosystem—is now generally attributed to the effects of the impact of the massive asteroid that produced the 180-kilometer-diameter (112-mile-diameter) Chicxulub Crater in Mexico and thus had an extraterrestrial cause, not an Earthly one.

The short-lived end-Ordovician Ice Age triggered a mass extinction of marine species but had little effect on life on land simply because there was not much life on land to be affected (McGhee 2013). My colleagues Peter Sheehan, Dave Bottjer, Mary Droser, and Matthew Clapham and I have conducted comparative paleoecological analyses that have revealed that the environmental degradation produced by the end-Ordovician glaciation precipitated a major loss of marine biodiversity, yet the extinction failed to eliminate any key taxa or evolutionary traits and was of minimal ecological impact (Droser et al. 2000; McGhee et al. 2004, 2012, 2013).

In terms of ecological severity, the end-Ordovician extinction had an impact equivalent to that of the end-Devonian extinction (table 1.3), and both extinctions were triggered by an intense glacial phase that was of a geologically short duration—1.9 million years in the end-Ordovician Hirnantian Age and 1.4 million years in the end-Devonian Famennian Age (McGhee et al. 2013; McGhee 2013).

In contrast to the end-Ordovician and end-Devonian biodiversity crises, the glaciation in the Early Carboniferous Serpukhovian Age triggered a precipitous drop in the speciation rate of marine species but only moderate diversity losses. However, the ecological impact of those diversity losses and ecosystem restructuring was an ecological level of magnitude larger than that seen in the end-Ordovician or end-Devonian extinctions (table 1.3) (McGhee et al. 2012). We will examine the sequence of Early Carboniferous evolutionary events and glaciation phases in more detail in chapter 2.

In contrast to all of the biodiversity crises considered thus far, the extinctions that occurred in the Capitanian Age at the end of the Middle Permian Epoch (tables 1.3 and 1.4) were associated with the waning of the Late Paleozoic Ice Age, not with its onset or expansion. The end of the Late Paleozoic Ice Age was triggered by the start of a major phase of global warming, and at the end of the Paleozoic Era occurred the end-Permian mass extinction, the most ecologically severe biodiversity crisis in the entire Phanerozoic—a truly global "Category 1" ecological crisis in both the Earth's marine and terrestrial biota (McGhee et al. 2004; Cascales-Miñana et al. 2015). We will consider the dire consequences of these two Permian biodiversity crises in greater detail in chapters 5 and 6.

This leaves the enigmatic Late Devonian biodiversity crisis—so called because it occurred *within* the Late Devonian Epoch at the end of the Frasnian Age (tables 1.3 and 1.4). The Late Devonian crisis was a very severe event, in terms of both the magnitude of biodiversity loss and the ecological impact. It has the highest ecological severity of any of the biodiversity crises associated with a Phanerozoic ice age—if it was associated with a Phanerozoic ice age! We will consider the relationship of the Late Devonian biodiversity crisis to the Late Paleozoic Ice Age in great detail in the section "The Onset of the Late Paleozoic Ice Age?" later in this chapter.

Finally, the reader will have noticed that the Cenozoic Ice Age (table 1.2) is not listed in the ecological-severity ranking of biodiversity crises in table 1.3. Does this mean that no extinction or loss of biodiversity occurred during the

Cenozoic Ice Age? No, significant extinctions are associated with the Cenozoic Ice Age, and we will also consider them in more detail in the "The Onset of the Late Paleozoic Ice Age?" section. In general, however, the biodiversity losses that occurred during the Cenozoic Ice Age were of much less magnitude than those associated with the Paleozoic crises. Why? Steve Stanley, a University of Hawaii paleontologist, has proposed an answer to that question: the biota of the Paleozoic world was more prone to extinction than the biota of the Cenozoic, a phenomenon revealed in the classic analyses of the University of Chicago paleontologists Dave Raup and Jack Sepkoski that demonstrate that the mean extinction rate of marine animals has declined through time; that is, Paleozoic marine species had a much higher rate of extinction than Cenozoic marine species (Raup and Sepkoski 1982; Stanley 2007). This phenomenon is not unexpected: the theory of natural selection would predict the evolution of increasing extinction resistance with time. In the Paleozoic, the biota was dominated by ancient species with higher extinction rates than modern species. For example, Stanley has shown that in the Late Devonian extinction, the older species lineages suffered extinction at a 20 percent higher rate than the more recently evolved species lineages, and as the majority of the Late Devonian species belonged to these older lineages, the total extinction rate was predictably high (Stanley 2007). In contrast, Cenozoic marine species are much more resistant to extinction, hence the total extinction rates seen in the Cenozoic Ice Age are predictably lower than those seen in the Late Paleozoic Ice Age. We will examine further evolutionary and ecological differences between the ancient Paleozoic species and the more modern Cenozoic species in more detail in chapter 6.

THE MYSTERY OF THE ICE AGES

What triggered these ice ages and snowball Earths in geologic time? Why did huge areas of the Earth become frozen in the past? Obviously, if the entire planet gets colder, it must somehow either not be *receiving* enough heat from the sun to maintain its previous temperature or be *losing* much more of its heat to space than previously or both.

The Earth does generate some of its own heat by the decay of radioactive minerals in its rocks, but the planet really depends upon electromagnetic radiation—light—from the nearest star, our sun, for heat. The sun is actually becoming hotter with time as it slowly burns more and more of its original hydrogen gas in the fusion production of helium. A cooler sun during the Proterozoic Eon may have contributed in part to the intensity of the snowball-Earth glaciations. In the past 541 million years, however, our models of the sun's energy production predict no major fluctuations that could have triggered the Phanerozoic ice ages.

If the production of energy by the sun has been relatively constant for the past 541 million years, then one way to change the amount of heat that the Earth receives is to change the distance of the Earth from the sun: a planet closer to the sun receives more energy per unit area than a planet farther away. Although it is not apparent to us on human timescales, the Earth's distance from the sun actually does vary on geologic timescales. The shape of the Earth's orbit around the sun changes with time, from being almost a perfect circle to being stretched out into an ellipse—a shape variation that is known as orbital eccentricity (table 1.5). When the Earth's orbit is nearly circular, the planet absorbs about the same amount of heat from the sun throughout the year, as the distance of the Earth from the sun—the radius of the circular orbit—does not change during the year. At the other extreme the Earth's orbit is highly elliptical, with a long axis and a short axis. Twice a year the Earth is located at the short axis of the ellipse and warms up as it is closer to the sun, and twice a year the Earth is located at the long axis of the ellipse and cools down as it is farther from the sun.

TABLE 1.5 Periodic orbital phenomena that affect the Earth's climate.

Phenomenon	Periodicity of phenomenon
1. *Orbital Eccentricity:* variation in the shape of the Earth's orbit, from near-circular at one extreme to elliptical at the other.	100,000 years and 400,000 years
2. *Rotational Axis Obliquity:* variation in the tilt of the Earth's spin axis, from a tilt of 22.1° at one extreme to 24.5° at the other.	41,000 years
3. *Rotational Axis Precession:* circular variation in the orientation of the Earth's spin axis, from pointing to Polaris as the north pole star at one extreme to pointing in the opposite direction, along an arc of 180°, at the other.	19,000 years and 23,000 years

The stretching of the Earth's orbit from a near circle to an ellipse and its rounding back to a circle again is a function of the gravitational influence of the other planets in the solar system, and it occurs periodically in cycles of 100,000 and 400,000 years of geologic time.

The Earth also experiences variation in solar radiation in two ways that are caused by its orientation relative to the sun, not its distance. The first of these, rotational axis obliquity, refers to the fact that the degree of tilt in the Earth's rotational axis is not constant. The spin axis of the Earth is not vertical; that is, the planet tilts over on its side as it rotates. This tilt is responsible for the four seasons that we experience. For example, when the Northern Hemisphere of the Earth is tilted toward the sun, it receives more energy per unit area and is thus hotter—the summer season—than when it is tilted away from the sun and is thus colder—the winter season. The tilt in the rotational axis of the Earth varies between 22.1° and 24.5°. When the spin axis of the Earth is tilted over at 24.5°, the seasonal differences in temperature are much more pronounced—the winters are colder and the summers are hotter—than when the planet is tilted at only 22.1°. The degree of tilt of the Earth's rotational axis oscillates in a 41,000-year cycle in geologic time. Finally, the direction in which the rotational axis of the Earth points varies with time—a phenomenon known as rotational axis precession. Simply expressed, the Earth wobbles with time like a toy top winding down, with its tilted-over rotational axis spinning around and around in a circle. The tilt of the Earth relative to the sun changes on 23,000- and 19,000-year cycles; thus, the Northern Hemisphere is now tilted toward the sun (summer) in June, but 23,000 years from now it will be tilted away from the sun (winter) in June.

To make things even more interesting, these three cycles—orbital eccentricity, axial obliquity, and axial precession—interact to magnify or diminish one another's effect. For example, when the Earth is located farthest from the sun on an elliptical orbit (aphelion) and its axis is tilted over at 24.5° and is positioned at one of the solstices, the planet will become quite cold in one hemisphere and even the equator will cool. The opposite effect occurs when the Earth's orbit is more circular, its axis is tilted at only 22.1°, and the equinoxes occur at aphelion. These orbital oscillations are called Milankovitch cycles, after the Serbian geophysicist Milutin Milankovitć (usually anglicized to Milankovitch in the English literature) who first proposed them. It has been demonstrated that the Earth's climate did indeed vary in Milankovitch cycles of 400, 100, 41, 23, and 19 thousand years during the Cenozoic Ice Age (Zachos et al. 2001), and the 400- and 100-thousand-year eccentricity cycles have been detected in Late Carboniferous strata during the Late Paleozoic Ice Age (Heckel 2008; Horton et al. 2012).

However, the Milankovitch cycles in climate change became apparent only *after* the Cenozoic Ice Age had started and glaciers had formed on the Earth. Thus it appears that some stronger cooling mechanism is necessary to chill the planet down to form glaciers, which then wax and wane on weaker Milankovitch thermal frequencies. The next suspect in the ice ages mystery is changes in the Earth's atmosphere—specifically, changes in the amount of the greenhouse gases carbon dioxide (CO_2) and methane (CH_4) in the atmosphere. These gases act much like the glass roof of a greenhouse (hence the name greenhouse gas): the glass roof allows higher-frequency sunlight to penetrate down into the greenhouse, where the light is absorbed by the plants and then radiated back as lower-frequency infrared light, but then the glass roof blocks the escape of the infrared light and thus causes the temperature within the greenhouse to become hotter and hotter. Likewise, given the same amount of energy received from the sun, the Earth will retain more of that energy as heat when the atmosphere of the planet is enriched with carbon dioxide and methane and will lose more of that energy back to space when the atmosphere contains little carbon dioxide and methane.

Thus, to cool the planet and initiate an ice age, the amount of greenhouse gases in the Earth's atmosphere must be reduced (Montañez and Poulsen 2013). Carbon dioxide is removed from the atmosphere by two principal mechanisms: biological photosynthesis[11] and chemical weathering of silicate rocks.[12] Of these two processes, the chemical weathering of silicate rocks appears to be the more important: it is estimated that of the carbon dioxide that was removed from the Earth's atmosphere during the Phanerozoic, about 80 percent was removed by silicate weathering and only about 20 percent by biological photosynthesis (Raymo and Ruddiman 1992). Carbon-dioxide removal by weathering is enhanced by the uplift and exposure of large surface areas of silicate rock to the Earth's atmosphere during mountain-building events (Raymo and Ruddiman 1992) and by the exposure of silicate rocks in the equatorial humid zone of the Earth where the combination of high precipitation and high temperature intensifies the process of chemical weathering (Kent and Muttoni 2008;

Irving 2008). Methane is removed from the atmosphere chiefly by oxidation, but the oxidation of methane then produces carbon dioxide,[13] so we are back to the mechanisms of atmospheric carbon-dioxide removal in order to trigger global cooling.

The role of tectonic-forcing mechanisms of widespread mountain-building events (crustal buckling and uplift caused by collisions of continental plates, accretion of island-arc terranes onto continental plates, and subduction of oceanic plates under the margins of continental plates) and continental positionings (large number of continental plates located in the equatorial humid zone) in removing carbon dioxide from the Earth's atmosphere via chemical weathering has been implicated in all of the known glaciations, from the four snowball-Earth glaciations in the Proterozoic (Melezhik 2006) to the end-Ordovician (Lenton et al. 2012), Cenozoic (Raymo and Ruddiman 1992; Zachos et al. 2001; DeConto and Pollard 2003; Kent and Muttoni 2008), and Late Paleozoic ice ages in the Phanerozoic. However, it has been argued that atmospheric carbon-dioxide depletion by biological evolutionary processes was also a significant contributing factor in triggering the Huronian Snowball Earth (the evolution of aerobic photosynthesis) (Kopp et al. 2005), the end-Ordovician Ice Age (the evolution of land plants) (Lenton et al. 2012), and the Late Paleozoic Ice Age (the evolution of forests) (Algeo et al. 1995, 2001).

The topic of this book is the Late Paleozoic Ice Age, so let us take a closer look at the proposed triggers of this climatic event. The Late Devonian was clearly a time of major tectonic activity. In the short time span (on geologic timescales) of only four million years during the Frasnian Age, major mountain-belt deformation and uplift spread across the Appalachian-Caledonian mountain chain in North America and Europe (then joined together in a single continent called Laurussia; see fig. 1.2) and further to the south in the Variscide mountain chain extending into northern Africa (on the giant southern supercontinent Gondwana; see fig. 1.2). These crustal deformations and mountainous uplifts were driven by the incipient collision of the southeastern margin of Laurussia (southeast North America) and the northwestern margin of Gondwana (northwest Africa). On the northeastern margin of Laurussia, mountainous uplifts occurred in the Ural mountain belt (western Russia), driven by the collision of the Kazakhstan crustal block with Laurussia, and to the east of that, the Central Asian mountain belt buckled and uplifted in the collision of the Siberian

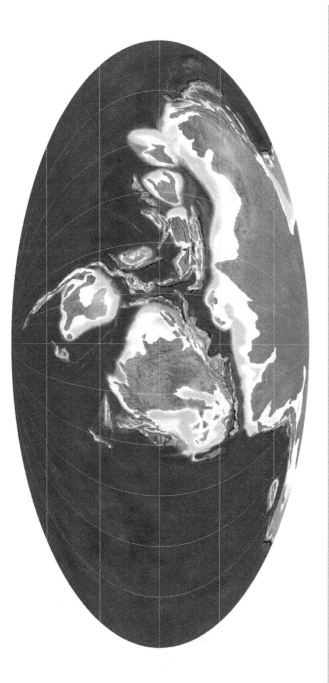

FIGURE 1.2 Paleogeography of the Earth in the late Frasnian Age. Lighter shaded areas surrounding the continents are shallow continental marine waters, and deep oceanic waters are black. The large continent straddling the equator in the left-center is Laurussia (modern-day North America and Europe), the smaller continent to the northwest of Laurussia is northern Asia, the islands east of Laurussia are pieces of eastern Asia, and the giant continent to the south of Laurussia is Gondwana (modern-day South America, Africa, India, Antarctica, and Australia all joined together).

Source: Global Paleogeography and Tectonics in Deep Time © 2016 Colorado Plateau Geosystems Inc. Reprinted with permission.

crustal block (eastern Russia) with the eastern margin of the Kazakhstan block. In terms of tectonic plate collisions, the Frasnian Age was a time of a multicar pileup on the geologic turnpike.

But that was not all. On the northwestern margin of Laurussia, crustal buckling and uplift occurred in the Antler mountain belt (western margin of the United States and Canada) eastward into the Ellesmerian-Svallbardian mountain belt (northern Canada). This deformation and uplift was triggered not by continental block collisions but rather by oceanic subduction and accretion of island-arc terranes along the northwestern margin of Laurussia. Nor was that all. To the south, on the giant continent Gondwana, similar oceanic-plate subduction and island-arc terrane accretion occurred in the Bolivianide mountain belt on the western margin of Gondwana (western South America) and in the Lachlan mountain belt on the eastern margin of Gondwana (eastern Australia and Antarctica). In summary, the Late Devonian Epoch was a period of intense tectonic activity on a global scale. (For a detailed discussion of the tectonic events that occurred during the Late Devonian, see McGhee 2013.)

Finally, there exists the possibility that yet another major tectonic mechanism might have been active in the Late Devonian—mantle-plume volcanism. In mantle-plume volcanism, a giant plume of magma rises from deep in the Earth's mantle and, when it intersects continental crust at the Earth's surface, erupts in huge volumes of basaltic lava—the weathering of which could extract huge volumes of carbon dioxide from the atmosphere. Possible evidence for mantle-plume volcanism comes from the Viluy igneous province in eastern Siberia (Courtillot and Renne 2003; Courtillot et al. 2010; Bond and Wignall 2014). The original size of the Viluy igneous province is unknown because it is highly eroded. Estimates suggest that the extent of these lava flows may have been as much as six million square kilometers (2.3 million square miles). Radiometric dating of the Viluy lavas is not precise, but it does place the eruptions in the interval of 377 to 350 million years ago—that is, sometime from the middle Frasnian Age in the Late Devonian to the late Tournaisian Age in the Early Carboniferous (table 1.4). Smaller igneous provinces, some with kimberlite[14] magmatism that demonstrates a mantle source for the volcanism, are also known in other areas of the ancient Laurussian continent from the same time interval; thus, there may have been several areas of mantle-plume volcanism active in the Late Devonian (Racki and Wignall 2005).

The uplift, deformation, and exposure to the atmosphere of huge surface areas of silicate rocks that occurred in the Late Devonian tectonic events should have triggered a massive drawdown of carbon dioxide from the atmosphere caused by the chemical weathering of these newly exposed rocks. In addition, the Laurussian continental block was positioned directly on the equator within the equatorial humid zone of the Earth in the Frasnian (fig. 1.2), thus enhancing the chemical weathering of silicate rocks located in this region. Thus, it is not unexpected that these potential carbon-dioxide-depleting tectonic mechanisms for triggering global cooling have been proposed as causes not only of the end-Famennian glaciation but also of a hypothesized glaciation that may have occurred at the end of the Frasnian as well (Averbuch et al. 2005). In addition, the University of Lille sedimentologist Olivier Averbuch and his colleagues have offered an independent geochemical line of evidence in support of the chemical-weathering hypothesis for global cooling in the Late Devonian: ratios of the isotopes strontium-87 and strontium-86 in seawater. Strontium-87 is radiogenic and is primarily found in continental rocks; strontium-86 is nonradiogenic and is primarily found in seafloor rocks. Thus higher ratios of strontium-87 to strontium-86 indicate higher rates of chemical weathering and erosion of crustal rocks on the continents, with the delivery of more strontium-87 to the oceans; lower ratios indicate the reverse (or increased seafloor hydrothermal activity). Strontium-87 to strontium-86 ratios increased during the Frasnian Age and spiked in the late Frasnian, indicating an intensification of continental weathering at this critical time. A second spike in strontium-87 to strontium-86 ratios occurred some five million years later in the early Famennian, indicating a second pulse of intense continental weathering at this time (Averbuch et al. 2005). Did these periods of intense weathering of continental silicate rocks deplete the Earth's atmosphere of enough carbon dioxide to trigger the formation of glaciers in the late Frasnian and early Famennian?

Another continental positioning (other than equatorial) also contributed to the formation of glaciers during the Late Devonian: the giant continent Gondwana was definitely positioned over the South Pole of the Earth in the late Famennian, 360 million years ago (fig. 1.3). Over the next 110 to 115 million years the geographic position of the South Pole on Gondwana shifted eastward across the landmass, from South America to Africa to Antarctica to Australia (fig. 1.3), as the giant tectonic plate holding Gondwana slowly moved westward

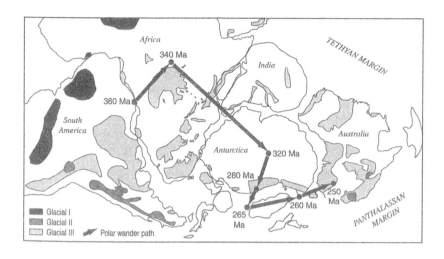

FIGURE 1.3 Polar wandering path across Gondwana from 360 million years ago to 250 million years ago. Also shown are the geologic outcrop areas of glacial strata, where Glacial I strata are Late Devonian to Early Carboniferous in age, Glacial II strata are late-Early to Late Carboniferous in age, and Glacial III strata are Early to Middle Permian in age.

Source: From *Geological Society of America Special Paper* 441, pp. 331–342, by T. D. Frank et al., "Paleozoic Climate Dynamics Revealed by Comparison of Ice-Proximal Stratigraphic and Ice-Distal Isotopic Records," copyright © 2008 Geological Society of America. Reprinted with permission.

over the South Pole. In the Cenozoic Ice Age the continent Antarctica was positioned over the South Pole (where it still is today); it has an ice-covered surface area of 13.72 million square kilometers (5.30 million square miles). In the Late Paleozoic, the giant continent Gondwana had a surface area of 74.23 million square kilometers (28.65 million square miles), almost five-and-one-half times bigger than Antarctica. Gondwana was so large that it was never entirely covered by ice during the Late Paleozoic Ice Age; rather, the centers of multiple glacial masses migrated across the giant continent in concert with its movement over the South Pole (fig. 1.3).

Independent biological weathering events also may have contributed to atmospheric carbon-dioxide depletion during the Devonian. A major event in the evolution of land plants occurred in the Givetian Age of the Middle Devonian—the first forests on Earth—and the evolution of forests certainly had to have added to the tectonic effects triggering global cooling. The Earth's oldest known forest is the famous Gilboa fossil forest of New York State; it consisted of trees that were large cladoxylopsids, extinct relatives of our modern-day ferns. (Tree ferns still exist today, such as the Australian tree fern *Cyathea cooperi.*) In the Givetian, the cladoxylopsid fernlike tree was the species *Wattieza (Eospermatopteris) erianus*, which stood more than eight meters (26 feet) tall and had trunks that were a half-meter to a meter (1.6 to 3.3 feet) in diameter. Yet these same large trees were anchored only by a broad, bulbous base that had numerous small, short roots.

The next step in the evolution of trees was the appearance of forests of giant *Archaeopteris* trees in the Frasnian Age of the Late Devonian. Unlike the fernlike cladoxylopsid trees, *Archaeopteris* trees were lignophytes—true woody trees with strong trunks; they towered 30 meters (100 feet) into the air and had deep, branched root systems that penetrated over a meter (3.3 feet) into the ground. These trees produced deep soil horizons by both chemical and mechanical weathering of the rocks into which they were rooted. However, the *Archaeopteris* trees still reproduced by spores, not seeds like our modern lignophyte trees, and they were generally restricted to the wetlands in the lowland areas of the Earth because they needed a reliable source of water for their reproduction by spore.

The final step in the evolution of trees was the appearance of the first seed-reproducing plants, the spermatophytes. The first spermatophyte lignophyte trees evolved in the Famennian Age and, not needing water in reproduction, colonized the dry highlands and mountains that were out of reach for non-spermatophyte *Archaeopteris* trees. Thus, huge areas of exposed silicate rock were now within reach of the seed plants for potential biological weathering.

The University of Cincinnati sedimentologist Thomas Algeo and his colleagues have argued that the evolution of the first woody forests on Earth was a major contributor to atmospheric carbon-dioxide depletion and a trigger for global cooling in the Late Devonian (Algeo et al. 1995). They proposed that the evolution of forests triggered three different mechanisms that acted in concert to remove significant amounts of carbon dioxide from the atmosphere. First, they argued that the photosynthetic productivity of the *Archaeopteris* trees that covered vast coastal and lowland areas of the Earth by the late Frasnian removed carbon from the atmosphere to form organic hydrocarbons. The increased burial rate of organic carbon on land began the process of depleting carbon dioxide in the atmosphere.

Second, they argued that the mechanical and chemical weathering of silicate rocks by *Archaeopteris* root systems produced deep soil horizons and removed carbon dioxide from the air by forming carbonates. Plant roots not only fracture rocks mechanically, exposing more rock surface area to contact with carbon dioxide and water and, thus, potential weathering, they also introduce organic acids that directly weather the rock chemically.

Third, they argued that enhanced soil formation by vascular plants "resulted in elevated fluxes of soil solutes (especially biolimiting nutrients) as a consequence of (1) enhanced mineral leaching, (2) fixation of nitrogen by symbiotic root microbes, and (3) shedding of plant-derived detrital carbon compounds . . . [and] elevated river-borne nutrient fluxes may have promoted eutrophication of semi-restricted epicontinental seas and stimulated algal blooms" (Algeo et al. 2001, 233). Eutrophic blooms of marine phytoplankton and algae in the seas would lead, in turn, to the depletion of oxygen in the shallow seas and the accumulation of unoxidized organic hydrocarbons in black shales, rather than returning the carbon to the atmosphere as carbon dioxide in the normal process of aerobic respiration by bacteria (Carmichael et al. 2016). The resultant carbon-dioxide drawdown from the atmosphere is proposed to have led to global cooling in a reverse greenhouse effect.

Algeo and colleagues thus proposed that major global cooling took place in the late Frasnian, that that cooling was driven by the extraction of large volumes of carbon dioxide from the atmosphere by the vast forests of the newly evolved *Archaeopteris* trees, and that that cooling was a trigger for the Late Devonian biodiversity crisis (table 1.3). This proposal received widespread attention in the popular press, with one science newsmagazine referring to the Earth's first widespread forests as "mass murderers of the Devonian" (Flangan 1995). However, the Algeo and colleagues model attributing the end-Frasnian extinctions to the spread of *Archaeopteris* forests on land drew immediate criticism from other Devonian workers. Tony Hallam, a University of Birmingham paleontologist, and his colleague, Paul Wignall, a University of Leeds paleontologist, commented: "Perhaps the only question arising from the Algeo model lies in the degree to which chemical weathering [by plants] increased in the Late Devonian. The *Archaeopteris* forests were restricted to floodplain environments, whereas more upland areas may not have been colonized until later in the Famennian, with the appearance of seed plants. Increased chemical weathering [by plants] may not therefore have become significant until the very end of the Devonian" (Hallam and Wignall 1997, 91).

Thus a much stronger case may be made for global cooling triggered by the biological weathering model of Algeo and colleagues for the late Famennian extinctions at the end of the Late Devonian than the late Frasnian ones within the Late Devonian (McGhee 2013). Evidence for the existence of glaciers on the Earth in the late Famennian Age is unequivocal (and will be discussed in detail in the next section of this chapter), but did glaciers form in the late Frasnian Age as well? If so, were the late Frasnian glaciers the result of carbon dioxide drawdown from the atmosphere by the chemical weathering of vast areas of silicate rocks exposed by numerous tectonic events in the late Frasnian? And were the late Famennian glaciers the result of further carbon dioxide depletion in the atmosphere exacerbated not only by the evolution of coastal and lowland spore-reproducing *Archaeopteris* forests but also by the spread of newly evolved highland and arid region seed-reproducing plants?

THE LATE PALEOZOIC ICE AGE

The University of Nebraska geologist Christopher Fielding and his colleagues have demonstrated that the main pulses of the Late Paleozoic Ice Age lasted for over 67 million years in eastern Australia (Fielding, Frank, Birgenheier, et al. 2008). Sixty-seven million years is a long time. The Phanerozoic Eon of geologic time is divided into three eras: the Paleozoic, Mesozoic, and Cenozoic, from oldest to youngest (table 1.1), and the main pulses of the Late Paleozoic Ice Age lasted longer than the entire Cenozoic Era! That is, the entire Age of Mammals—the period of time in which we, the mammals, have dominated the terrestrial ecosystems of the Earth following the termination of the Age of Dinosaurs at the end of the Cretaceous—has existed only for 66 million years, not quite as long as the Late Paleozoic Ice Age. If we include the Late Devonian glacial episodes that were the harbingers of the main glaciation interval (as will be argued in the next section of the chapter), then the entire duration of the Late Paleozoic Ice Age is even longer—some 115 million years in total.[15] This 115-million-year duration estimate for the total span of the Late Paleozoic

Ice Age is more in accord with the fact that the giant continent Gondwana was situated over the South Pole for more than 110 million years (fig. 1.3) (Frank et al. 2008), just as the glaciated continent Antarctica has been situated over the South Pole during the Cenozoic Ice Age. With a duration of 115 million years, the Late Paleozoic Ice Age persisted for more than 20 percent of the entire Phanerozoic Eon.

Not only was the ice age of enormously long duration, but many strange things happened during the Late Paleozoic Ice Age as well. About 90 percent of all of the coal-bearing strata on Earth were deposited during the Late Paleozoic Ice Age; that is, almost all of the fossil-fuel coal reserves of the entire world were deposited in an interval of time that constitutes less than 2 percent of the total age of the planet (Lane 2002).[16] Giant plants grew in and around the coal swamps of the Late Paleozoic Ice Age. Our modern little lycophytes (club mosses) and equisetophytes (horsetails) grow to be only about 12 to 25 centimeters (five to ten inches) tall, respectively; ancient relatives of the lycophytes towered 50 meters (164 feet) into the sky in the Late Paleozoic Ice Age, and the equisetophytes produced trees 20 meters (65 feet) high.

In the Late Paleozoic Ice Age, giant animals roamed the grounds beneath the unusual giant plants and flew in the skies above the forests—dragonfly-like insects with wingspans like a seagull's, salamander-like four-legged vertebrates and multilegged millipedes as long as alligators, slithering silverfish as big as grasshoppers, spiders as large as lobsters. In general, many of the ancient animals that lived during the Late Paleozoic Ice Age were five to seven times larger than their living relatives—and sometimes 12 times larger!

How did such a world, such an Earth so unlike our own, come to be? The initial environmental signals—harbingers of what was to come—first began to show up in the Late Devonian world.

THE ONSET OF THE LATE PALEOZOIC ICE AGE?

When did the Late Paleozoic Ice Age begin? Curiously, that is not an easy question to answer. Let us begin by considering the events that occurred at the onset of an ice age much closer in time to us—the Cenozoic. Three million years ago, in the late Pliocene (table 1.6), glaciation of the Earth had become bipolar: in the

TABLE 1.6 The timing of geologic and biotic events at the onset of the Cenozoic Ice Age.

Geologic Time Period	Epoch	Ice Sheets			Biotic Events
		Ma	SH	NH	
Quaternary	Holocene	00	ICE	ICE	
	Pleistocene	01	ICE	ICE	← Land extinctions
		02	ICE	ICE	← Marine extinctions
Neogene	Pliocene	03	ICE	ICE	← Marine extinctions
		04	ICE		
		05	ICE		
	Late	06	ICE		
	Miocene	07	ICE		
		08	ICE		
		09	ICE		
		10	ICE		
		11	ICE		
		12	ICE		
	Middle	13	Cold		
	Miocene	14	Cold		← Marine + land extinctions
		15			
		16			
	Early	17			
	Miocene	18			
		19			
		20			
		21			
		22			
		23			← Mi-1 Glacial Pulse
Paleogene	Late	24			
	Oligocene	25			
		26			
		27	ICE		
		28	ICE		
	Early	29	ICE		
	Oligocene	30	ICE		
		31	ICE		
		32	ICE		← Land extinctions
		33	ICE		← Marine extinctions
		34	ICE		← Oi-1 Glacial Pulse
	Late	35			
	Eocene (pars)	36			

Source: Cenozoic timescale modified from Walker and Geissman (2009); geologic and biotic events modified from McGhee (2013).

Note: Ma = millions of years before the present; SH = Southern Hemisphere; NH = Northern Hemisphere.

Southern Hemisphere, Antarctica was totally covered by the Antarctic ice sheet; in the Northern Hemisphere, northern North America was under the Laurentide ice sheet and northern Europe was under the Fennoscandian ice sheet. The onset of glaciation in the Northern Hemisphere coincided with extinction pulses first in the marine realm in the Pliocene (table 1.6) and then later on land in the Pleistocene (Hayward 2002). While significant, the Plio-Pleistocene extinctions were smaller in magnitude than the Paleozoic extinctions, in terms of the amount of biodiversity loss and the ecological impact, and are not in the list of the eight most severe biodiversity crises in the Phanerozoic (table 1.3). (For further discussion, see McGhee 2013, 203–212.)

So the Cenozoic Ice Age began in the late Pliocene when glaciation became bipolar, right? The answer is no, as the giant Antarctic ice cap is much older than the late Pliocene. The question of the timing of the onset of the Cenozoic Ice Age thus becomes: when did the Antarctic ice cap form? Twelve million years ago, in the late Miocene (table 1.6), glaciation of the Earth was unipolar; massive ice sheets had formed in Antarctica that persist to the present day. So the Cenozoic Ice Age began in the late Miocene when the Southern Hemisphere ice cap formed, right? The answer is no, as the onset of continental glaciation in Antarctica is much older than the late Miocene. The Earth began to cool rapidly 14 million years ago, in the middle Miocene, and a pulse of extinction occurred in both the marine and terrestrial realms at this same time (Sepkoski 1996; Lewis et al. 2008). However, the onset of Miocene glaciation in Antarctica was even older than that; it is dated from the Mi-1 glacial pulse in the earliest Miocene, some 23 million years ago (table 1.6), which lasted for about 200,000 years (Zachos et al. 2001).

So the Cenozoic Ice Age began in the earliest Miocene, right? The answer is no, as the onset of glaciation in Antarctica is even older than that. The first pulse of continental glacier formation in Antarctica, and the onset of the Cenozoic Ice Age, is dated to the Oi-1 glacial pulse in the earliest Oligocene, some 34 million years ago (table 1.6). The Oi-1 glacial pulse initiated the formation of ice sheets that lasted some eight million years in eastern Antarctica, and it is estimated that the ice-sheet coverage of the Oligocene glaciers was about 7.0 to 11.9 million square kilometers (2.7 to 4.6 million square miles).[17] The onset of the early Oligocene glaciations coincided with two separate pulses of extinctions that occurred in the oceans about 33 million years ago and with a third pulse of extinction that occurred on land about 32 million years ago (table 1.6) (McGhee 2001).

Now the story becomes even more interesting: where is the sedimentary proof that Antarctica was glaciated in the Oligocene? The answer is that there is precious little sedimentary evidence of Oligocene glaciation on the continent of Antarctica itself; possibly only two sites preserve evidence of Oligocene glacial sediments (Strand et al. 2003; Ivany et al. 2006). The problem is that the sedimentary evidence of the smaller Oligocene glaciers was removed by the erosive action of the much larger glaciers that formed in the Miocene. The Oligocene glaciers are estimated to have covered between 7.0 and 11.9 million square kilometers (2.7 to 4.6 million square miles) of Antarctica, whereas it is estimated that the Miocene glaciers covered 14 to 16.8 million square kilometers (5.4 to 6.5 million square miles).[18] The much larger Miocene glaciers not only totally covered the area once occupied by the Oligocene glaciers but also extended much farther across the surface of Antarctica, eroding and stripping away the glacial sedimentary deposits of the older Oligocene glaciers in the process.

For this reason, sedimentary evidence for the existence of the Oi-1 glaciers is found offshore from Antarctica rather than on the mainland itself. This evidence comes in the form of glacially derived, ice-rafted debris found in offshore marine sediments. University of Michigan geologist James Zachos and colleagues have documented the presence of layers of angular quartz sands and heavy minerals at the Oi-1 stratigraphic level on the Kerguelen Plateau in the southern Indian Ocean. These layers contain medium-size[19] and larger sand debris, debris that was too large to have been transported offshore from Antarctica by wind and thus must have been transported by ice (Zachos et al. 1992). Werner Ehrmann and Andreas Mackensen, geologists at the Alfred Wegener Institute, have reported the presence of gravel with pebbles at the same stratigraphic horizon containing the ice-rafted sand deposits on the Kerguelen Plateau. The presence of gravel in offshore marine deposits is unequivocal evidence of ice rafting, and the presence of ice-rafted debris as far north as 61°S on the Kerguelen Plateau is evidence of either a high frequency of icebergs in the area or a few large debris-containing icebergs, both of which evidence large-scale continental Oi-1 glaciation rather than small-scale local glaciation in the Antarctic (Ehrmann and Mackensen 1992).

Other evidence for the Oi-1 glaciations on Antarctica comes from models of glacioeustatic sea-level changes and from geochemical analyses of Oligocene strata for evidence of changes in the temperature of sea-surface waters. Sequence stratigraphic reconstructed sea-level curves estimate a 67-meter (220-foot) drop in sea level at the Oi-1 stratigraphic level (Katz et al. 2008), and independent models of ice-volume changes with time give an average estimate of a 70-meter (230-foot) drop in sea level (Pusz et al. 2011). Clearly the drop in sea level in the early Oligocene was driven by the removal of water from the sea by the formation of frozen-water deposits—glaciers—on land in Antarctica. Proof of a drop in sea-surface temperature at the Oi-1 stratigraphic horizon is hampered by the fact that the critical strata are missing in the rock record, further evidence of a drop in sea level. Strata that have been preserved indicate that a drop of 3 to 4°C (5 to 7°F) in sea-surface temperatures occurred immediately before the Oi-1 glacial pulse, but the total decline in temperature in the glacial pulse itself remains as yet unknown (Wade et al. 2012).

In summary, the onset of the Cenozoic Ice Age is dated to the earliest Oligocene, the Oi-1 glacial pulse (table 1.6). Furthermore, the development of glaciers in the Cenozoic Ice Age was not a continuous global cooling process but rather took place in three glaciation steps in geologic time—the early Oligocene, the middle Miocene, and the late Pliocene—and each of these steps was associated with extinctions and biodiversity losses (table 1.6).

I have proposed that the onset of the Late Paleozoic Ice Age was similar in timing to that of the Cenozoic Ice Age, and that it also occurred in three glaciation steps—the late Frasnian, the late Famennian, and the early Serpukhovian (McGhee 2013, 203–212). It has been argued that glaciation in the Early Carboniferous had become bipolar by the early Serpukhovian Age: in the Southern Hemisphere, western Gondwana was covered in ice sheets that would become comparable in size to those present in the maximum phase of the Cenozoic Ice Age; in the Northern Hemisphere, glaciers had formed on the Siberian crustal block in northern Asia (González-Bonorino and Eyles 1995; Stanley and Powell 2003). Major extinctions in the marine realm were also associated with the onset of the Serpukhovian freezing; these glacial and biological events will be considered in detail in chapter 2.

However, just as in the onset of the Cenozoic Ice Age, the beginning of the Late Paleozoic Ice Age is not dated to the time when glaciation of the Earth became bipolar, as the massive glaciers in Gondwana are much older than the early Serpukhovian. Just as in the onset of the Cenozoic Ice Age, the question of the timing of the onset of the Late Paleozoic Ice Age becomes: when did the ice sheets form on Gondwana? In the late Famennian Age of the Late Devonian, 363 million years ago (table 1.7), massive ice sheets had already formed in Gondwana (Caputo et al. 2008). These ice sheets covered at least 16 million square kilometers (6.2 million square miles) of land in western Gondwana—present-day South America and Africa (fig. 1.4) (Isaacson et al. 2008). And, as discussed

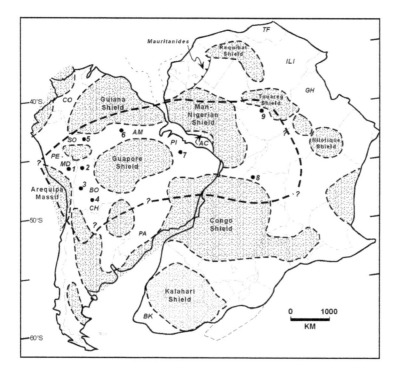

FIGURE 1.4 Paleogeographic extent of the Famennian ice sheet (heavy dashed line) on the continent of Gondwana, which covered parts of present-day South America and Africa. Lighter dashed lines indicate continental shield regions, and numbered points indicate sedimentary basins containing glacial strata.

Source: From *Palaeogeography, Palaeoclimatology, Palaeoecology*, volume 268, pp. 126–142, by P. E. Isaacson et al., "Late Devonian–Earliest Mississippian Glaciation in Gondwanaland and Its Biogeographic Consequences," copyright © 2008 Elsevier. Reprinted with permission.

TABLE 1.7 The timing of geologic and biotic events in the proposed onset of the Late Paleozoic Ice Age.

Age	Geologic Time (Ma)		Geologic and Biotic Events
TOURNAISIAN	347		
	348		
	349	Cold	Tournaisian Gap
	350	ICE	Tournaisian Gap
	351	ICE	Tournaisian Gap
	352	ICE	Tournaisian Gap
	353	ICE	Tournaisian Gap
	354	Cold	Tournaisian Gap
	355	Cold	Tournaisian Gap
	356	Cold	Tournaisian Gap
	357	Cold	Tournaisian Gap
	358	Cold	Tournaisian Gap
FAMENNIAN	359	ICE	← Land extinctions
	360	ICE	← Marine extinctions
	361	ICE	
	362	ICE	
	363	ICE	
	364		
	365		
	366		
	367	Cold	Famennian Gap
	368	Cold	Famennian Gap
	369	Cold	Famennian Gap
	370	Cold	Famennian Gap
	371	Cold	Famennian Gap
	372	ICE?	← Marine extinctions, Famennian Gap
FRASNIAN (pars)	373	ICE?	← Marine extinctions
	374	ICE?	← Marine + land extinctions
	375	ICE?	
	376		
	377		

Source: Timescale modified from Gradstein et al. (2012); Famennian and Tournaisian glacial data are from Caputo et al. (2008); biotic events modified from McGhee (2013).

Note: Ma = millions of years before the present.

previously, the Famennian glaciation coincided with the seventh most ecologically severe biodiversity crisis in the Phanerozoic (table 1.3).

At 16 million square kilometers, the late Famennian glaciers are comparable in size to those that formed in the Miocene phase of the onset of the Cenozoic Ice Age, which have been estimated to have been in the range of 14 to 16.8 million square kilometers (5.4 to 6.5 million square miles), as discussed earlier in this chapter.[20] The sedimentary evidence of glacial striated pavements and glacial tillites are still present in terrestrial strata in South America and South Africa, and decimeter- to meter-size (four-inch- to yard-size) ice-rafted glacial dropstones are found in offshore marine strata of late Famennian age (Streel, Vanguestaine, et al. 2000; Isaacson et al. 2008; Brezinski et al. 2010). The formation of frozen water masses on land the size of the Famennian glaciers should have produced a major drop in sea level. When sea level falls, the coastlines of the oceans retreat from the land. Land that was once under water is now exposed and, more important, the mouths of rivers that used to empty into the ocean are now far from the new coastline and at an altitude higher than the new sea level. As a result, the rivers cut downward in their river beds until they reach the new sea level, producing deeply incised valleys. In the maximum phase of the Cenozoic Ice Age, valley incision depths approached 100 meters (328 feet), indicating, of course, that the level of the sea fell by 100 meters during the period of maximum ice buildup on the land.

The late Famennian glaciation occurred over 359–363 million years ago (table 1.7), but geological evidence of valley incision still exists in isolated regions of the world. In both North America and Europe there are ancient incised valleys with incision depths ranging from 75 to 90 meters (246 to 295 feet), almost as deep as those seen in the Pleistocene Epoch of the Cenozoic Ice Age (Brezinski et al. 2010). Thus the severity of global cooling in the late Famennian approached that seen in the maximum phase of the Cenozoic Ice Age in terms of the amount of water frozen on the land.

So the Late Paleozoic Ice Age began in the late Famennian Age of the Late Devonian, right? Many paleoclimatologists today would answer that question with a "yes," but many also have their doubts and think that the onset of the Late Paleozoic Ice Age is even older than that. Sedimentary evidence for glaciation and glacially driven changes in sea level in the older Frasnian Age of the Late Devonian has increased steadily over time. Around the world, cyclic patterns

FIGURE 1.5 Frasnian sedimentary cycles at Devils Gate, Nevada, that may be the result of fluctuations in sea level triggered by the expansion and contraction of glacial ice sheets in Gondwana. The rock hammer spanning one such cycle in the center of the figure is 30 centimeters (12 inches) tall.

Source: Photograph courtesy of Dr. Peter Isaacson, Department of Geological Sciences, University of Idaho.

of sedimentation are known to exist in Frasnian strata (fig. 1.5) that appear to represent patterns of sea-level rise and fall that occur in cycles of around 10,000 to 100,000 years (Sandberg et al. 1988; Montañez and Isaacson 2013; Becker et al. 2016; see also Elrick and Witzke 2016).

More recently, Jonathan Filer, a geologist at Towson University, has demonstrated the existence of numerous sedimentary cycles in late Frasnian subsurface strata that are continuous over 700 kilometers (435 miles) in eastern North America, revealed in the analysis of data from over 600 hydrocarbon test wells in the Appalachian Basin, and has argued that these cycles are evidence of sea-level changes driven by the expansion and contraction of ice sheets on land in Gondwana (Filer 2002). Wilson McClung, a geologist at Chevron, and his colleagues have traced 12 of Filer's Frasnian sedimentary cycles from the subsurface to surface outcrops and have argued that the ability to correlate these cycles in both outcrop and subsurface, both parallel and perpendicular to the ancient shoreline in the Appalachian Basin, confirms that the sedimentary cycles are the product of global rises and falls in sea level produced by glacial cycles of melting and freezing on land (McClung et al. 2013). Moreover, McClung and colleagues have estimated that these 12 sedimentary cycles had a temporal periodicity of around 375,000 years, and they have further measured over 70 smaller-scale sedimentary cycles in outcrop with an estimated periodicity of around 65,000 years, cyclic frequencies that are similar in magnitude to the long-term orbital-eccentricity and rotational-axis-obliquity climatic cycles (table 1.5) seen in the Cenozoic glaciations (Zachos et al. 2001). Finally, McClung and colleagues have also demonstrated the existence of incised-valley fills some 35 to 45 meters (115 to 148 feet) thick in outcrop, evidence of a late Frasnian glacioeustatic sea-level fall of about half the magnitude of the sea-level fall that produced the incised valleys in the late Famennian glaciation. (For an extensive discussion of the evidence for a major drop in sea level during the late Frasnian—particularly the karstification of carbonate platforms around the world—see McGhee 2013, 132–135.)

As yet, evidence for late Frasnian glaciation is found only outside of Gondwana. On Gondwana itself, all efforts to find late-Frasnian glacial striated pavements and tillites, similar to those found in the late Famennian, have failed. The western edge of the landmass of Gondwana was positioned over the South Pole in the Frasnian as well as in the Famennian (figs. 1.2–1.4), so where were the Frasnian glaciers? I have argued that a closer examination of the pattern of glacial onset seen in the Cenozoic Ice Age may explain the enigma. In essence, I argue that the absence of late Frasnian glacial sediments on Gondwana is to be expected, and that Gondwana is the wrong place to search for such evidence (McGhee 2014a, 2014b).

I predict that glacial sediments of late Frasnian age will never be discovered on the landmass of Gondwana because they were removed by the erosive action of the much larger glacier that formed in the late Famennian, analogous to the removal of the Oi-1 glacial sediments on Antarctica by the much larger Miocene glacier. For the same reason that sedimentary evidence for the existence of the Oi-1 glaciers is found offshore from Antarctica rather than on the mainland, I predict that sedimentary evidence for the late Frasnian glaciers will be found offshore from Gondwana rather than on the mainland. That evidence will come in the form of glacially derived ice-rafted debris found in offshore marine sediments (McGhee 2014a, 2014b).

The erosive effects of the successive glaciations of the Late Paleozoic Ice Age can be clearly seen in figures 1.3 and 1.4. Note that the existing glacial strata of late Famennian age (Glacial I in fig. 1.3) are found only in northern Africa and northern South America, yet the position of the South Pole in the late Famennian (360 Ma in fig. 1.3) was to the *south* of those outcrop areas. Clearly, glacial strata of late Famennian age once existed all across middle and southern Africa and South America, a mirror image about the South Pole to the existing glacial strata in northern Africa and South America, yet those strata no longer exist—they were removed by successive later phases of glaciation (Glacial II and Glacial III in fig. 1.3). A more detailed map of the distribution of late Famennian glacial strata is shown in figure 1.4; note the question marks along the southern and eastern margins of the heavy dashed line indicating the geographic extent of the late Famennian ice sheet. The University of Idaho geologist Peter Isaacson and his colleagues are clearly indicating here that the Famennian ice sheet had to have extended further to the south and east than the current existing glacial outcrops, but that those southern and eastern Famennian glacial strata are no longer to be found.

As discussed above, McClung and his colleagues have demonstrated the existence of Frasnian incised-valley fills that suggest a late Frasnian glacioeustatic sea-level fall about 50 percent of the sea-level fall that produced the incised valleys in the late Famennian. The minimum size of the late Famennian glaciers has been measured to have been 16 million square kilometers (6.2 million square miles) in western Gondwana (Isaacson et al. 2008), and I have proposed that glaciers approximately 50 to 71 percent of the size of the Famennian ice sheet, or eight to 11 million square kilometers (3.1 to 4.2 million square miles), were present in western Gondwana in the late Frasnian (McGhee 2014a, 2014b). This estimate is based on the scaling of the size range of the first-step Oi-1 glaciers to the size range of the second-step Miocene glaciers at the onset of the Cenozoic Ice Age, and the assumption that that scaling was similar in the size ranges of the first-step late Frasnian glaciers to the second-step Famennian glaciers in the proposed onset of the Late Paleozoic Ice Age (McGhee 2014a, 2014b).

Subsequent fieldwork by McClung and his colleagues in the Appalachian Mountains in North America has yielded further empirical stratigraphic data that have allowed them to produce an independent calculation for the possible size of the ice sheet proposed to have existed in Gondwana during the late Frasnian.

Their calculation falls in the range of 5.6 to 7.4 million square kilometers (2.2 to 2.9 million square miles), values that they noted are "similar to McGhee's (2014) estimate for the area of late Frasnian glaciation" (McClung et al. 2016, 139).

Based on either my larger estimate or the somewhat smaller estimate of McClung and his colleagues, the late Frasnian ice sheet still would have been totally covered by the much larger late Famennian glaciers. As in the case of the Oi-1 glaciation in Antarctica, the late Famennian glaciers would have erased the trace of the initial late Frasnian glaciers on Gondwana—hence my prediction that glacial sediments produced by the late Frasnian glaciers will only be found in marine sediments offshore from the Gondwana landmass.

To test the hypothesis that glaciers formed in the late Frasnian, a worldwide search should be initiated for the presence of ice-rafted debris in late Frasnian marine strata deposited offshore from Gondwana. Rather than the decimeter- to meter-size (four-inch- to yard-size) Famennian ice-rafted debris, a search should be initiated for the presence of Frasnian ice-rafted smaller debris of medium to coarse sand grains and gravels with pebbles, similar to the ice-rafted debris found in the Oi-1 marine sediments (Zachos et al. 1992; Ehrmann and Mackensen 1992).

Aside from the sedimentary evidence for glaciation, biological evidence of global cooling during the late Frasnian biodiversity crisis has been amassed by paleontologists and paleobotanists for over 36 years (table 1.8). Biological evidence also exists for a six-million-year-long cold interval following the late Frasnian biodiversity crisis in the early Famennian, an interval of time I have called the Famennian Gap (table 1.7) (McGhee 2013, 104–107).[21] The late Famennian biodiversity crisis was followed by a similar ten-million-year protracted cold interval in the Early Carboniferous known as the Tournaisian Gap, which was punctuated with an additional period of glaciation 350 to 353 million years ago (table 1.7).[22] The names of both the Famennian and Tournaisian Gaps refer to gaps in the fossil record in which very low diversity existed in both land animal and plant species, and in marine species as well (McGhee 2013).

As discussed above, the onset of the Cenozoic Ice Age was stepwise, with the first step of glaciation taking place in the early Oligocene, followed by a warming period, and the second step of glaciation taking place in the middle Miocene (table 1.6). If the onset of the Late Paleozoic Ice Age also took place in a stepwise fashion, with the first step of glaciation taking place in the late Frasnian and the second step of glaciation taking place in the Famennian (table 1.7), then

TABLE 1.8 Empirical biological observations in both marine and terrestrial ecosystems that have been used to argue for global cooling in the late Frasnian.

A. MARINE ECOSYSTEMS

1. Differential survival of high-latitude marine faunas:
   - Brachiopods (Copper 1977, 1986, 1998)
   - Microbial reef biota (Copper 2002)

2. Differential survival of deepwater marine faunas:
   - Glass sponges (McGhee 1996)
   - Rugose corals (Oliver and Pedder 1994)
   - Tornoceratid ammonoids (House 1988)

3. Migration of deepwater marine faunas into shallow waters:
   - Glass sponges (McGhee 1996)
   - Tornoceratid ammonoids (House 1988)

4. Blooms in cold-water plankton:
   - Prasinophytes (Streel, Vanguestaine, et al. 2000)
   - Radiolarians (Racki 1998, 1999; Copper 2002)
   - Chitinozoans (Paris et al. 1996; Streel, Vanguestaine, et al. 2000; Grahn and Paris 2011)

5. Differential survival of freshwater versus marine species:
   - Acanthodian fishes (Dennison 1979)
   - Placoderm fishes (Dennison 1978; Long 1993)

6. Latitudinal contraction of geographic range in surviving equatorial marine faunas:
   - Foraminifera (Kalvoda 1990)
   - Stromatoporoid and coral reefs (Stearn 1987; Copper 2002)
   - Tentaculitoids (Wei et al. 2012)
   - Trilobites (Morzadec 1992)

B. TERRESTRIAL ECOSYSTEMS

1. Differential survival of high-latitude terrestrial biota:
   - Land plants (Streel, Caputo, et al. 2000)

2. Latitudinal contraction of geographic range in surviving equatorial terrestrial biota:
   - Land plants (Streel, Caputo, et al. 2000)
   - Tetrapod vertebrates (McGhee 2013)

Source: Modified from McGhee (2014a).

the timing of the onset of the glaciations and associated extinction pulses in the Paleozoic and Cenozoic Ice Ages is strikingly similar (table 1.9). In the onset of the Cenozoic Ice Age, the time interval between the first-step extinctions in the Oligocene and the second-step extinctions in the Miocene was about 19 million years; in the proposed onset of the Late Paleozoic Ice Age, the time interval between the onset of the Frasnian and Famennian extinctions was about 14 million years. The sequential severity of the extinctions was also the same: in the Cenozoic Ice Age, the first-step Oligocene extinctions were much more severe than the second-step Miocene (Prothero 1994; Prothero et al. 2003), and in the proposed onset of the Late Paleozoic Ice Age, the first-step Frasnian extinctions were much more severe than the second-step Famennian (table 1.3). Finally, Cenozoic glaciers persisted for five million years following the Oligocene extinctions, and the Famennian Gap cold interval persisted for six million years following the Frasnian extinctions. If the Earth cooled at a similar rate in the Cenozoic and Late Paleozoic, then the temporal spacing and magnitudes of extinction seen in these two time intervals may not be coincidental (table 1.9).[23]

TABLE 1.9 Biological similarities between the stepwise onset of the Cenozoic Ice Age and proposed stepwise onset of the Late Paleozoic Ice Age

Cenozoic Ice Age	Late Paleozoic Ice Age
A. Glacial Step 2	
Miocene extinction pulses 13–14 million years ago	Famennian extinction pulses 359–360 million years ago
B. Glacial Step 1	
Oligocene extinction pulses 32–33 million years ago	Frasnian extinction pulses 372–374 million years ago
C. Severity of extinction pulses	
Oligocene (Step 1) > Miocene (Step 2)	Frasnian (Step 1) > Famennian (Step 2)
D. Time interval between onset of extinction pulses 1 and 2	
19 million years	14 million years
E. Duration of Antarctic glaciers following Oligocene extinction pulses	**E. Duration of Famennian Gap cold interval following Frasnian extinction pulses**
five million years	six million years

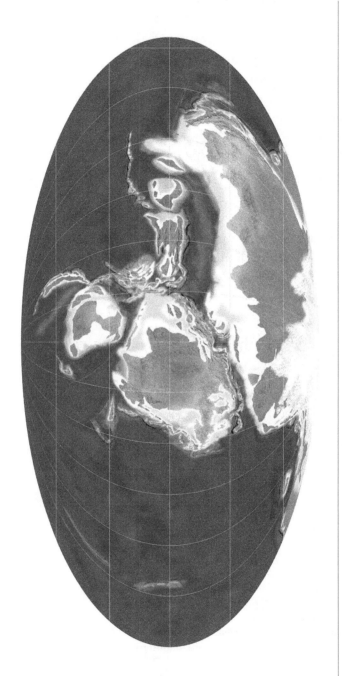

FIGURE 1.6 Paleogeography of the Earth in the late Famennian Age. Note the similarity of the positioning of the continents in the Famennian world to that of the Frasnian (fig. 1.2), but the presence of an ice cap at the South Pole in the Famennian world.

Source: Global Paleogeography and Tectonics in Deep Time © 2016 Colorado Plateau Geosystems Inc. Reprinted with permission.

Did the Late Paleozoic Ice Age begin in the late Frasnian or the late Famennian? Current paleogeographic reconstructions of the Earth in the late Famennian show a large ice cap at the South Pole, on the western margin of Gondwana (fig. 1.6), caused by the presence of glacial deposits in South American and Africa that are of late Famennian in age (figs. 1.3 and 1.4). The western margin of Gondwana was also positioned on the South Pole in the late Frasnian (fig. 1.2), but no ice cap is shown there at present—will one be demonstrated to have existed there in the future? Abundant circumstantial evidence exists for late Frasnian glaciation (fig. 1.5, tables 1.6–1.9); what remains to be discovered is a few remnant glacial strata of late Frasnian age on Gondwana or ice-rafted glacial debris in marine sediments offshore from the Gondwana landmass, similar to that found for the Oligocene glaciation in Antarctica.

In either event, the glaciations of the Late Devonian were unipolar, confined to the giant southern continent of Gondwana (fig. 1.6), and thus were only harbingers of the bipolar glaciation that was to come with the formation of a polar sea-ice cap in the Northern Hemisphere of the Earth. That freezing came in the Serpukhovian Age of the Early Carboniferous, and we will explore its consequences in the next chapter.

2 | The Big Chill

The long duration of the LPIA [Late Paleozoic Ice Age] sets it apart from earlier Paleozoic intervals of massive glaciation. . . . Glaciation was bipolar: glacial marine deposits of Serpukhovian age provide the earliest evidence of late Paleozoic glaciation in Siberia, although ice volume was greatest in the Southern Hemisphere, where glaciers expanded over broad areas of Gondwana.

—Stanley and Powell (2003, 877–878)

THE STRANGE WORLD OF THE TOURNAISIAN GAP

The world at the dawn of the Early Carboniferous Epoch was very different from that of the Late Devonian.[1] On land, the great forests of the Late Devonian were gone—the largest land plants stood only two meters (6.6 feet) high, in stark contrast to the towering 30-meter-high (100-foot-high) *Archaeopteris hibernica* trees that were so widespread before the end-Devonian extinctions (we will consider *Archaeopteris* trees in more detail in chapter 3). The land plants that did survive into the Early Carboniferous were mostly shrubby—the dense stands of trees and shady forests of the Late Devonian had vanished from the Earth. Only in the Visean Age of the Early Carboniferous would tall trees once again evolve, chiefly the 40-meter-tall (131-foot-tall) woody seed-tree *Pitus primaeva*.[2]

In the oceans, the great reefs were gone. The Late Devonian reefs were the largest, most widespread reef tracts ever to exist in Earth history—covering some five million square kilometers (almost two million square miles) of shallow marine seafloor. They were hit hard in the Late Devonian extinctions, and it took 130 million years before the corals recovered and once again became major reef-building organisms in the Middle Triassic. Even the composition of the shellfish in the seas had changed: the Devonian had been the golden age of the ancient brachiopods,[3] the lampshells, but now the more modern shellfish species of the molluscs,[4] the clams and oysters, had diversified following the extinction of over three-quarters of all brachiopod species. The giant predatory armored fishes of the Devonian world were all extinct. Never again would the oceans be populated by huge fish as big as modern-day killer whales but with strange bony armor around their heads and no teeth! Instead of teeth, they had sharp bone blades in their mouths that, similar to the sharp bone beaks of modern-day snapping turtles, functioned very efficiently in slicing up prey animals (fig. 2.1).

Welcome to the world of the Tournaisian Gap, a peculiar world of depauperate ecosystems and a gap in evolutionary innovation that was to last some ten million years in the Tournaisian Age.[5] For most of the Tournaisian Gap, fossils of our ancestors, the tetrapod vertebrates, are missing in the rock record. Only in the very latest Tournaisian Age, and continuing into the following Visean Age, would fossils of the tetrapod vertebrates once again be found in the rock record. Why? The answer appears to be that the tetrapods

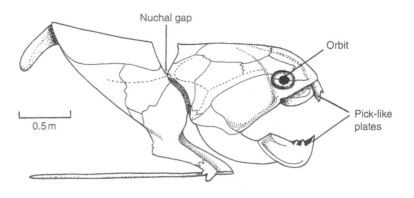

FIGURE 2.1 Bone shield of the Late Devonian armored fish *Dunkleosteus*.

Source: Redrawn from Moy-Thomas and Miles (1971) in *Invertebrate Palaeontology*, by M. J. Benton, copyright © 2015 Michael J. Benton. Reprinted with permission.

existed in such small populations during the Tournaisian Gap that the probability that any individual specimen would be preserved in the fossil record was very low. The tetrapods had been hit very hard by the Late Devonian extinctions and barely survived into the Early Carboniferous.[6] However, in the late Tournaisian and early Visean Ages, the tetrapods began to recover and once again to spread across the land areas of the Earth—and to show up in the fossil record.

The Tournaisian Gap world was the product of a classic evolutionary phenomenon known as an evolutionary bottleneck—a period of time in which major losses of biodiversity occur and only a small fraction of the individuals and species that existed before the "bottleneck" survive. An evolutionary bottleneck is produced when the population sizes of a given species shrink almost to the critical minimum level from which a species cannot recover. One of the immediately observable consequences of a species' surviving an evolutionary bottleneck is a sharp reduction of genetic diversity in the population of survivors. Numerous examples of this phenomenon are evident in nature. The cheetahs in Africa survived a severe evolutionary bottleneck, but the remaining cheetahs have such low genetic diversity that they are still endangered with extinction. They came very close indeed to the minimum-viable-population limit. European and Asian populations of humans, *Homo sapiens*, also show the characteristic low genetic diversity that results from passing through an evolutionary bottleneck. It is estimated that all European and Asian humans could be the descendants of as few as 160 people, possibly a single clan of hunter-gatherers, who managed to exit Africa by crossing the Bab al-Mandab (Gate of Grief) of the Red Sea about 50,000 years ago (Wade 2006).

The Tournaisian Gap world was actually the product of two bottlenecks, which I call the End-Frasnian Bottleneck and the End-Famennian Bottleneck (McGhee 2013). Two bottlenecks are worse than one, as we will see. For example, the giant armored fish of the Late Devonian (fig. 2.1) were severely affected by the End-Frasnian Bottleneck and lost half of their phylogenetic lineages and genetic diversity—yet they managed to survive. They survived in a weakened condition, however, and the End-Famennian Bottleneck proved fatal. None survived that diversity constriction, and the genetic lineage of the great armored fishes was terminated forever.

The End-Frasnian Bottleneck sharply reduced the morphological diversity that was previously present in the tetrapod lineages. The low morphological diversity seen in the survivors reflects their low genetic diversity, and only the more conservative or plesiomorphic (close to the ancestral condition) tetrapods survived the bottleneck constriction. In particular, the diversity of sizes seen in Frasnian tetrapods was lost, as only the midsize forms survived into the Famennian. The worldwide geographic distribution of the Frasnian tetrapods was also lost in the bottleneck constriction, as only the Laurussian tetrapods managed to survive. The same evolutionary bottleneck phenomena of reduction in morphological variance and geographic range are seen in tetrapod species that survived the End-Famennian Bottleneck—but now evolutionary reversals occurred in the survivors as well during the Tournaisian Gap. The Famennian tetrapods had made major evolutionary advances in their step-by-step emergence from the rivers and lakes of the Earth and their invasion of the land: they had lost their fishlike teeth and lateral-line systems—used by fish to "hear" in water, a trait useless on dry land—and they had lost their internal gills and could no longer breathe in water but relied exclusively on their lungs to breathe air. When the tetrapods begin to show up again in the fossil record in the late Tournaisian and the Visean, following the Tournaisian Gap, it is shocking to see that many of the tetrapod groups had abandoned dry land entirely and returned to an aquatic mode of life. What could have happened during the span of the Tournaisian Gap that would trigger a period of reverse evolution? What caused many tetrapod species to retreat from the land and go back into the water?[7]

Only the tetrapods of the family of the whatcheeriids seem to have managed to begin to diversify in the late Tournaisian Age. Starting with the only known Tournaisian species of the family, *Pederpes finneyae* from Scotland (fig. 2.2), the whatcheeriids diversified and by the mid-Visean had achieved a worldwide distribution.[8] By the early Visean Age, other tetrapod groups were recovering their population numbers, were producing new evolutionary innovations and clades, and were geographically on the move. No less than 23 new tetrapod species appeared in the fossil record by the late Visean (see McGhee 2013, table 7.4), proof that the harsh climatic conditions—whatever they may have been—of the Tournaisian Gap world had ended.

FIGURE 2.2 Reconstruction of the whatcheeriid tetrapod *Pederpes finneyae* from Scotland. Note the presence of six toes on the forefoot and the laterally placed eyes in the head. The animal was about one meter (3.3 feet) long.

Source: Illustration by Kalliopi Monoyios © 2013. Reprinted with permission.

THE RETURN OF THE GLACIERS

Evidence for the existence and timing of the glaciation in the Famennian Age of the Late Devonian (table 1.7) is undisputed, but once we cross into the Tournaisian Age of the Early Carboniferous, the data are more open to question (table 2.1). In a comprehensive 2003 summary of the geologic evidence for glaciation, University of Wisconsin geologist John Isbell and his colleagues argued that three distinct glacial phases could be recognized in the Late Paleozoic Ice Age: Phase GI, the time span from the Famennian Age in the Late Devonian through the Tournaisian Age in the Early Carboniferous; Phase GII, the time span from the Serpukhovian Age in the Early Carboniferous through the Bashkirian Age of the Late Carboniferous; and Phase GIII, the time span from the late Moscovian Age in the Late Carboniferous to the middle of the Sakmarian Age in the Early Permian (see table 1.4 for the geologic timescale in this interval of time). Isbell and colleagues (2003) argue that glacial strata of the Famennian and Tournaisian ages, their proposed glacial phase GI, are present in the northwestern Gondwana basins in Brazil and Bolivia in South America and in the Tim Mersoï basin in north-central Africa. Thus in the first column in table 2.1, the glaciation that started in the Famennian Age of the Late Devonian is shown to continue throughout the entire Tournaisian Age of the Early Carboniferous, and the entire Visean Age of the Early Carboniferous is shown to be free of continental glaciers.

TABLE 2.1 The timing of geologic and biotic events in the Tournaisian and Visean Ages of the Early Carboniferous.

Age	Geologic Time (Ma)	Geologic Events (1)	(2)	(3)	Biotic Events
VISEAN	331				
	332				
	333			ICE	
	334			ICE	
	335		ICE	ICE	
	336		ICE	ICE	
	337		ICE	ICE	
	338		ICE	ICE	
	339		ICE	ICE	
	340		ICE	ICE	
	341				
	342				
	343				
	344				
	345			ICE	
	346			ICE	
TOURNAISIAN	347	ICE		ICE	
	348	ICE		ICE	
	349	ICE		ICE	Tournaisian Gap
	350	ICE	ICE	ICE	Tournaisian Gap
	351	ICE	ICE	ICE	Tournaisian Gap
	352	ICE	ICE	ICE	Tournaisian Gap
	353	ICE	ICE	ICE	Tournaisian Gap
	354	ICE		ICE	Tournaisian Gap
	355	ICE			Tournaisian Gap
	356	ICE			Tournaisian Gap
	357	ICE			Tournaisian Gap
	358	ICE		ICE	Tournaisian Gap
FAMENNIAN (pars)	359	ICE	ICE	ICE	← Land extinctions
	360	ICE	ICE	ICE	← Marine extinctions

Source: Geologic events column (1) from Isbell et al. (2003), column (2) from Caputo et al. (2008), and column (3) from Frank et al. (2008). Biotic events from table 1.7; timescale modified from Gradstein et al. (2012).

In contrast, the University of Pará geologist Mário Vicente Caputo and his colleagues (2008) argue that glacial strata only from the middle Tournaisian Age can be reliably dated biostratigraphically,[9] and that the early Tournaisian and late Tournaisian time intervals may have been cold but free of continental ice sheets. A major fall in sea level did occur in the middle Tournaisian in North America on the Laurussian continent to the north of the giant continent Gondwana, leading to erosional valley incisions as deep as 60 meters (200 feet) in some areas (Kammer and Matchen 2008), and it has been argued that this sea-level drop was triggered by the formation of glacial ice sheets on the land.[10]

Furthermore, Caputo and colleagues report the existence of glacial strata in Brazil and Argentina that have been biostratigraphically dated to the middle Visean Age of the Early Carboniferous,[11] an interval of time that was paleoclimatically reconstructed as glacier-free by Isbell and colleagues (see table 2.1). The biostratigraphic dating of these glacial strata to the middle Visean has been corroborated by radiometric dating of volcanic strata that occur with the glacial strata in Argentina. Uranium-lead dating gives an age of 335.99 ± 0.06 million years ago, a date that falls in the middle of the Visean Age (Gulbranson et al. 2010).[12] Milo Barham, a geologist at the National University of Ireland, and his colleagues have presented further geochemical evidence of a sharp 4.5°C (8.1°F) drop in sea-surface temperatures that occurred 336 million years ago, consistent with the onset of global icehouse conditions (Barham et al. 2012). The sedimentary strata in scattered regions around the world also begin to show cycles in sedimentary deposition that have been argued to have Milankovitch periodicities in the middle to late Visean. It is argued that cyclic-deposited sedimentary strata in Europe and North America were driven by periodic fluctuations in global sea level as continental ice sheets waxed and waned on Gondwana in the Visean (Schwarzacher 1989; Smith and Read 2000; Wright and Vanstone 2001; Gastaldo, Purkyňová, Šimůnek, and Schmitz 2009; Bishop et al. 2009; Giles 2009). Thus, in the second column of table 2.1, only the middle of the Tournaisian Age is shown to be glaciated, but, unlike in column 1, the middle of the Visean Age is also shown to be glaciated.

Finally, the University of Nebraska geologist Tracy Frank and colleagues (2008) have conducted geochemical analyses of oxygen-isotope ratios in ancient brachiopod shells in an effort to reconstruct paleoclimatic fluctuations in the Early Carboniferous (see McGhee 2013, 196–199). Collecting brachiopod isotopic data from North America, Europe, Iran, and China—all offshore from the giant continent Gondwana—they reconstructed the earliest Tournaisian to have been glacial in Gondwana, followed by a warming episode and ice-free conditions until the middle Tournaisian, global cooling and a return to glaciation from the middle Tournaisian through the earliest Visean, another warming episode and ice-free conditions in the early Visean, and a return to glacial conditions in the middle Visean (column 3 in table 2.1).

In summary, hard empirical data from dated glacial strata in Brazil and Argentina demonstrate that Gondwana was glaciated in the middle Tournaisian and in the middle Visean. Geochemical analyses suggest an expansion in the duration of the middle Tournaisian glaciation from four to ten million years (extending into the earliest Visean), and an expansion in the duration of the middle Visean glaciation from six to eight million years (table 2.1). However, regardless of the precise pattern of the development of the Tournaisian and Visean glaciers, those glaciers were still unipolar, confined to the giant southern continent of Gondwana.

THE ONSET OF BIPOLAR CONTINENTAL ICE?

The freezing of the Earth began in earnest in the Serpukhovian Age. Following a two- to four-million-year warmer interval in the late Visean (table 2.1), the ice returned to western and central Gondwana with a vengeance in the Serpukhovian and persisted through the remainder of the Early Carboniferous and through the Bashkirian Age in the following Late Carboniferous—a time span of some 15 million years of continental glaciation (Frank et al. 2008). This time, however, the glaciers were not confined to western and central Gondwana but spread to eastern Gondwana—present-day Australia—as well. The University of Nebraska geologist Christopher Fielding and his colleagues (Fielding, Frank, Birgenheier et al. 2008) document four distinct pulses of glacial advance and retreat in eastern Gondwana, which they have labeled C1 (for first Carboniferous glacial pulse, the oldest) through C4 (for fourth Carboniferous glacial pulse, the youngest; see column 2 in table 2.2). The first glacial pulse, C1, consisted of alpine glaciers that formed in highland areas of eastern Gondwana. It was short-lived, but the next glacial pulse, C2 in the Serpukhovian, was much colder, and continental glaciers formed for the first time in eastern Gondwana (table 2.2).

TABLE 2.2 The possible onset of bipolar glaciation during the Late Paleozoic Ice Age.

Age	Geologic Time (Ma)	South Pole Climatic Events		North Pole
		West Gondwana	East Gondwana	Siberia
MOSCOVIAN	308		ICE	
	309		ICE	
	310		ICE	
	311		ICE	ICE?
	312		ICE	ICE?
	313		ICE	ICE?
	314		ICE	ICE?
	315		ICE	ICE?
BASHKIRIAN	316	ICE	ICE-C4	ICE?
	317	ICE		
	318	ICE	ICE	
	319	ICE	ICE	
	320	ICE	ICE	
	321	ICE	ICE-C3	
	322	ICE		
	323	ICE		
SERPUKHOVIAN	324	ICE	ICE	ICE?
	325	ICE	ICE	ICE?
	326	ICE	ICE-C2	Cold
	327	ICE		
	328	ICE		
	329	ICE	ICE	
	330	ICE	ICE-C1	
VISEAN (pars)	331			
	332			

Source: Gondwana data from Isbell et al. (2003), Frank et al. (2008), and Fielding, Frank, Birgenheier et al. (2008); Siberian data from Epshteyn (1981a), Chumakov (1994), and Raymond and Metz (2004). Timescale modified from Gradstein et al. (2012).

Note: Bold type indicates the existence of continental glaciers and normal type indicates the presence of alpine glaciers, where glaciations in eastern Gondwana are designated C1 through C4, from oldest to youngest.

Independent geochemical analyses of oxygen isotope ratios in brachiopod shells from the Russian Platform on the continent of Laurussia—far to the north of Gondwana—reveal a major positive anomaly[13] in the late Serpukhovian, an anomaly that is argued to be evidence of a buildup of ice volume on the Earth that was equivalent "to a full Pleistocene glaciation" (Mii et al. 2001, 144). The analyses of the Universidad Nacional de Salta glacial geologist Gustavo González-Bororino and his University of Toronto colleague Nicholas Eyles agree with that glacial magnitude assessment and indicate that, in the late Serpukhovian cooling of the Earth, an ice cap formed on Gondwana that covered some 21 million square kilometers (eight million square miles) of land, just a little less than the 23.5-million-square-kilometer (9.1-million-square-mile) ice cover that developed on the Earth in the Last Glacial Maximum (LGM) phase of the Cenozoic Ice Age (González-Bororino and Eyles 1995).

But did the glaciers become truly bipolar in the Serpukhovian freezing event—did ice sheets exist on the continental landmasses both in the giant southern continent of Gondwana and in the high-latitude, Siberian region of the northern continent Laurussia? In our present-day world, only two large areas of continental glaciation still exist (and they are rapidly dwindling in size). These are the glacial ice cap in Antarctica, which covers 13,720,000 square kilometers (5,297,322 square miles) of land, and the glacial ice cap in Greenland, which covers 2,166,097 square kilometers (836,335 square miles). Together they total some 15,886,097 square kilometers (6,133,657 square miles) of the Earth that is covered by ice today. But in the Pleistocene maximum glaciation phase of the Cenozoic Ice Age, much of northern Europe was covered by the Fennoscandian continental glacier, and much of northern North America was covered by the Laurentide continental glacier. When these two giant glaciers are added to the mix, then a total of 23.5 million square kilometers (9.1 million square miles) of the land-masses of the Earth were covered by ice in the LGM of the Cenozoic Ice Age (González-Bororino and Eyles 1995). Was the continental coverage of ice on the Earth in the Serpukhovian similar to that seen in the LGM in the Cenozoic, with a continental ice cap at both poles?

Many paleontologists think that continental glaciation became truly bipolar in the Serpukhovian, as evidenced by the epigraph at the beginning of this chapter from the work of the paleontologists Steve Stanley and Matthew Powell (see also Fielding, Frank, and Isbell 2008). The analysis of faunal migration patterns in marine species by Anne Raymond and Cheryl Metz, paleontologists at Texas A&M University, indicate that the North Pole had cooled significantly in the late Serpukhovian, and it is probable that massive sea ice had formed offshore from Siberia (column 3 in table 2.2) (Epshteyn 1981a; Chumakov 1994; Raymond and Metz 2004). By the end of the Bashkirian Age, it has been argued that continental ice sheets were present in Siberia in the Northern Hemisphere, evidenced by the presence of ice-rafted debris and glacial-marine sediments in the rock record.[14] Thus by late Bashkirian and into the early Muscovian (table 2.2), the Late Paleozoic Ice Age had definitely become bipolar—or had it?

The data have been called into question by the University of Wisconsin geologist John Isbell and his colleagues (2016), who argue that a sea-ice-cap coverage of the North Pole (like today's) is still not a continental, land-ice coverage, and that the Bashkirian-Muscovian data are evidence only for the presence of sea ice. They argue that continental glaciation on the Earth was never truly bipolar in the late Paleozoic, and that all continental glaciers were confined to the Southern Hemisphere. We will return to Isbell and colleagues' arguments when we discuss the Permian, where the bipolarity question again arises. In light of this ongoing debate, I have placed question marks on the distribution of Northern Hemisphere continental ice in table 2.2.

Whatever the distribution of continental ice on the Earth in the Serpukhovian, the effect of the Serpukhovian "Big Chill" was devastating to marine life—precipitating the sixth most severe ecological crisis in the Phanerozoic, greater even than that of the glacially triggered end-Famennian extinctions (table 1.3). The biological effects of that crisis will be examined in detail in the next two sections of this chapter.

THE SERPUKHOVIAN CRISIS IN THE OCEANS

The Serpukhovian biodiversity crisis marks the boundary between the Early and Late Carboniferous Epochs (or the Mississippian and Pennsylvanian Epochs in the United States; see table 1.4) (Gradstein et al. 2012). Although the biodiversity crisis was large enough to allow a major subdivision in the geologic timescale to be recognized, it has received relatively little analysis until recently. Indeed, the University of Birmingham paleontologist Tony Hallam and his colleague, the

University of Leeds paleontologist Paul Wignall, have commented (1997, 92): "The importance ascribed to the event varies from author to author: Saunders and Ramsbottom (1986) called it a major extinction, and others (Ziegler and Lane 1987; Weems 1992), with undoubted exaggeration, ranked it as a mass extinction comparable to the big five." The "big five" that Hallam and Wignall refer to are the end-Ordovician, Late Devonian (end-Frasnian), end-Permian, end-Triassic, and end-Cretaceous mass extinctions (see table 1.3), first recognized by the University of Chicago paleontologists Dave Raup and Jack Sepkoski in biodiversity analyses they conducted back in 1982. These five extinction events triggered the five largest losses of biodiversity in the Earth's oceans in Phanerozoic history, although the ecological impact of the end-Ordovician event was relatively minor in contrast to the end-Cretaceous event, which triggered the least loss of biodiversity of the big five but precipitated the second most severe ecological disruption in the entire Phanerozoic, as discussed in chapter 1 (table 1.3) (McGhee et al. 2004).

My colleagues Peter Sheehan, Dave Bottjer, and Mary Droser and I have conducted comparative paleoecological analyses that have revealed that the ecological impact of the Serpukhovian biodiversity crisis was greater than that of the end-Ordovician (one of Raup and Sepkoski's big five) and end-Devonian (end-Famennian) extinctions (table 1.3)—both of which were triggered by brief but intense glaciations, as discussed in chapter 1. Yet the magnitude of the biodiversity loss that occurred in the Serpukhovian crisis was less than in any of the big five—much less than in the end-Ordovician crisis—so why did the Serpukhovian crisis cause a larger ecological disruption? There is a significant difference in the mechanism by which biodiversity was lost in the Serpukhovian crisis as opposed to the end-Ordovician crisis. In the end-Ordovician, as well as the end-Permian and end-Cretaceous crises, diversity loss was driven by a sharp major increase in the extinction rate (Bambach et al. 2004). In contrast, the Serpukhovian biodiversity loss was driven by a precipitous drop in the speciation rate at the beginning of the Serpukhovian during a period of elevated extinction rates (Bambach et al. 2004, fig. 6; Stanley 2007), as shown in figure 2.3. In this, the Serpukhovian biodiversity crisis is more similar to the Late Devonian (end-Frasnian) biodiversity crisis (McGhee 1988, 1996) and the end-Triassic biodiversity crisis (Bambach et al. 2004) than to the end-Ordovician, end-Permian, and end-Cretaceous mass extinctions.

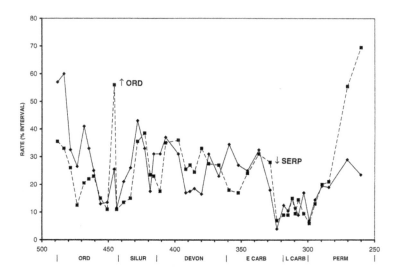

FIGURE 2.3 The magnitude of Late Paleozoic marine extinction rates (dashed line and square data points) and origination rates (solid line and diamond data points) is measured in percent change in generic diversity per time interval (vertical axis). Geologic timescale abbreviations (horizontal axis): ORD, Ordovician; SILUR, Silurian; DEVON, Devonian; E CARB, Early Carboniferous; L CARB, Late Carboniferous; PERM, Permian; and SERP, Serpukhovian (within the figure); see text for discussion.

Source: From McGhee et al. (2012), modified from Stanley and Powell's (2003) original figure.

The University of Hawaii paleontologist Steve Stanley and his colleague, the University of California (Santa Cruz) paleontologist Matthew Powell, have argued that the Serpukhovian crisis initiated "a new state of the global marine ecosystem" (Stanley and Powell 2003, 877) and "reordered the marine ecosystem to one dominated by widespread genera, which had intrinsically low macroevolutionary rates" (that is, low speciation and extinction rates) (Powell 2008, 525). This phenomenon can be seen clearly in figure 2.3: both extinction and speciation rates plummet at the end of the Serpukhovian and remain low for the entire duration of the major phase of the Late Paleozoic Ice Age, from the Early Carboniferous through the Middle Permian. There was no significant recovery of biodiversity in marine ecosystems after the Serpukhovian crisis. Instead, both speciation rates and extinction rates remained anomalously low for 50 million years (fig. 2.3), and the marine fauna of the Late Paleozoic Ice Age were characterized by "faunal persistence, not origination or extinction" (Bonelli and Patzkowsky 2011, 14) in a protracted period of ecological and evolutionary stagnation in paleotropical regions.[15]

The ecological dynamics of the global marine ecosystem have been characterized by a series of evolutionary faunas during the Phanerozoic, first recognized in the biodiversity analyses of Jack Sepkoski (1984, 1990). We will examine these evolutionary faunas in great detail in chapter 6, but here let us briefly consider some characteristics of two of them: the Paleozoic evolutionary fauna and the modern evolutionary fauna. In general, the older Paleozoic evolutionary fauna was characterized by a higher rate of extinction than the younger modern evolutionary fauna, which is more extinction resistant, a phenomenon we considered briefly in chapter 1. Steve Stanley's analyses have revealed that the Serpukhovian crisis was the most ecologically selective of the Paleozoic biodiversity crises in that it preferentially eliminated species of the Paleozoic evolutionary fauna, and spared the species of the modern evolutionary fauna, to a much greater extent than any of the other Paleozoic crises (Stanley 2007). For example, the Serpukhovian crisis eliminated genera of the Paleozoic evolutionary fauna relative to the genera of the modern evolutionary fauna in a ratio of 2.4 to 1, in contrast to the end-Ordovician mass extinction, which had a Paleozoic-to-modern evolutionary fauna extinction ratio of only 1.4 to 1. The modern evolutionary fauna experienced a biodiversity loss of only 15.4 percent in the Serpukhovian crisis, the smallest modern evolutionary fauna diversity loss in the any of the Paleozoic biodiversity crises (Stanley 2007). In essence, the Serpukhovian biodiversity crisis was a harbinger of the demise of the dominant Paleozoic evolutionary fauna and its ecological replacement by the modern evolutionary fauna in the Mesozoic; we will consider that demise in great detail in chapter 6, "The End of the Paleozoic World."

The Serpukhovian crisis was also ecologically selective in that it preferentially spared marine genera with broad latitudinal geographic ranges and eliminated those with narrow latitudinal distributions (Powell 2005, 2008). Matthew Powell has further argued that the key selective factor underlying this ecological differentiation was thermal tolerance; no other ecological trait, such as niche breadth, geographic dispersal range, species richness, body size, or habitat position in an onshore-offshore gradient, played a significant role in survival in the Serpukhovian crisis (Powell 2008). Because these other ecological traits are taken as selective in times of background extinction (Jablonski 1986), as opposed to mass extinction, Powell has proposed that the Serpukhovian biodiversity crisis "adhered to a mass-extinction, not background, regime" (Powell 2008, 526). Thus, contrary to the assessment of Tony Hallam and Paul Wignall (quoted at the beginning of this section of the chapter) that to consider the Serpukhovian crisis as comparable to the big five extinctions was an exaggeration (Hallam and Wignall 1997, 92), Powell does consider the Serpukhovian biodiversity crisis to have possessed the qualities of a mass extinction comparable to the big five.

Finally, although the great Late Devonian reefs—dominated by ancient and now extinct stromatoporoid sponges, tabulate corals, and rugose corals—had been destroyed in the Devonian extinctions, the rugose corals (or horn corals, since they looked like the horns of a bull) had survived into the Early Carboniferous. But those rugose coral survivors were strongly impacted by the Serpukhovian biodiversity crisis: the ecologically dominant rugose species of the Early Carboniferous were decimated and were ecologically replaced by entirely new rugose coral species in the following Late Carboniferous. This ecological-replacement event has been called the "mid-Carboniferous rugose coral evolutionary event" (Wang et al. 2006, 339) in which the Serpukhovian crisis triggered a major ecological shift from a coral fauna dominated by large, solitary corals to one of massive colonial corals that, it is argued, have been adapted to the changed marine conditions associated with the onset of glaciation.

It is interesting that both the end-Ordovician mass extinction and the Serpukhovian biodiversity crisis were triggered by glaciations. Other than that common trigger, the two events were very different. Glaciation in the Ordovician triggered an enormous jump in the extinction rate of marine organisms (fig. 2.3) and major biodiversity losses, yet the ecological impact of those extinctions was minimal. Glaciation in the Serpukhovian triggered a precipitous drop in the speciation rate (fig. 2.3) but only moderate biodiversity losses, yet the ecological impact of those diversity losses and ecosystem restructuring was an order of ecological magnitude larger (table 1.3).

Why were these two events so different? First, in contrast to the Serpukhovian glaciations, the primary Ordovician glaciations ended abruptly, although there were significant brief glacial advances and accompanying sea-level changes during the early Silurian (Finnegan et al. 2011; Harris and Sheehan 1998). These latter glacial events may have set back recovery for brief intervals, but in general faunas recovered more or less continuously during the early Silurian. Thus, in the post-Ordovician, glacial changes were still ongoing but not of sufficient strength to reset to a new state in the global marine ecosystem as it did in the post-Serpukhovian.

Second, unlike the post-Ordovician, the Serpukhovian glaciations struck the descendants of a fauna that had already survived a series of glaciation events (tables 1.7 and 2.1) in the end-Famennian, mid-Tournaisian, mid-Visean Ages—and possibly in the end-Frasnian Age as well. Thus, in the Serpukhovian, global cooling did not trigger a sharp jump in extinction rates of marine organisms, as it did in the Ordovician (fig. 2.3). Rather, it was the loss of habitat diversity in the new globally homogeneous marine environments that triggered a drop in speciation rate and subsequent loss of biodiversity. Major glaciation continued in the post-Serpukhovian, and the cool and globally similar marine environments (only cool-water niches were available to be filled, warm tropical niches did not exist; see Powell 2005) perpetuated the reset in global marine ecosystems triggered by the Serpukhovian biodiversity crisis. Ultimately, the vast expansion of forests in the Carboniferous may have been an ecological trigger for some of the differences between the Serpukhovian and Ordovician glaciations. Without abundant land plants, carbon dioxide could not be drawn down in the Silurian as easily as in the Carboniferous, and global temperatures rebounded in the Silurian but not in the Carboniferous (Sheehan 2001; Stanley 2007). We will examine the vast—and strange—Carboniferous forests in great detail in the next chapter.

THE SERPUKHOVIAN CRISIS ON LAND

Paleontologists Ann Raymond and Cheryl Metz have compared the climatic effects of the Cenozoic and Late Paleozoic Ice Ages and have noted that climatic similarities of the two ice ages "include glacial onset in the Southern Hemisphere and subsequent extension of glaciation to the Northern Hemisphere; cyclic advance and retreat of continental glaciers governed by Milankovitch orbital cycles, resulting in cyclothems [cyclic sedimentary layers]; formation of an ever-wet, ever-warm climate zone centered on the paleoequator; and strengthening of the pole-to-equator temperature gradient" (Raymond and

Metz 2004, 657). The formation of an ever-wet, ever-warm equatorial climate zone produced a stable environment in which "equatorial land-plant diversity increased dramatically relative to mid-latitude and high-latitude diversity." On the other hand, the high-latitude poles became much colder with massive ice buildup, and the mid-latitude regions centered on the 30°N and 30°S latitudinal belts developed "strongly expressed arid climatic zones" (Raymond and Metz 2004, 658).

The C1 glacial pulse at the beginning of the Serpukhovian was followed by a warming period of about two million years before the much colder C2 glacial pulse began (table 2.2). During this warm interval, a major extinction in land plants occurred in the mid-latitude regions of Laurussia, in the present-day Silesian Basin of the Czech Republic. The Colby College paleobotanist Robert Gastaldo and his colleagues report that nearly all plant clades suffered extinctions—species of groundcover plants, climbing vines, understory shrubs, and trees all were affected, whether they reproduced by spores or by seeds. This extinction and loss of land-plant diversity occurred about 327 million years ago (table 2.3), "several million years prior to the onset of maximum glaciation and sea-level downdraw at the Mississippian-Pennsylvanian [Serpukhovian-Bashkirian] boundary," and appeared to be due to a climatic "shift toward greater seasonality, with an increased number of dry months" (Gastaldo, Purkyňová, and Šimůnek 2009, 351).

In great contrast, Gastaldo and colleagues report that the wetland species were unaffected. Here we begin to see a major difference in response of marine and land-dwelling species to the great climatic shift that took place in the Serpukhovian. For marine species the news was uniformly bad, as discussed in the previous section of the chapter, but among land species some were driven to extinction while others thrived. The winners during the Serpukhovian crisis on land were the wetland species—and some of these winners were very strange indeed.

The strangest were the lycophyte trees. The lycophytes are still alive today, although most people have never seen one. A fairly large living lycophyte plant belongs to the stag's horn clubmoss species *Lycopodium clavatum*, which stands about 12 centimeters high—about five times as tall a quarter (fig. 2.4).[16] In contrast, its ancient relatives during the Carboniferous formed gigantic lycophyte trees that towered 50 meters (164 feet) into the sky. Small lycophyte trees—thin, pole-like trees covered with tiny leaves all over their trunks as well as their short

TABLE 2.3 Biotic events during the Serpukhovian "Big Chill".

Age	Geologic Time (Ma)	Climatic Events			Biotic Events
		(1)	(2)	(3)	
MOSCOVIAN	308		ICE		
	309		ICE		
	310		ICE		
	311		ICE	ICE?	
	312		ICE	ICE?	
	313		ICE	ICE?	
	314		ICE	ICE?	
	315		ICE	ICE?	
BASHKIRIAN	316	ICE	ICE-C4	ICE?	
	317	ICE			
	318	ICE	ICE		
	319	ICE	ICE		
	320	ICE	ICE		
	321	ICE	ICE-C3		
	322	ICE			
	323	ICE			
SERPUKHOVIAN	324	ICE	ICE	ICE?	← Marine biodiversity minimum
	325	ICE	ICE	ICE?	
	326	ICE	ICE-C2	Cold	← First lycophyte rainforests
	327	ICE			← Extinction in land plants
	328	ICE			
	329	ICE	ICE		
	330	ICE	ICE-C1		← Marine biodiversity loss begins
VISEAN (pars)	331				
	332				

Source: Timescale modified from Gradstein et al. (2012).

Note: Column (1) is West and Central Gondwana and column (2) is East Gondwana (Australia), both in the Southern Hemisphere; column (3) is Siberia (northeastern Russia) in the Northern Hemisphere.

FIGURE 2.4 The Stag's horn clubmoss *Lycopodium clavatum* is a living lycophyte plant that stands about 12 centimeters (4.7 inches) high.

Source: Modified and redrawn from Lecointre and Le Guyader (2006).

branches at the top of the trunk—are found as fossils in Earth's oldest forest, the famous Middle Devonian Gilboa fossil forest of New York State (Stein et al. 2012) that we briefly considered in chapter 1. While other plant lineages evolved adaptive innovations like woody tissue, which allowed the construction of large trunks, and seed reproduction, which freed the plants from wetland habitats, the ancient lycophytes remained behind in the wetlands. Their time would come—and it came in the C2 freezing of the planet (table 2.3). The first forests that were actually dominated by species of lycophyte trees appeared on the Earth about 326 million years ago in Central Asia (Cleal and Thomas 2005). These strange trees flourished in the Karaganda Basin in present-day northeastern Kazakhstan, spreading across marshy land and river deltas formed as the sea level began to fall with the freezing of large masses of ice at the poles. The lycophyte trees would eventually form huge rainforests that spanned the globe in the ever-wet, ever-warm equatorial climate formed by the Late Paleozoic Ice Age. We will examine these rainforests in great detail in the next chapter.

THE MYSTERY OF THE SERPUKHOVIAN LAND ANIMALS

How did the Serpukhovian crisis affect the land animals—was there a crisis at all? Or, like the wetland species of land plants, did the land animals thrive during the chilling of the Earth during the Serpukhovian? We have no definitive answer to that question, at least not yet.

First, we have no known evidence of a major extinction event in land-animal species during the Serpukhovian crisis.[17] Second, we do know that major evolutionary innovations occurred in the mid-Carboniferous land-animal lineages—both tetrapod and arthropod—but it remains unclear whether these innovations occurred in the Bashkirian Age at the beginning of the Late Carboniferous or earlier, during the Serpukhovian crisis period at the end of the Early Carboniferous. For example, the tetrapods evolved the amniote egg and related morphological changes that freed them from water in their reproduction. This was a crucial step in the invasion of land by the vertebrate animals—prior to this innovation, the tetrapods still laid their eggs in water, much like fish. Many amphibians, which are not amniotes, still do so. In addition, the insect lineage of the arthropods evolved the first wings and capability of flight among land animals. This also was a crucial step in the invasion of land by the arthropods—prior to this innovation, insects were constrained to walking on the ground, their dispersal rates limited by their walking speed and small size. With the evolution of flight, the insects could take to the skies and disperse over huge areas of land in a very short time.[18]

There are two evolutionary lineages of amniote animals: the reptilian amniotes, the ancestors of living reptiles; and the synapsid amniotes, the ancestors of living mammals (table 2.4). The evolutionary split between the reptiles and the synapsids had already occurred in the Bashkirian as the oldest known (as yet) skeletal fossils of a reptilian amniote, the species *Hylonomus lyelli*, and the oldest known skeletal fossils of a synapsid amniote, the species *Protoclepsydrops haplous*, are dated to about 317 million years ago in the late Bashkirian Age (table 2.4).[19] Thus the amniote ancestor to the reptilian and synapsid lineages had to be older than the late Bashkirian.

TABLE 2.4 Amniote tetrapod lineages and species in the Early to Late Carboniferous transition.

TETRAPODA (limbed vertebrates)
– basal tetrapods
– Neotetrapoda
– – BATRACHOMORPHA (ancestors of living amphibians)
– – – basal batrachomorphs ("temnospondyls")
– – REPTILIOMORPHA (ancestors of amniote tetrapods)
– – – basal reptiliomorphs ("anthracosaurs")
– – – – Seymouriamorpha
– – – – AMNIOTA (amniote tetrapods, conquerors of the land)
– – – – – basal amniotes
– – – – – – *Casineria kiddi?* ← **Visean**
– – – – – SYNAPSIDA (ancestors of living mammals)
– – – – – – basal synapsids
– – – – – – – *Protoclepsydrops haplous* ← **Bashkirian**
– – – – – REPTILIA (ancestors of living reptiles)
– – – – – – basal reptiles
– – – – – – – *Hylonomus lyelli* ← **Bashkirian**

Source: Phylogenetic classification modified from Benton (2015).

Note: The geologic age of species is in bold; older paraphyletic tetrapod group names are in quotation marks; major clades are in capitals.

The late Bashkirian species *Protoclepsydrops haplous* is a basal member of the clade of the synapsid amniotes—but we have trace fossil evidence that much more highly evolved, derived species of the synapsid amniote clade also existed in the Bashkirian! This evidence comes from a fossil trackway in Germany that preserves the footprints of a large animal of an as yet unknown species, but the trackway (which was given the trace-fossil species name *Dimetropus*) is very similar to those produced by more highly derived ophiacodontid, edaphosaurid, or sphenacodontid synapsids, all of which were large animals (Voigt and Ganzelewski 2010; McGhee 2013; we will examine all of these synapsid groups in detail in chapter 4). If highly derived synapsid species were also present in the Bashkirian Age, then the ancestors of the synapsid amniote lineage must be older than the Bashkirian.

The oldest known (as yet) possible candidate for the first amniote is a small animal known only from its postcranial skeleton; that is, its body skeleton is pretty well preserved but the fossil is missing its head. This species is *Casineria kiddi*, and its body possesses several traits that are usually taken to be amniote traits, but it cannot be proved that the animal is definitely an amniote because its skull is missing. The *Casineria kiddi* fossil is from Cheese Bay, Scotland, and is dated to the mid-Visean Age (see the detailed discussion of *Casineria kiddi* in McGhee 2013). Even though it is impossible to prove that *Casineria kiddi* is the first amniote, the phylogenetic analyses of the University of Bristol vertebrate paleontologist Michael Benton have led him to predict that the amniotes evolved about 340 million years ago, in the middle of the Visean (Benton 2015, 123–124 and text box 5.1). If the amniotes evolved 340 million years ago, in the Visean (table 2.1), why did they diversify and attain population numbers large enough to appear in the fossil record only in the Bashkirian, some 17 million years later? That is, did the chilling climatic shift in the intervening Serpukhovian Age have a negative impact on the evolution of the earliest amniotes?

A similar puzzle exists concerning the evolution of the first winged insect. It was long thought that the oldest known member of the clade of the winged insects, the Pterygota (table 2.5), was the species *Eugeropteron lunatum* from Bashkirian-aged strata in Argentina (Grimaldi and Engel 2005). *Eugeropteron lunatum* is an odonatopteran, the clade of the dragonflies (table 2.5), which is almost the basal clade of the winged insects—but not quite. That honor belongs to the mayflies, the ephemeropterans (table 2.5), so *Eugeropteran lunatum*

TABLE 2.5 Flying insect lineages and species in the Early to Late Carboniferous transition.

INSECTA (insects)

– basal insects

– PTERYGOTA (winged insects, conquerors of the land)

– – Ephemeroptera (mayflies)

– – Metapterygota

– – – Odonatoptera (dragonflies, damselflies)

– – – – *Eugeropteron lunatum* ← **Bashkirian**

– – – Palaeodictyopterida

– – – – *Delitzschala bitterfeldensis* ← **Serpukhovian**

– – – NEOPTERA (folding-wing insects)

– – – – Protoptera

– – – – – *Kemperala hagensis* ← **Serpukhovian**

– – – – Polyneoptera

– – – – – Anartioptera

– – – – – – Orthopterida

– – – – – – – Orthoptera

– – – – – – – – Archaeorthoptera

– – – – – – – – – *Ampeliptera limburgica* ← **Serpukhovian**

Source: Phylogenetic classification modified from Grimaldi and Engel (2005) and Lecointre and Le Guyader (2006).

Note: The geologic age of species is in bold; major clades are in capitals.

cannot have been the first flying insect, but it was close to the origin of flight. However, the University of Exeter biologist Robin Wootton and his colleagues (1998) noted that *Eugeropteron lunatum* was surprisingly advanced in its wing morphology and suggested that there had to be earlier, less highly adapted, odonatopteran flying species that were the ancestors of *Eugeropteron lunatum*.

Indeed, in 2005, the Czech Charles University zoologist Jakub Prokop and his colleagues announced the discovery of four winged-insect fossil species from Serpukhovian-age strata in the Czech Republic (Prokop et al. 2005). These four species are *Delitzschala bitterfeldensis* (a palaeodictyopteran), *Kemperala hagensis* (a protopteran), *Ampeliptera limburgica* (an archaeorthopteran), and fragments of another archaeorthopteran species as yet unidentified. These fossils thus provide definitive corroboration that the winged insects evolved on Earth in the Serpukhovian Age of the Early Carboniferous. Or do they? The protopteran and archaeorthopteran species are highly derived neopteran insects (table 2.5)—insects that had evolved the capability of folding their wings, an adaptation not present in the basal clades of the mayflies and dragonflies. Thus the presence of highly derived neopteran species in Serpukhovian strata at the end of the Early Carboniferous argues for the evolution of flying insects even earlier in the Early Carboniferous. Yet even if the winged insects evolved in the Visean Age—along with the first amniote, *Casineria kiddi*, if it is an amniote— they did not become numerous enough, or have population sizes large enough, to appear in the abundance seen in the fossil record until the Bashkirian Age at the start of the Late Carboniferous. Why? What delayed the diversification of species possessing these new adaptive innovations?[20] Did the Serpukhovian crisis negatively affect the land animals after all?

Regardless of whether the evolution of the crucial adaptive innovations of the tetrapod amniote egg or the insect wing occurred in the dawn of the Late Carboniferous or the twilight of the Early Carboniferous, it is undisputed that a major diversification of both tetrapod and insect species started in the Bashkirian Age. Five new species of basal reptiliomorphs (table 2.4) appeared in the fossil record in the Bashkirian Age at the beginning of the Late Carboniferous, seven more new species in the following Moscovian Age, two more in the Kasimovian Age, and one in the Gzhelian Age at the close of the Late Carboniferous (see McGhee 2013, table 7.6). In the synapsid amniotes, ancestors to the mammals, the Bashkirian species *Protoclepsydrops haplous* (table 2.4) was joined by three new species in the Moscovian, four more in the Kasimovian, and three in the Gzhelian. And, in addition to the Bashkirian species *Hylonomus lyelli* (table 2.4), five new species of reptilian amniotes appeared in the Moscovian, one in the Kasimovian, and two in the Gzhelian.

The flying insects and other arthropod groups show a similar pattern of diversification beginning in the Bashkirian Age (see McGhee 2013, 237–249). Fourteen new species of millipedes, centipedes, and arachnids (spiders, scorpions, and kin) appeared in the Late Carboniferous—including the gigantic anthropleurid millipedes, the largest arthropods ever to exist on land. In the flying insects, the gigantic griffenflies (ancient relatives of modern dragonflies) also evolved in the Late Carboniferous. These were the largest flying insects ever to inhabit the Earth; we will consider the evolution of giant Carboniferous arthropods in detail in chapter 4. The flying insects also experienced a burst of ecological and morphological innovation in the Late Carboniferous, evolving the ability to fold their wings and the ability to eat and digest living plants—the evolution of the first herbivorous insects.

All of the newly evolved tetrapods and arthropods appeared on an Earth very different from our modern-day world. Viewed from space, the Carboniferous Earth had brilliant white ice caps at both poles, patches of brown and reddish arid deserts scattered around the planet in the temperate latitudes, and a broad green band of lush rainforests in the equatorial zone. The trees that inhabited those rainforests were fantastically different from our modern rainforest flora, as we will see in chapter 3.

3 | The Late Carboniferous Ice World

The Late Paleozoic Ice Age is the only time in Earth history, other than the Neogene, when vegetated land masses were subjected to climate fluctuations associated with extended intervals of polar glaciation.

—Gastaldo, Purkyňová, Šimůnek, and Schmitz (2009, 336)

THE STRANGE CARBONIFEROUS TREES

Our modern rainforests are dominated by flowering plants (angiosperms) and conifers (pinophytes), both of which are spermatophyte lignophyte euphyllophytes—that is, "seed-reproducing woody leafy trees" in the precise language of modern evolutionists (table 3.1). Like the rainforests in the cool Carboniferous world, our modern rainforests evolved in a cool Neogene world, where the South Pole has been glaciated for 12 million years, since the late Miocene, and the North Pole has been variously glaciated for three million years, since the late Pliocene (table 1.6). However, in stark contrast to the trees in our modern rainforests, the giant trees of the Carboniferous rainforests did not reproduce with seeds (they were not spermatophytes), did not have woody trunks (they were not lignophytes), and did not have large leaves (they were not euphyllophytes). In addition, there were no flowers of any kind in the Carboniferous rainforests—the angiosperms, the flowering plants, would only evolve some 140 million years later in the Early Jurassic.[1] To understand just how strange the giant Carboniferous rainforest trees were, we need first to examine the sequence of the evolution of plants on land (table 3.1).

TABLE 3.1 Land-plant lineages of the Late Paleozoic Ice Age.

EMBRYOPHYTA (land plants)
– Marchantiophyta (liverworts)
– STOMATOPHYTA (stomate plants)
– – Anthocerophyta (hornworts)
– – Hemitracheophyta
– – – Bryophyta *sensu stricto* (mosses)
– – – POLYSPORANGIOPHYTA (branched-sporophyte plants)
– – – – Horneophyta†
– – – – TRACHAEOPHYTA (vascular plants)
– – – – – Rhyniopsida†
– – – – – Eutracheophyta
– – – – – – LYCOPHYTA (microphyll-leafed plants)
– – – – – – – Zosterophyllopsida†
– – – – – – – Asteroxylales
– – – – – – – – Drepanophycales†
– – – – – – – – Lycopodiales
– – – – – – – – – Protolepidodendrales†
– – – – – – – – – Lycopodiaceae (club mosses)
– – – – – – – – – – Selaginellales (spike mosses)
– – – – – – – – – – – Lepidodendrales† (scale trees)
– – – – – – – – – – – – Lepidodendraceae†: *Lepidodendrid scale trees*
– – – – – – – – – – – – Sigillariaceae†: *Sigillarian scale trees*
– – – – – – – – – – – Selaginellaceae
– – – – – – – – – – Isoetales (quillworts)
– – – – – – – – – – – Chaloneriaceae†
– – – – – – – – – – – Isoetaceae
– – – – – – EUPHYLLOPHYTA (megaphyll-leafed plants)
– – – – – – – Moniliformopses
– – – – – – – – Cladoxylopsida†
– – – – – – – – – Pseudosporochnaceae†: *Wattieza (Eospermatopteris) erianus trees*
– – – – – – – – Equisetophyta (horsetails)
– – – – – – – – – Calamitaceae†: *Calamitean horsetail trees*
– – – – – – – – – Sphenophyllaceae†
– – – – – – – – – Equisetaceae
– – – – – – – – Filicophyta (true ferns)
– – – – – – – – – Marattiales: *Marattialean tree ferns*
– – – – – – – LIGNOPHYTA (woody plants)
– – – – – – – – Aneurophytales†
– – – – – – – – Archaeopteridales†
– – – – – – – – – Archaeopteridaceae†: *Archaeopterid spore trees*

```
– – – – – – – – SPERMATOPHYTA (seed plants)
– – – – – – – – – Lyginopteridales†
– – – – – – – – – – Lyginopteridaceae†: *Lyginopterid seed-fern trees*
– – – – – – – – – Medullosales†
– – – – – – – – – – Medullosaceae†: *Medullosan seed-fern trees*
– – – – – – – – – Gigantopteridales†
– – – – – – – – – – Gigantopteridaceae†: *Gigantopterid seed-fern trees*
– – – – – – – – – Core seed plants
– – – – – – – – – – Ginkgophyta (ginkgos)
– – – – – – – – – – Pinophyta (conifers)
– – – – – – – – – – – Cordaitales†
– – – – – – – – – – – – Cordaitaceae†: *Cordaitean conifer trees*
– – – – – – – – – – Cycadophyta (cycads)
– – – – – – – – – – Gnetophyta (gnetophytes)
– – – – – – – – – – [ANGIOSPERMAE (flowering plants) not present in Paleozoic!]
```

Source: Phylogenetic classification modified from Kenrick and Crane (1997a, 1997b), Donoghue (2005), and Lecointre and Le Guyader (2006).

Note: Important tree groups are in bold italics; extinct lineages are marked with a dagger (†); major clades are in capitals. The classification given in this table is a phylogenetic one, where the taxa listed are monophyletic clades. In the older literature the reader will encounter non-phylogenetic classifications of plants containing groupings of plants that are now recognized as paraphyletic. To help the reader translate the older plant classifications, here is a helpful list of the major paraphyletic groups that have been used in the past: "Pteridophytes" = non-spermatophyte tracheophytes, "Progymnosperms" = non-spermatophyte lignophytes, "Gymnosperms" = non-angiosperm spermatophytes, "Pteridosperms" = non-core seed plants ("seed ferns"); for discussions, see Niklas (1997) and Donoghue (2005).

Land plants, the Embryophyta, evolved from freshwater filamentous green algae, and it appears that the first tiny land plants emerged from the aquatic world in the Middle Ordovician (Lecointre and Le Guyader 2006; McGhee 2013). The most primitive of the land plants are the marchantiophytes, the liverworts (table 3.1). The oldest known spores and cuticular fragments that are believed to be from liverworts are found in strata dated to the Dapingian Age of the Middle Ordovician (see table 1.4 for the Ordovician Age divisions of the geologic timescale), around 470 million years ago (Rubinstein et al. 2010; Wellman 2010). About 15 million years later, definite proof that the liverworts had evolved comes from fossils of Katian Age of the Late Ordovician, about 453 million years ago (Wellman et al. 2003). These fossils include not just isolated spores but also tissue fragments of the sporangia, the structures that held the spores together in life. Some of these fossil sporangia hold as many as 7,450 spore tetrads together, and the shape of the spherical sporangia fragments suggest that they could have held as many as 95,000 spore tetrads in the living plant.

The ability to produce large numbers of spores is considered to be an adaptation to living in harsh terrestrial environments and, apart from the terrestrial strata in which they are found, is further evidence that these fossils were from land plants. Finally, the microscopic ultrastructure of the walls of the spores reveals parallel-arranged lamellae of a type that is found only in the spores of living liverworts, providing further evidence that the plants that produced them were indeed liverworts, the most primitive of the land plants (Wellman et al. 2003; Wellman 2010). Today, these simple liverwort land plants resemble lichen, the terrestrial algal-fungal symbionts that are not really plants, and like lichen, they can encrust rocks or tree trunks.

The liverworts have no roots, no vascular system, no true stomata (these plant structures will be discussed shortly), and are confined to humid environments. The next step in the evolution of land plants was the evolution of the Stomatophyta (table 3.1), plants so named because they possess stomata to regulate gas exchange in dry air. Stomata are openings in the wall tissues of the plant that can be opened and closed by two guard cells; they allow the plant to regulate gas and water vapor exchange between it and the atmosphere around it. It is critical to a land plant (as opposed to a water plant) to prevent as much water loss as possible to the surrounding dry atmosphere of the land regions, and the stomata are key anti-dehydration structures in this effort. The most primitive or least derived members of the stomatophyte clade are the anthocerophytes, the hornworts, and fossil *Stomatophytes* hornwort spores demonstrate that the stomatophytes had evolved by the Katian Age of the Late Ordovician, about 453 million years ago (Lecointre and Le Guyader 2006).

The bryophytes, the mosses, are more advanced stomatophytes of the hemitracheophyte clade (table 3.1). Both the hornworts and the mosses have upright sporophytes; that is, the sporangium is held up in the air on a thin vertical stem. Upright sporophytes are another adaptation to life in dry air, but it is debated whether this adaptation evolved once in the hornworts and was simply inherited by the mosses, or whether the mosses independently evolved upright sporophytes and thus this adaptation is convergent in the two groups (Lecointre and Le Guyader 2006; for an extensive discussion of convergent evolution in plants, see McGhee 2011). The next step in the evolution of land plants was the evolution of plants with sporophytes that were branched—that is, vertical stems that supported more than one sporangium. These plants are the Polysporangiophyta (table 3.1), which evolved in the Early Silurian. The most primitive polysporan-giophytes were the now extinct horneophytes and include the fossil species *Horneophyton lignieri* and *Aglaophyton major* from the Early Devonian Emsian Age (for more information about these fossil species, see McGhee 2013).

Following the evolution of plants with erect, multibranched stems, the next step in the evolution of land plants was the evolution of the vascular plants—that is, plants possessing water-conducting tubes that transport water up the stem against the force of gravity. These tubes are called tracheids, hence the name Tracheophyta (table 3.1) for these plants. The tracheophytes also evolved in the Early Silurian, and the most primitive members of the tracheophyte clade are another extinct group, the rhyniopsids, such as the fossil species *Rhynia gwynne-vaughanii* from the Early Devonian Pragian Age (for more information about these fossil species, see McGhee 2013).

The basal, leafless tracheophytes gave rise to the more derived Eutracheophyta, the first land plants to evolve leaf-like structures. The eutracheophytes are divided into two major clades: the microphyll-leafed vascular plants, the Lycophyta, which have numerous tiny leaves; and the megaphyll-leafed vascular plants, the Euphyllophyta, which have less numerous big leaves (table 3.1). The clade of the tiny-leafed lycophyte plants includes modern-day small club mosses, spike mosses, and quillworts. It also includes the gigantic Carboniferous trees of the Late Paleozoic Ice Age. The evolutionary split between the lycophytes and the euphyllophytes appears to have occurred by the Middle Silurian Homerian Age, as fossils of the basal lycophyte species *Cooksonia pertoni* and *Baragwanathia longifolia* are found in strata of that age (for more information about these fossil species, see McGhee 2013).

The clade of the large-leafed plants, the euphyllophytes, includes most of the land plants that are familiar to us in our modern world. This large clade is itself divided into two major subclades: the Moniliformopses and the Lignophyta (table 3.1). The clade of the moniliformopses includes the cladoxylopsids, an extinct group of fernlike plants; the equisetophytes, the modern-day horsetails; and the filicophytes, the modern-day true ferns. All of these plants still reproduce by spores and have very little, if any, woody tissue.

The other clade of the euphyllophytes, the Lignophyta, are known as the woody plants as they contain substantial amounts of wood tissue. The lignophytes also evolved "bifacial cambium," an advanced trait that will be discussed in more detail below. The most primitive or least derived members of the ligno-phyte clade, the Aneurophytales and the Archaeopteridales (table 3.1), still repro-duced by spores, and both groups of plants are now extinct. The more derived lignophytes, the Spermatophyta, evolved the first seeds in the Late Devonian Famennian Age. The evolution of seed-reproducing plants was a major inno-vation in the geologic history of plants; the implications of this evolutionary innovation in reproductive type will be discussed in more detail below. The last major innovation in plant evolution was the evolution of the flowering plants, the Angiospermae (table 3.1), but that did not occur until the early Jurassic in the Mesozoic Era. Life in the ancient Paleozoic world saw no flowers.

Now let us consider the evolution of trees. Trees are a characteristic adap-tation of plants to life on land (McGhee 2011, 2013). Simple plant growth in a two-dimensional plane soon leads to crowding and overgrowth, with one plant shading out another in the competition for light from the sun, the energy source for plant survival. When crowding occurs in the two-dimensional plane of the land surface, the solution is to move into the third dimension above the land surface, to evolve plants with single vertical stems, then branched stems, and then branched stems with leaves—hence the evolution of the Stomatophyta, the Polysporangiophyta, and the Euphyllophyta (table 3.1).

Even now, however, crowding will again become a problem when these types of plants cover the surface of the land. And, just as in human cities when crowd-ing occurs in a region of low buildings, the solution is to move even higher into the third dimension above the land surface—to construct towering skyscraper buildings or, in the case of plants, to evolve trees. The force of gravity has to be reckoned with, and new support structures had to be evolved in order to create a massive central structure, a tree trunk, that rises vertically from the land surface. At some distance above the ground, branches extend out from the tree trunk in order to capture as much sunlight as possible for the survival of the tree and, in order to ensure the survival of the species, to facilitate fertilization and dispersal of the tree's offspring. All of these structures are heavy, and the tree trunk must be strong enough to support them without breaking or bending. Given these difficulties, one might think that only one advanced clade of plants would suc-cessfully evolve a tree. Thus it is astonishing that not one, but no less than nine separate phylogenetic lineages of plants independently, convergently, evolved the tree form, each with its own different solution to the problem of growing a sufficiently strong central trunk structure (Niklas 1997; McGhee 2011). The lycophytes, cladoxylopsid moniliformopses, and lignophytes all separately evolved the tree form in the Middle Devonian Givetian Age. Then the equise-tophyte and filicophyte moniliformopses independently evolved the tree form in the Late Devonian and Early Carboniferous. The other four convergently evolved tree forms occurred later in geologic time, after the end of the Paleozoic Era (twice more in the filicophytes, and twice more in the lignophytes) (Niklas 1997; McGhee 2011).

The oldest tree fossils yet known are of the extinct cladoxylopsid *Wattieza* (*Eospermatopteris*) *erianus* trees (table 3.1), found in the famous Gilboa fossil forest in New York State (Stein et al. 2007). These ancient trees stood eight meters (26 feet) tall and looked like thin, tall poles topped with a shaving brush (fig. 3.1)! They had no horizontal branches and no woody tissue—in contrast, the trunk of the tree was probably hollow like a reed. More significantly, down in the understory of the Gilboa forest were other smaller, thin, pole-like trees covered with numerous tiny leaves—the first lycophyte trees, ancestors of the giant Carboniferous trees (Stein et al. 2012).

The Gilboa fossil forest is of Givetian Age in the Middle Devonian, and in the later Givetian, the woody lignophytes also convergently evolved a tree form—the archaeopteridalean *Archaeopteris* (table 3.1). An *Archaeopteris* tree was much more like the trees that are familiar to us in the modern world (fig. 3.2) in that it had a trunk with a core of heartwood, and it had numerous horizontal branches with leaves. However, it was unlike any modern lignophyte trees in that it still reproduced with spores, not seeds. Unlike the short-lived *Wattieza*-type cladoxylopsid trees, the *Archaeopteris* spore trees could tower 30 meters (100 feet) into the sky. *Archaeopteris* trees became very numerous in the Late Devonian, and by the middle Frasnian Age vast areas of the Earth were covered with *Archaeopteris* forests. Then, mysteriously, all of the *Archaeopteris* trees died out in the end-Devonian biodiversity crisis (McGhee 2013).

In the Early Carboniferous Tournaisian Gap (table 2.1), following the End-Famennian Bottleneck, the tallest trees on the Earth were only two meters (6.6 feet) high (DiMichele and Hook 1992; McGhee 2013). The lignophytes had not lost the ability to produce trees, however, and in the later Visean Age of the Early Carboniferous, the first of the lignophyte seed-fern trees evolved (table 3.1), such as the 40-meter-tall (131-foot-tall) *Pitus primaeva*, discussed

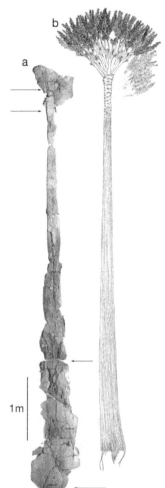

FIGURE 3.1 The extinct cladoxylopsid-lineage fernlike tree *Wattieza (Eospermatopteris) erianus*: (a) the reassembled tree fossils; (b) a reconstruction of the living tree. The living tree was about eight to ten meters (26 to 33 feet) tall.

Source: From Macmillan Publishers Ltd: *Nature* (Stein et al., 2007), copyright © 2007. Reprinted with permission.

in chapter 2. The lycophytes also would begin to produce large trees—the gigantic tropical scale trees—starting in the Serpukhovian Age of the Early Carboniferous (table 2.3).

Having summarized the evolutionary events that led to the origin of trees on Earth, we now turn specifically to the types of trees that were present in the great rainforests of the Carboniferous. Five types of trees were dominant: the lycophyte scale trees, the equisetophyte horsetail trees, the filicophyte tree ferns,

FIGURE 3.2 The extinct archaeopterid-lineage spore tree *Archaeopteris* was about 30 meters (100 feet) tall, here compared to a four-meter- (13-foot-) tall male African elephant.

Source: Illustration by Mary Persis Williams.

the medullosan seed-fern trees, and the cordaitean pinophyte trees (table 3.1) (Pfefferkorn et al. 2008). In that order, let us examine each of these five types of tree in detail. Only the last two tree types were seed plants with woody core tissue like our modern trees—and the first two were unlike any trees alive on Earth today.

The Lycophyte Scale Trees

The strangest trees in Carboniferous rainforests—from our modern perspective—were the towering, pole-like lepidodendrid lycophytes (fig. 3.3). First, their gigantic size is difficult to understand. The lycophytes are still alive today, but they are quite small plants. A fairly large modern lycophyte, the stag's horn club moss *Lycopodium clavatum*, stands only 12 centimeters (five inches) high (fig. 2.4). In contrast, an average *Lepidodendron* tree could tower 50 meters (160 feet) into the sky (Cleal and Thomas 2005). How could a lycophyte plant in the Carboniferous grow to be over 400 times taller than any living lycophyte on the Earth today?

Second, the lycophyte trees had determinate growth, a mode of plant growth that is unusual in plants in our modern world. Consider, for example, the mode of growth of a modern dandelion weed, *Taraxacum officinale*, in your lawn. For many days the dandelion plant has bladelike leaves that are low to the ground, and eventually it develops a bright yellow flower that is also located low to the ground. Then, suddenly, the plant shoots up a long stem as much as 15 centimeters (six inches) high, crowned with a fuzzy white globe of tufted seeds. This is the seed-dispersal phase in the reproductive cycle of the dandelion in which it elevates its seeds above ground level in order to more efficiently expose the seeds to wind currents that will carry the tufted seeds considerable distances away from the parent plant.

The mode of growth of the *Lepidodendron* tree was even stranger than that of the modern dandelion: after it reproduced, it died. Like a young dandelion plant, the young *Lepidodendron* tree grew low to the ground—but for many years, not just for a few weeks. During this time it resembled a low stump, but it was green rather than bark-brown as it was covered with the tiny scalelike leaves of the lycophytes. Below ground, however, it was constantly expanding its peculiar rootlike stigmarian system: in *Lepidodendron* trees even the stigmarian rootlets could photosynthesize, and they often protruded above ground in order to catch sunlight.

When the *Lepidodendron* tree reached its reproductive phase, similar to the development of a stem by a dandelion plant, it began to grow a pole-like trunk that stretched into the sky (fig. 3.3). This trunk could reach heights of 50 meters (160 feet) but was quite slender, only about one meter (3.3 feet) in diameter—in

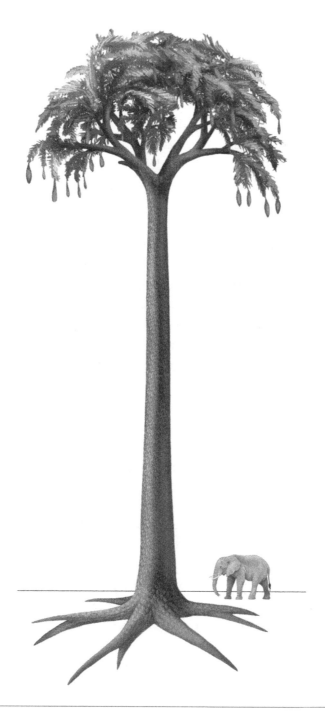

FIGURE 3.3 The lycophyte scale tree *Lepidodendron* was about 50 meters (160 feet) tall, here compared to a four-meter- (13-foot-) tall male African elephant.

Source: Illustration by Mary Persis Williams.

essence a thin, towering pole covered with tiny, green, scalelike leaves. As the trunk grew taller and taller, it would begin to shed the leaves in its lower reaches, leaving only the leaf-cushion bases while thickening the trunk. Fossil trunks of these trees are covered with leaf-cushion bases that look like the pattern of scales seen on the body of a snake, hence these trees are often given the common name of "scale trees." But the scalelike leaf-cushion bases could still photosynthesize—even without leaves—and hence produce food locally on the trunk.

Only when the trunk reached its maximum height would it then begin to produce branches in pairs, drooping on either side of the central trunk, and to develop cones on the branches. Like the upper trunk, these branches were covered with sleeves of leaves, but they were smaller than those found on the trunk. The *Lepidodendron* tree would then rapidly produce spores in its cones, which would be dispersed by winds at a height of 50 meters above ground level. The tree's microspores were small and lightweight; thus they could easily be carried by the wind and did not need structures like the tufts of the much larger dandelion seed, which increase the surface area available to catch wind current for support in transporting the seed through the air. After its spores were all shed, the *Lepidodendron* tree would die. The towering dead trunks of the *Lepidodendron* trees would remain standing for a while before they began to rot and then, one by one, the great trunks would fall to the swamp waters below.

Astonishingly, the trunks of the gigantic lycophyte trees were not supported by woody tissue like our modern giant trees, such as the giant redwood *Sequoiadendron giganteum*, which can stand 90 meters (300 feet) high. Most modern trees belong to the clade of the lignophytes, the woody plants (table 3.1). Woody trees possess a bifacial cambium, with secondary xylem tissue toward the center of the trunk and secondary phloem tissue toward the outside (Donoghue 2005). The secondary phloem of a modern woody tree transports food produced by the photosynthesizing leaves in the crown of the tree down to the trunk and underground root regions, and the woody tissue supports the mass of the trunk, branches, and photosynthetic needles or leaves against the downward pull of gravity.

In contrast, the trunks of the lycophyte trees were constructed in a most peculiar manner. They had only a unifacial cambium with no secondary phloem. Without secondary phloem, the tree could not transport food and nutrients for long distances; thus the tree developed photosynthesizing structures—tiny leaves and even the bases holding the leaves—all over its branches and trunk to provide living tissues with food that was locally produced. Even the peculiar stigmarian rootlike tissues of the tree could photosynthesize food. Thus, unlike modern woody trees with food production confined to the crown of the tree with its photosynthesizing leaves, the lycophyte tree produced food over just about its entire surface (Donoghue 2005).

The unusual growth mode of the trunk of the lycophyte tree was probably due to the absence of secondary phloem and lack of sufficient woody tissue to support the trunk against the force of gravity. The weight of the tree was held up by an outer layer, called the periderm, of cortex and leaf bases—in essence, the external bark of the tree, rather than an internal wood core, was its structural support. The trunk of the tree was produced only when it was needed, in the reproductive phase, when the spores were developed high above the surface of the Earth to be dispersed by wind. Once the spores of the tree were dispersed, the trunk was no longer needed, and it died and fell to the Earth. In contrast, modern-day woody trees go through decades of reproductive cycles before dying.

The Equisetophyte Horsetail Trees

The second strange type of trees present in the Carboniferous rainforests that do not exist today were the equisetophyte trees (fig. 3.4). Like the lycophytes, equisetophyte plants still exist today, but most are quite small. The modern field horsetail *Equisetum arvense* stands 25 centimeters (ten inches) high and is abundant in many moist environments today, such as roadside ditches (Lecointre and Le Guyader 2006). In contrast, Carboniferous horsetail trees were gigantic— some *Calamites* trees stood 20 meters (65 feet) tall—80 times taller than our modern small horsetail rushes (fig. 3.4) (Taylor and Taylor 1993).

Although the calamitean horsetail trees belonged to the clade of the euphyllophytes (table 3.1), they still possessed microphylls—small leaves that were whorled about jointed branches attached to trunks that were also jointed. These jointed trunks grew upwards from large rhizomes buried in the soil, rather than from rootlike stigmarian systems as in the lycophyte trees (fig. 3.4). Like the modern-day field horsetail, these underground rhizomes had rootlet clusters at the nodes where the erect stems, or trunks as in the case of *Calamites*, were produced.[2]

Like the lycophyte trees, the ancient equisetophyte trees had a unifacial cambium and no secondary phloem; also like the peculiar lycophyte trees, they possessed the unusual trait of determinate growth (Donoghue 2005). Unlike the

FIGURE 3.4 The equisetophyte horsetail tree *Calamites* was about 20 meters (65 feet) in height, here compared to a four-meter- (13-foot-) tall male African elephant.

Source: Illustration by Mary Persis Williams.

lycophyte trees, the segmented trunks of the equisetophyte trees were structurally supported by internal wedges of wood (Niklas 1997). Although the equisetophyte trees possessed more woody support than the lycophyte trees, the inward-pointing wedges of wood that were arranged around the circular interior of the trunk of the equisetophyte trees were very different from the massive amount of heartwood and sapwood found in the core of the trunk of a lignophyte tree.

The Filicophyte Fern Trees

The third major type of tree found in Carboniferous forests, the tree ferns, are not so strange to us as the extinct lycophyte and equisetophyte trees because some tree ferns, such as the Australian tree fern *Cyathea cooperi*, still exist today. However, most living ferns are small herbaceous plants, or more rarely shrub sized. They are easily recognizable by their characteristic large fronds of leaves that unfurl from fiddlehead-shaped spirals and the clusters of dark spores that periodically develop on the undersides of their leaves. In contrast, the Carboniferous tree ferns were large: the marattialean tree fern *Psaronius* (fig. 3.5) stood ten meters (33 feet) high (Taylor and Taylor 1993).

The true ferns, the filicophytes (table 3.1), reproduce by spores and are not lignophyte plants. Thus the living tree ferns still exhibit those traits that

FIGURE 3.5 The marattialean true tree fern *Psaronius* was about ten meters (33 feet) tall, here compared to a four-meter- (13-foot-) tall male African elephant.

Source: Illustration by Mary Persis Williams.

they share with their extinct Carboniferous relatives, the equisetophyte trees. Unlike the equisetophytes, which evolved only a single type of tree, the filico-phytes have convergently evolved tree forms no less than three separate times: *Psaronius*-type tree ferns in the Carboniferous, *Tempskya*-type tree ferns in the Cretaceous, and the tree ferns that survive today (McGhee 2011). Each group invented a different type of trunk in their evolution: the Carboniferous tree ferns had trunks supported by an outer mantle of adventitious roots; the Creta-ceous tree ferns had trunks consisting of interwoven stems bound together by adventitious roots; and living tree ferns have a lower columnar base of adventi-tious roots and an upper trunk supported by an outer layer of cortex and exter-nal layer of leaf bases (Niklas 1997). Thus all three types of tree ferns construct their trunks in a very different manner than either the lycophyte or equiseto-phyte trees: they modify their root-growth systems for usage as support struc-tures above ground.

The Medullosan Seed-Fern Trees

Only two of the five major tree types found in the great Carboniferous rain-forests were spermatophyte lignophytes—seed plants with woody core tissue like our modern trees. These were the seed-fern trees and the pinophyte conifer trees (table 3.1). The evolution of the lignophyte plants in the Middle Devonian introduced a typical tree-trunk construction—having a structural support core of heartwood in the center of the trunk—that is very familiar to us in the mod-ern world, but it could be argued that this construction is less mechanically effi-cient than the tree-trunk construction of the ancient non-lignophyte lycophyte, equisetophyte, and filicophyte trees in the Carboniferous forests. Karl Niklas, a paleobotanist at Cornell University, notes: "Engineers have long known that the best location for mechanically supportive materials is just beneath the surface of a vertical column, where mechanical bending and torsional forces reach their maximum intensities. This strategy is particularly important when the quantity of the stiffest building material is limited, perhaps for economy in design. It is no coincidence that the stiffest tissues in a variety of plants tend to develop just beneath or very near the external surface of vertical stems . . . the stiff lycopod periderm produced by the lepidodendrids . . . and the pertinacious mantle of adventitious roots around the stems of *Psaronius* are but a few examples of this mechanical strategy" (Niklas 1997, 331–333).

On the other hand, the evolution of bifacial cambium—one in which both secondary xylem and phloem tissues are produced—in the lignophytes was an adaptive innovation (Donoghue 2005). The secondary phloem in lignophytes makes possible the long-distance transport of food from the leaves high up in the crown of the tree down to the roots located far underground at the other end of the tree. The lycophyte and equisetophyte trees had a unifacial cambium and did not possess secondary phloem (Donoghue 2005). The growth consequences of possessing only a unifacial cambium for the lycophyte and equisetophyte trees have been noted previously.

The evolution of the seed plants, the spermatophyte lignophytes, in the Late Devonian Famennian Age was a major adaptive innovation in the geological history of plants. The key adaptation of the seed plants is their freedom from needing water in reproduction, unlike the spore-reproducing plants, and the spermatophytes could now colonize the dry highlands and mountains that were out of reach for non-spermatophyte plants. The evolution of the seed in plants was the ecological equivalent of the evolution of the amniote egg in vertebrates (Niklas 1997), a key adaptation that also freed the synapsid and reptilian amniotes (ancestors of modern-day mammals and reptiles, table 2.4) from needing water in their reproduction. These two groups, the seed plants and the amniote vertebrates, would become the victorious plant and vertebrate conquerors of the terrestrial realm of the Earth in the Carboniferous (McGhee 2013).

The ancient seed-fern trees were not true ferns (table 3.1), the spore-reproducing filicophytes (fig. 3.5), but rather were seed-reproducing plants that possessed foliage that looked very similar to that of a true fern. Seed ferns were also not a monophyletic clade in that three separate clades of spermatophyte lignophytes are often called "seed ferns"—the lyginopteridales, the medullosales, and the gigantopteridales (table 3.1)—all three of them extinct. The lyginopterids produced some fairly large seed-fern trees in the Early Carboniferous, such as *Pitus*, but the most important seed-fern trees in the Late Carboniferous were the medullosans, trees such as *Medullosa* and *Sutcliffia*. *Medullosa* was a small seed-fern tree that stood only about five meters (16 feet) tall, about half as tall as the true fern tree *Psaronius* (fig. 3.5). Smaller seed ferns like *Neuralethopteris* and *Alethopteris* also contributed to ground cover in the Carboniferous rainforests, as will be discussed shortly.

The Cordaitean Conifer Trees

The only other modern-type lignophyte trees found in the great Carboniferous rainforests were the pinophyte conifer trees (table 3.1). It is in the clade of the pinophytes that we finally find modern trees that are even more gigantic than the Carboniferous giants. The giant redwood trees of western North America, such as *Sequoiadendron giganteum*, can reach a height of 90 meters (300 feet)—almost twice the 50-meter (160-foot) height of the peculiar Carboniferous *Lepidodendron* scale trees.

The important conifer trees in the Late Carboniferous were the cordaitean pinophytes, trees such as *Cordaites* and *Walchia*. *Cordaites* was a small conifer tree that stood only about five meters (16 feet) tall, about half as tall as *Psaronis* (fig. 3.5).[3] Some cordaitean trees had a tree-on-stilts appearance much like a modern mangrove tree, such as *Rhizophora mangle*, in that the trunk of the tree was elevated above ground by an extensive root system, the upper part of which was subaerial. Also like a modern mangrove tree, *Cordaites* trees are thought to have lived in wet environments on the margins of swamps and in coastal areas.

The Peculiar Carboniferous Rainforests

In summary, walking or boating in most of the great Carboniferous rainforests of the ancient Earth would have been quite a different experience than walking or boating in a modern rainforest. In a modern rainforest, the forest floor is in perpetual gloom as sunlight is shaded out by the canopy of leaves of the giant woody seed trees. A distinct canopy ecosystem exists up in the air of the modern rainforest, an ecosystem inhabited by vertebrate and arthropod carnivores and herbivores that live their lives in the sunlit canopy of leaves high above the Earth's surface.

In contrast, in most of the Carboniferous rainforests the forest floor was illuminated, not shrouded in gloom, as the towering lycophyte and equisetophyte spore trees did not have large leaves that would shade out the forest floor below (fig. 3.6). The lycophytes also developed their canopies of branches only in the reproductive phase of their life cycle, so that any shading effect produced by

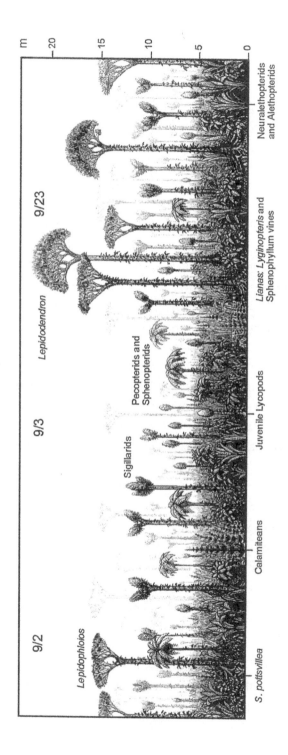

FIGURE 3.6 In situ reconstruction of a lycophyte tropical rainforest of Bashkirian Age, early Late Carboniferous, in present-day Alabama in North America. Characteristic flora are labeled, and the vertical scale in meters (right margin of the figure) gives the height of the plants in the foreground; see text for discussion.

Source: From Gastaldo, R. A., I. M. Stevanović-Walls, W. N. Ware, and S. F. Greb, "Community Heterogeneity of Early Pennsylvanian Peat Mires," *Geology* vol. 32, pp. 693–696, copyright © 2004 Geological Society of America. Reprinted with permission.

their canopies of branches existed only for a fraction of the lifetime of the tree. Thus, much of the time, the rainforest may have had the peculiar appearance of being filled with many tall, pod-topped green poles without branches.

Figure 3.6 illustrates a painstaking, labor-intensive reconstruction of a Late Carboniferous, Bashkirian Age, North American rainforest in Alabama (Gastaldo et al. 2004). The Colby College paleobotanist Robert Gastaldo and his colleagues descended into old coal mines in Alabama and took a careful census of in situ trees and plants embedded in the roofs of the mines, located immediately above the mined-out coal layer. This sampling procedure not only allowed the paleobotanists to census the species composition of the rainforest plants that existed at a single instant in time, but it also allowed them to spatially map the geographic distribution of the plants within the rainforest.

Rainforest reconstructions from three of their 17 field sites are shown in figure 3.6, labeled "9/2", "9/3", and "9/23". Site 9/3, in the middle of the figure, shows the rainforest region with the least number of trees in canopy stage and greatest number of juvenile pole-like trees (94 percent). Taller *Sigillaria* lycophyte scale trees and lower *Psaronis* Marattialean tree ferns (*Pecopteris* leaf fossils) and *Sphenopteris* seed-fern trees are labeled in this region of the figure. Site 9/2, on the left in the figure, shows the rainforest with taller, mature *Lepidophloios* scale trees that have gone into canopy stage, with branches at the top of the pole-like trunk that itself is covered with a sleeve of leaves. Below the *Lepidophloios* and *Sigillaria* lycophyte trees are shown lower *Calamites* equisetophyte horsetail trees and a ground cover of *Sphenopteris pottsvillea* seed ferns. On the right in figure 3.6, at site 9/23, the rainforest is shown with a 53 percent canopy cover with still taller, mature *Lepidodendron* lycophyte trees—note again the fuzzy appearance of the trunk of the tree, which is covered in a sleeve of leaves. Below the canopy of *Lepidodendron* and *Lepidophloios* lycophyte trees—a canopy that covers only half of the sky in the rainforest— vinelike *Lyginopteris* seed ferns and *Sphenophyllum* equisetophytes are shown on the left, and ground cover *Neuralethopteris* and *Alethopteris* seed ferns are shown on the right.

This Alabama rainforest was overwhelmingly dominated by juvenile trees, with 95 percent juveniles in the youngest forests and 89 percent juveniles in the field sites with the greatest number of mature lycophyte trees. The youthfulness of the forest may have been a function of its origin: the forest formed when the mire basin subsided, water and sediment flooded in, and the new trees took root directly on top of the older, buried forest debris that would become the massive coal bed that was mined out by humans some 320 million years later. The lycophyte trees in this Alabama rainforest are also smaller than those in many other lycophyte-dominated forests, as the *Lepidodendron* scale trees are only 23 meters (75 feet) high—in other rainforests, these lycophyte trees could be twice that height. Unlike in a modern rainforest, there was no vertical stratification of specific lycophyte and equisetophyte tree species adapted to capturing light at different height zones in the ancient Carboniferous rainforests (Cleal 2010). Given their determinate mode of growth, the differing heights of the lycophyte and equisetophyte trees within the forest reflected the differing ages of the trees instead (fig. 3.6).

It has generally been argued that geographic variation in the composition of the lycophyte and equisetophyte trees species within the Carboniferous rainforests were a function of the areal distribution of different soil types and conditions such as waterlogged regions, muddy-margin regions along riverbanks, better-drained regions between rivers, dry upland regions, and so on (Cleal 2010). In particular, it has been argued that *Lepidophloios* trees preferred waterlogged regions, *Lepidodendron* trees preferred better-drained regions, and *Sigillaria* trees preferred drier upland regions. In contrast, the Alabama rainforest contained species of all three of these lycophyte tree genera coexisting in the same area (fig. 3.6). Gastaldo and colleagues suggest that the previously argued soil-moisture gradient for lycophyte tree species could be a sampling artifact, as previous reconstructions of Carboniferous rainforests were made from coal-ball assemblages—assemblages that are time averaged, with floras that "represent the resistant biomass contribution from several plant generations to the peat," so that "resistant plant parts that accumulate and are buried may represent a century or more of biomass concentration in any one coal ball" (Gastaldo et al. 2004, 693). Alternatively, Gastaldo and colleagues suggest that the absence of a soil-moisture gradient in lycophyte tree distributions in the Alabama rainforest may be a function of the early geologic age of the forest in the Bashkirian, at the dawn of the Late Carboniferous. They suggest that a soil-moisture preference by the lycophyte trees may have evolved later, a result of "increasing habitat specialization as peat mires became more extensive in the Late Pennsylvanian" (Gastaldo et al. 2004, 696).

THE SPREAD OF THE GREAT RAINFORESTS

The lycophytes first evolved tree forms in the Middle Devonian Givetian Age, but it was not until the Early Carboniferous Serpukhovian Age that the first of the lycophyte-dominated tropical rainforests formed (table 2.3). The most extensive of these early tropical rainforests formed in what is now central Asia and northwest China, with smaller rainforests scattered in northern Europe and in North America. Best estimates of the original geographic extent of these rainforests indicate that they probably covered almost a half-million square kilometers (almost 200,000 square miles) of the tropical region of the Early Carboniferous Earth (table 3.2). In the Late Carboniferous Bashkirian and Moscovian Ages, the lycophyte-dominated rainforests progressively expanded in the Earth's tropical region until they reached a maximum geographic extent of almost 2.5 million square kilometers (almost a million square miles; table 3.2) in the late Moscovian (fig. 3.7) (Cleal and Thomas 2005).

TABLE 3.2 Climatic events, area of rainforest coverage, and biological events during the Carboniferous phase of the Late Paleozoic Ice Age.

Age	Geologic Time (Ma)	Climatic Events (1)	(2)	(3)	Areal Extent of Rainforests	Biological Events
(Permian)	298	ICE	ICE-P1		1,255,000 km²	← Recovery in the rainforests
GZHELIAN	299				1,087,000 km²	
	300				1,095,000 km²	
	301				1,110,000 km²	
	302				1,111,000 km²	
	303				1,120,000 km²	
KASIMOVIAN	304				1,131,000 km²	
	305				1,131,000 km²	
	306				1,131,000 km²	
	307				1,131,000 km²	← Extinction in the rainforests

		(1)	(2)	(3)		
MOSCOVIAN	308		ICE		2,395,000 km²	
	309		ICE		2,170,000 km²	
	310		ICE		1,945,000 km²	
	311		ICE	ICE?	1,721,000 km²	
	312		ICE	ICE?	1,840,000 km²	
	313		ICE	ICE?	1,822,000 km²	
	314		ICE	ICE?	1,804,000 km²	
	315		ICE	ICE?	1,786,000 km²	
BASHKIRIAN	316	ICE	ICE-C4	ICE?	1,456,000 km²	
	317	ICE			1,258,000 km²	
	318	ICE	ICE		1,060,000 km²	
	319	ICE	ICE		826,000 km²	
	320	ICE	ICE		664,000 km²	
	321	ICE	ICE-C3		467,000 km²	
	322	ICE			467,000 km²	
	323	ICE			467,000 km²	
SERPUKHOVIAN	324	ICE	ICE	ICE?	450,000 km²	
	325	ICE	ICE	ICE?	450,000 km²	
	326	ICE	ICE-C2	Cold	450,000 km²	← First rainforests
	327	ICE				
	328	ICE				
	329	ICE	ICE			
	330	ICE	ICE-C1			
VISEAN (pars)	331					
	332					

Source: Gondwana data from Isbell et al. (2003), Frank et al. (2008), and Fielding, Frank, Birgenheier et al. (2008); Siberian data from Epshteyn (1981a), Chumakov (1994), and Raymond and Metz (2004). Rainforest data are from Cleal and Thomas (2005); timescale modified from Gradstein et al. (2012).

Note: Climatic events column (1) is West and Central Gondwana and column (2) is East Gondwana (Australia), both in the Southern Hemisphere; column (3) is Siberia (northeastern Russia) in the Northern Hemisphere. Bold type indicates the existence of continental glaciers, and normal type indicates the presence of alpine glaciers; glaciations in eastern Gondwana are designated C1 through C4, from oldest to youngest.

In the late Moscovian Age, huge expanses of lycophyte-dominated tropical mires stretched across the continental-interior and Appalachian-basin regions of the present-day United States (fig. 3.7). To the east, lycophyte rainforests stretched continuously from eastern Canada in North America across southern Ireland, through Wales and England, across northern Europe, to vast

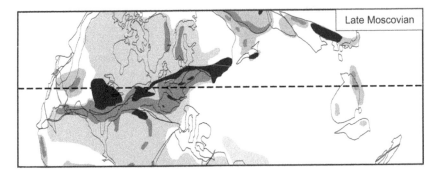

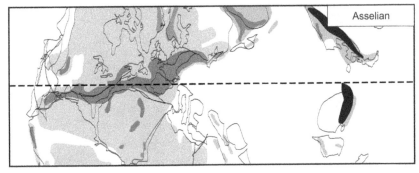

FIGURE 3.7 Paleogeographic distribution of lycophyte rainforests in the Late Carboniferous late Moscovian Age (upper figure) contrasted with the Early Permian Asselian Age (lower figure). Rainforest areas are shown in black, highland areas are shown in dark gray, lowland areas are shown in light gray, and areas under marine waters are shown in white; see text for discussion.

Source: Thanks to Dr. Christopher Cleal, Department of Biodiversity and Systematic Biology, National Museums and Galleries of Wales, for creating this figure for this book (modified from Cleal and Thomas, 2005).

tropical mires in Kazakhstan. Along the way, large isolated basins filled with lycophyte-dominated mires existed in northern Africa, Spain, and middle Europe. Further to the east, lycophyte rainforests existed in basins in western Asia and in vast mires on the Sino-Korean continental block (present-day Korea and North China; fig. 3.7).

The spread of the great rainforests is clearly correlated with the onset of the major Carboniferous phase of the Late Paleozoic Ice Age. Massive continental ice sheets formed first in western and central Gondwana in the early Serpukhovian (glacial phase GII of Frank et al. 2008), with alpine ice sheets in the highlands of eastern Gondwana (table 3.2) (glacial phase C1 of Fielding, Frank, Birgenheier et al. 2008). By the late Serpukhovian, lowland continental ice sheets had formed in eastern Gondwana (glacial phase C2 of Fielding, Frank, Birgenheier et al. 2008), and the northern hemisphere of the Earth had become glaciated as well, with a polar sea-ice cap in Laurasia (off modern-day Siberia, northeastern Russia) (Epshteyn 1981a; Frakes et al. 1992; Stanley and Powell 2003; Raymond and Metz 2004).

Why was the spread of the great rainforests correlated with the onset of the massive-ice-formation phase of the Carboniferous glaciation? The huge accumulation of frozen water on land in massive glaciers had the effect of lowering global sea levels, and as the seas drained off of the land, vast areas of once flooded shallow marine shelf were exposed to subaerial conditions in the tropics. Thus vast areas of flat, miry, wetland habitats were created, in what the Kentucky Geological Survey geologist Stephen Greb and colleagues call the "largest tropical peat mires in Earth history" (Greb et al. 2003, 127)—perfect habitats for the lycophyte trees in particular.

Not only did the falling sea level expose vast areas of land that was once under water, it also had the effect of lowering the base level of erosion in the uplands. Rivers that once had been near sea level were now at an elevation considerably higher than sea level, and they began to incise their valleys, cutting downward toward the new sea level. This increased erosion transported huge amounts of sediment outward over the newly exposed flatlands, creating large river deltas like our modern-day Amazon delta—again, perfect habitats for the lycophyte trees in particular.

Plants are photoautotrophic organisms; that is, they produce their own food by using energy from the sun. In this process, they extract carbon dioxide from the atmosphere, split hydrogen atoms from water molecules in their leaves, and combine the hydrogen and carbon dioxide molecules to produce hydrocarbon molecules like sugar. Once the hydrogen is split from water molecules, the oxygen in the water molecules is left over as a waste product and the plants simply dump it into the atmosphere.[4]

The spread of the great rainforests had profound consequences for the Earth's atmosphere. Huge amounts of carbon dioxide were being extracted from the atmosphere to form food for the land plants in the forests. At the same time, huge amounts of oxygen were being produced by the plants and dumped into the atmosphere. Hour after hour, day after day, year after year, for millions and millions of years, this process took place in the great rainforests. The carbon-dioxide content of the Earth's atmosphere must have declined as a result, and the oxygen content of the Earth's atmosphere must have increased—but how can we measure these changes in the atmosphere of the Earth in the Carbonifer-ous, over 300 million years ago?

THE GREAT RAINFORESTS AND ATMOSPHERIC OXYGEN

We know that the concentration of oxygen in the Earth's atmosphere was very low in the Frasnian Age of the Late Devonian but that the amount of oxygen in the atmosphere began to increase following the Famennian Gap. How do we know this? The empirical data that support this conclusion come from the distribution of charcoal deposits in the sedimentary rock record. Charcoal is produced by wildfires, and wildfires do not occur unless quite specific levels of oxygen are present in the atmosphere. Despite extensive stratigraphic searches, the Frasnian Age is known for the rarity of charcoal deposits in its strata. The absence of wildfires in the Frasnian is quite striking, given the spread of Earth's first forests during this same time interval. The geologists Andrew Scott, of the University of London, and Ian Glasspool, of the Field Museum of Natural His-tory, comment: "*Archaeopteris*, the first large woody tree, evolved in the Late Devonian and spread rapidly. By the Mid-Late Frasnian, monospecific archae-opterid forests dominated lowland areas and coastal settings over a vast geo-graphic range. Despite this extensive wood biomass, charcoal occurrences are rare with only isolated fragments of charred *Callixylon* (archaeopterid) wood reported from this interval and small amounts of inertodetrinite (microscopic charcoal fragments) preserved in early Late Devonian Canadian coals" (Scott and Glasspool 2006, 10863). In fact, the entire interval of the Eifelian Age through the Frasnian Age has been termed the "charcoal gap," as there is very lit-tle evidence of any wildfires anywhere during this period (Scott and Glasspool 2006, 10862, 10864).

In contrast, the Famennian Age is well known for numerous charcoal deposits in its strata, deposits that demonstrate that wildfires were common. The oldest Famennian charcoal deposits yet discovered come from strata in

Pennsylvania dated to about 362.7 million years ago.[5] Charcoal data from around the world indicate that frequent wildfires occurred in widely separated regions of the Earth in the last 3.5 million years of the Famennian (Scott and Glasspool 2006, fig. 1; Marynowski and Filipiak 2007; Marynowski et al. 2010). An oxygen level of at least 13 percent has to be present in the atmosphere in order for wildfires to ignite (Scott and Glasspool 2006). Thus hard empirical data exist to demonstrate that the atmosphere of the Earth contained 13 percent oxygen, or more, in the last 3.5 million years or so of the Famennian. The absence of charcoal in strata in the Famennian Gap, and in the mid to late Frasnian, can be used to argue that oxygen levels present in the Earth's atmosphere were below 13 percent during this span of time. Woody plant material was abundant in the Frasnian; if wildfires occurred, then charcoal should be present in Frasnian strata—but it is not.

In addition to the empirical charcoal data, another type of evidence supports the previous conclusion concerning oxygen levels in the Earth's atmosphere during the Late Devonian: the "empirical model," a model whose predictions depend not only on the mathematics of the model's assumptions but also on input from empirical data. The geochemist Robert Berner of Yale University and his colleagues have constructed two empirical models, Rock-Abundance and Geocarbsulf, in an attempt to predict atmospheric oxygen concentrations throughout the span of the Phanerozoic on Earth (Berner et al. 2003; Berner 2006). Both models predict that the concentration of oxygen in the Earth's atmosphere increased from the Frasnian into the Famennian—thus matching the charcoal data—and then continued to increase throughout the Carboniferous into the Permian (fig. 3.8). Specifically, both models predict minimum oxygen concentrations in the atmosphere in the early Frasnian (380 million years ago): 17 percent in the Rock-Abundance model and 13 percent in the Geocarbsulf model, and higher oxygen concentrations in the Famennian: 21–23 percent in the Rock-Abundance model and 16–17 percent in the Geocarbsulf model.

However, Scott and Glasspool urge some caution in accepting either model's predictions: "Predictions of the degree to which O_2 fluctuated [in the Late Devonian] are based on data-driven models. However, the complexity of the feedback mechanisms that govern O_2 levels results in a large degree of uncertainty in these models, as is evident in their frequent refinement. Additional data sets are invaluable to these refinements" (Scott and Glasspool 2006, 10861). They note that the

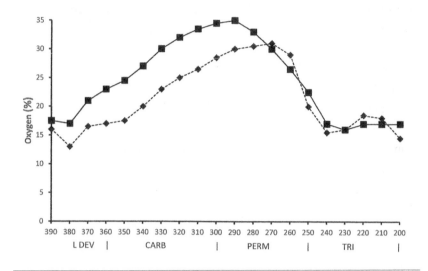

FIGURE 3.8 Modeled fluctuations in atmospheric oxygen content from the Late Devonian through the Triassic. The square data points and solid line show the predictions of the Rock-Abundance model, and the diamond data points and dotted line show the predictions of the Geocarbsulf model; see text for discussion. Geologic timescale abbreviations: L DEV, Late Devonian; CARB, Carboniferous; PERM, Permian; TRI, Triassic.

Source: Data from Berner et al. (2003) and Berner (2006).

empirical models' predictions of lower oxygen concentrations in the Frasnian atmosphere fit with the evidence of the "charcoal gap" (Scott and Glasspool 2006, 10862) in the strata of this time period. However, given the frequency of charcoal occurrences in late Famennian strata, they suggest that oxygen concentrations in the atmosphere in the late Famennian may have been higher than the 17 percent predicted by the Geocarbsulf model and more in accord with the 23 percent concentration predicted by the Rock-Abundance model. The discrepancy between the charcoal data set and the Geocarbsulf model becomes even more marked in the Carboniferous: "Collectively, these data suggest levels of O_2 modeled for this interval rising from 17 percent to 23.5 percent [in the Geocarbsulf model] are inappropriate and instead favor prior, higher levels modeled at ≈23–31.5 percent [in the Rock-Abundance model], values further supported by the occurrence of very large arthropods at this time" (Scott and Glasspool 2006,

10863). The appearance of gigantic arthropods on Earth in the Carboniferous will be considered in detail in chapter 4.

Thus we have two lines of evidence that oxygen was in short supply in the atmosphere of the Frasnian world: the charcoal-distribution data and the empirically based geochemical models. This evidence leads immediately to the question: why was oxygen in short supply? One obvious way to address this question is to consider the organisms that were producing the oxygen in the first place—the plants. As outlined in chapter 1, it is known that land plants suffered a major loss in diversity in the end-Frasnian biodiversity crisis. The number of known macrofossil genera dropped to 13, a minimum diversity level that persisted unchanged through the Famennian Gap. Taking a look at longer timescales, Anne Raymond and Cheryl Metz, paleobotanists at the University of Texas, have shown that land plants were losing diversity through the entire Middle Devonian into the Frasnian (Raymond and Metz 1995), though the loss was not as severe as the diversity loss that occurred in the late Frasnian. Land-plant macrofossil diversity was at its maximum in the Early Devonian Emsian, with 38 genera known. Diversity dropped to 31 genera in the Middle Devonian Eifelian, dropped further to 24 genera in the middle Frasnian, and then dropped precipitously to a low of 13 genera in the late Frasnian. In summary, the decline in land-plant diversity seen in the fossil record roughly parallels the decline in oxygen levels in the atmosphere predicted by the geochemical models for the Frasnian, suggesting a causal relationship between land-plant diversity and atmospheric oxygen levels.

The reverse of this argument is also true: the increase in land-plant diversity and the geographic spread of the great rainforests in the Carboniferous and their persistence in the Permian should have produced a major increase in atmospheric oxygen levels on the Earth. And indeed, the geochemical models predict that just such an increase did in fact occur (fig. 3.8). But, as Scott and Glasspool have cautioned, the two geochemical models give different predictions: the Rock-Abundance model predicts that a maximum concentration of 35 percent oxygen in the atmosphere occurred about 290 million years ago, in the Artinskian Age of the Early Permian, whereas the Geocarbsulf model predicts that a maximum concentration of 31 percent oxygen in the atmosphere occurred much later, about 270 million years ago, in the Wordian Age of the Middle Permian. More important, the Geocarbsulf model predicts that the oxygen content of the Earth's atmosphere did not increase to 30 percent or higher until the Artinskian Age of the Early Permian—atmospheric oxygen concentrations for the entire Late Carboniferous are predicted to have been below 30 percent (fig. 3.8).[6] The biological implications of these model differences in the predicted oxygen content of the Earth's atmosphere in the Carboniferous will be explored in detail in chapter 4, where we will consider the evolution of animal gigantism in the Carboniferous.

THE EARTH UP IN SMOKE?

Wildfires are common in our present-day world with its atmosphere of 21 percent oxygen—we are all familiar with scenes of firefighters battling blazes in California or Arizona in western North America, or in Spain or Greece in Mediterranean Europe. However, truly massive wildfires are confined to these generally arid regions of the Earth. In contrast, in a world with an atmosphere of 25 percent oxygen, wildfires will become widespread even in wet climatic regions (Scott and Glasspool 2006). And in a world with an atmosphere of 30 percent oxygen, almost every single tree struck by lightning during a storm will catch fire—regardless of how wet it may be. A large thunderstorm with multiple lightning strikes in a localized area would result in a raging wildfire, burning even swamp foliage down to the water level.

The Rock-Abundance model predicts that the oxygen content of the Earth's atmosphere (fig. 3.8) exceeded 25 percent around 340 million years ago in the Visean Age of the Early Carboniferous, and the Geocarbsulf model predicts a somewhat later date of 320 million years ago in the Bashkirian Age of the Late Carboniferous. Moreover, the Rock-Abundance model further predicts that oxygen in the Earth's atmosphere reached the 30 percent level around 330 million years ago, in the Serpukhovian Age of the Early Carboniferous, and remained above 30 percent until 260 million years ago, in the Late Permian.

Were wildfires almost constantly present in the great rainforests of the Late Carboniferous Earth? Two lines of empirical evidence suggest that they were. First, the University College London biochemist Nick Lane notes that over 15 percent of the volume of some Late Carboniferous coals is charcoal, "an extraordinary amount if we consider that coal beds are formed in swamps, which under modern conditions virtually never catch fire" (Lane 2002, 95).[7] The abundant presence of charcoal in Carboniferous coal strata is hard fossil evidence that the wet, tropical mires of the Earth were frequently on fire. At present atmospheric levels of 21 percent oxygen, our modern rainforests and swamps in Indonesia and Malaysia are constantly wet and virtually fire free; thus the atmosphere of the Earth must have contained much more oxygen in order for tropical mires to catch fire in the Carboniferous.

Second, Lane notes that empirical evidence exists that Late Carboniferous wildfires burned exceedingly hot compared to our present-day wildfires, a point that further explains how wet foliage could burn so frequently:

> Coals that formed during periods of hypothetically high oxygen, such as the Carboniferous and Cretaceous, contain more than twice as much charcoal as the coals that formed during low-oxygen periods like the Eocene (54 to 38 million years ago). This implies that fires raged more frequently in time of high oxygen and were not related to climate alone. Some of the properties of the charcoal support this interpretation. The shininess of charcoal depends on the temperature at which it was baked. Charcoals formed at temperatures about 400°C are shinier than those that cooked at lower temperatures, and so reflect back more of the light directed at them. The difference can be detected with great accuracy using a technique known as reflection spectroscopy. The shininess of fossil charcoals from both the Carboniferous and Cretaceous implies that they formed at searing temperatures, almost certainly above 400°C and perhaps as high as 600°C, in fires of exceptional intensity. (Lane 2002, 95–96)

Fueled by a rich oxygen atmosphere, flames burning at 400°C (752°F) to 600°C (1,112°F) could sweep through even the soggiest rainforest foliage in the Late Carboniferous. These two empirical data—the amount of charcoal present in Carboniferous coals and the high temperature at which that charcoal was formed—strongly suggest that the Rock-Abundance model for the oxygen content of the Earth's atmosphere in the Carboniferous and Permian is more accurate than the Geocarbsulf model (fig. 3.8).

The Late Carboniferous sky was probably almost always yellow-beige in color, somewhat similar to the present-day orange-beige color of the sky on Mars that we see in the numerous photographs taken by our robot rovers on the

Martian surface. The Martian sky is colored by dust; the Carboniferous sky was colored by smoke. If we could have viewed the Earth from space in the Late Carboniferous, we would have seen thick black plumes of smoke rising from numerous spots in the green band of the great rainforests, smearing out to long streaks of brown where the rising smoke plumes encountered higher-level atmospheric winds. The faint smell of smoke was probably almost constantly detectable in the air around the world, even in higher-latitude localities far away from the tropical band straddling the equator that contained the great rainforests and their numerous wildfires.

WHY DID SO MUCH COAL FORM IN THE LATE CARBONIFEROUS?

Nick Lane points out that 90 percent of the Earth's coal strata were deposited in the time span from the Serpukhovian Age in the Early Carboniferous to the Wuchiapingian Age in the Late Permian, and marvels at the "riddle posed by this 70-million-year period, which lasted from around 330 to 260 million years ago. [This means that] 90 percent of the world's coal reserves date to a period that accounts for less than 2 percent of the Earth's history. The rate of coal burial was therefore 600 times faster than the average for the rest of geological time" (Lane 2002, 84).

Why did so much coal form in the Late Carboniferous? One key to this mystery may be the type of coal that is found in these strata—namely, its unusually high charcoal content. Lane points out that "charcoal is virtually indestructible by living organisms, including bacteria. No form of organic carbon is more likely to be buried intact" (Lane 2002, 79) and that some Carboniferous coal beds "contain over 15 per cent fossil charcoal by volume. . . . The closest modern equivalents to Carboniferous coal swamps, the swamps of Indonesia and Malaysia, are almost charcoal-free" (Lane 2002, 95). Scott and Glasspool further point out that we now know that the mineral inertinite is fossil charcoal, and that inertinite is "a constituent component of many coals" (Scott and Glasspool 2006, 10861) in the Paleozoic. Thus part of the "riddle posed by this 70-million-year period, which lasted from around 330 to 260 million years ago" (Lane 2002, 84) in which the rate of coal burial was astoundingly high, may be attributable to the high oxygen content of the Earth's atmosphere during this same period of time. If the Rock-Abundance model is correct, then the oxygen content of the Earth's atmosphere was above 30 percent for the entire period from 330 to 260 million years ago (fig. 3.8). The resulting wildfires in the tropical mires of the Earth produced enormous amounts of nearly indestructible charcoal to be buried in the rock record.

A second key to the mystery of the massive amounts of coal found in Carboniferous strata may have been the general absence of wood consumers in Carboniferous ecosystems. Even if 15 percent of Carboniferous coal is near-indestructible charcoal, what accounts for the preservation of the other 85 percent that is not? Scott Richard Shaw, a paleoentomologist at the University of Wyoming, asks: "Why does most of our coal and much of our petroleum date from the Carboniferous period? The orthodox view is simply that the Carboniferous swamps provided optimal conditions for fossil fuel formation. After that time, the world became drier, and conditions were not as favorable. But is that all there is to it? Forests didn't go away after the Carboniferous. If anything, there were even more trees. . . . Something other than a change in the weather must have occurred" (Shaw 2014, 74). Shaw argues that that "something other" was the evolution of numerous wood consumers in post-Carboniferous forests: "In the Early Carboniferous, most of today's macroscopic and microscopic consumers of dead wood had not yet evolved. There were no birds, mammals, bees, wasps, bark beetles, wood-boring beetles, bark lice, termites, or ants." Thus, "the Late Devonian and Carboniferous really were special for their excess production of plant materials . . . because the plants were able to produce more biomass than the herbivores could consume, for millions of years. The first important insect wood consumers—the wood roaches—did not appear until the Late Carboniferous. They were followed by the appearance of bark lice and the diversification of wood-boring beetles in the Permian" (Shaw 2014, 75). In contrast, in the Early Carboniferous, the fungus consumers of trees were joined only by the oribatid mites and a few detritus-feeding millipedes and primitive wingless insects (Shaw 2014, 205, note 5).

A third key to the mystery of the Carboniferous coal strata might be found in the peculiar growth mode of the ancient lycophyte and equisetophyte trees. Did these trees simply produce more biomass than existing Carboniferous wood consumers could destroy, or was there another component to their excess biomass production? One possibility is that the lycophyte trees in particular had extremely fast determinate growth. It has been estimated that a 50-meter-high (160-foot-high) mature tree could be produced in ten years—a tree that would then promptly die after producing its spores (Phillips and DiMichele 1992). However, this argument has been challenged by the Stanford University paleobotanist Kevin Boyce and reconsidered by the Smithsonian paleobotanist William DiMichele, who maintain that "such rapid growth would violate all known physiological mechanisms" and that the giant lycophyte trees grew at a pace comparable to that of a modern-day palm tree (Boyce and DiMichele 2016). In any event, a single dead *Lepidodendron* tree falling down into the surrounding mire could contain as much as 3.2 tonnes (3.5 tons) of carbon (estimated by Cleal and Thomas, 2005, using the carbon-biomass estimates of Baker and DiMichele, 1997).

In a Carboniferous rainforest containing 500 to 1,800 such giant, fast-growing lycophyte trees per hectare (2.5 acres) (DiMichele et al. 2001), Cleal and Thomas calculate that these trees could extract between 160 and 578 tonnes (176 and 636 tons) of carbon from the atmosphere per year. Unlike in modern lignophyte trees, this fixed carbon did not then continue to be present in living biomass as the trees lived on for decades or even centuries. The mature lycophyte trees died in only ten years or so, and the carbon contained in the dead tree would be taken out of the living ecosystem and transferred to the swamp waters below. The water in the tropical swamps appears to have been acidic and with low fungal activity (DiMichele et al. 1985; Robinson 1990), and Cleal and Thomas calculate that only about 25 percent of the carbon in the dead trees in the mires was returned to the atmosphere by fungal consumption of the trees. Another 5 percent of the carbon was probably lost in runoff water from the swamp, resulting in a net accumulation of from 108 to 390 tonnes (119 to 429 tons) of carbon in the swamp sediments per year.

These estimates of carbon accumulation in the rock record for the Late Carboniferous rainforests are radically different from those for modern rainforests. The lignophyte trees in a modern rainforest grow slowly and live for decades or even centuries, and most of the carbon in the dead vegetation is recycled back into the atmosphere by aerobic-respiring fungal and bacterial consumers. In contrast to the estimated 108 to 390 tonnes (119 to 429 tons) of carbon stored in the sediments per year in the lycophyte-dominated rainforest, it is estimated that in a modern lignophyte-dominated rainforest, between two and ten tonnes (2.2 and 11 tons) of carbon goes into long-term sequestration per year (Cleal and Thomas 2005).

Further evidence that the peculiar growth mode of the lycophyte trees was a major factor in the accumulation of such major amounts of coal in the rock record comes from comparing the amount of coal produced by the Bashkovian-Moscovian rainforests, which were dominated by lycophyte trees, with that of the Permian rainforests, which were dominated by tree ferns. Cleal and Thomas note that both rainforests produced a similar tonnage of coal—about 200 billion tonnes (220 billion tons)—but that that amount was produced in about 11 million years by the lycophyte-dominated rainforests in the Bashkovian-Moscovian span of time, whereas it took the tree-fern-dominated rainforests about 35 million years in the Permian to produce the same amount of coal. That is, the lycophyte-dominated rainforests, with their fast-growing, determinate-growth-mode trees, produced coal at a rate three times higher than the slower-growing, non-determinate-growth-mode rainforests dominated by tree ferns (Cleal and Thomas 2005).

A fourth key to the mystery of the Carboniferous coal deposits might be found in the atmosphere of the Carboniferous world: there simply may have been much more carbon dioxide in the atmosphere that could be extracted and fixed in plant tissues than in our modern world. That is, net plant productivity may have been much higher in the Carboniferous world than in our modern world, where plants are carbon deprived in our atmosphere containing 0.04 percent carbon dioxide. Even with a hyperoxic atmosphere, atmospheric modeling by the University of Sheffield botanist David Beerling and geochemist Robert Berner of Yale University suggests that in an atmosphere with a carbon-dioxide content of 0.06 percent, the "CO_2 fertilization effect is larger than the cost of photorespiration, and ecosystem productivity increases leading to the net sequestration of 117 Gt C [117 billion tonnes of carbon] into the vegetation and soil carbon reservoirs. In both cases, the effects result from the strong interaction between pO_2 [partial pressure of oxygen in the atmosphere], pCO^2 [partial pressure of carbon dioxide], and climate in the tropics" (Beerling and Berner 2000, 12428). And the tropics were where the great rainforests were located in the Carboniferous.

In summary, all of these factors—excess wildfires and charcoal production in a world with a hyperoxic atmosphere, the lack of numerous wood consumers in

Carboniferous ecosystems, rapid determinate growth in the peculiar lycophyte and equisetophyte trees, and more carbon dioxide in the atmosphere for the Carboniferous plants to fix into more plant biomass—acted in concert to produce 90 percent of the world's coal reserves in a period of time that accounts for less than 2 percent of the history of the Earth.

THE KASIMOVIAN CRISIS IN THE GREAT RAINFORESTS

Something happened to the great rainforests in the Kasimovian, something that would have been noticeable even from space. The rainforests still existed in the Kasimovian, but the area of the Earth covered by their green expanse had shrunk by 53 percent (table 3.2); that is, over half the area of the green band that circled the Earth in the Moscovian was gone in the Kasimovian (Cleal and Thomas 2005).

Descending into the rainforests from space, we would immediately notice that the trees were different—the towering, peculiar lepidodendrid scale trees that were so common in the Bashkirian and Moscovian rainforests were gone. Instead, the majority of the trees surrounding us were now tree ferns, marattialean filicophytes (table 3.1). What happened to the great lycophyte trees?

The University of Pennsylvania paleobotanist Hermann Pfefferkorn and colleagues argue that the fossil record shows this marked changeover in the great Carboniferous rainforests occurred in three steps: First came the abrupt disappearance of the towering, pole-like lepidodendrid scale trees. Species of the once-abundant lycophyte tree genera *Lepidodendron* (fig. 3.3), *Lepidophloios*, *Paralycopodites*, *Hizemodendron*, *Diaphorodendron*, and *Synchysidendron* all vanish from the fossil record. Second, there was a short interval of time in which the wetlands were populated by smaller understory seed-reproducing trees— medullosan seed-fern trees and cordaitean pinophyte trees—and a ground cover consisting of spore-reproducing plants—sphenophylleacean horsetails and chaloneriacean quillworts (table 3.1), both of which groups are now extinct. This brief interval is sometimes represented by a single coal bed in the fossil record. Third, the wetlands then began to be repopulated with spore-reproducing trees—large marattialean tree ferns (fig. 3.5) and the surviving sigillarian lycophyte scale trees (fig. 3.9). The sigillarian scale trees were smaller than the

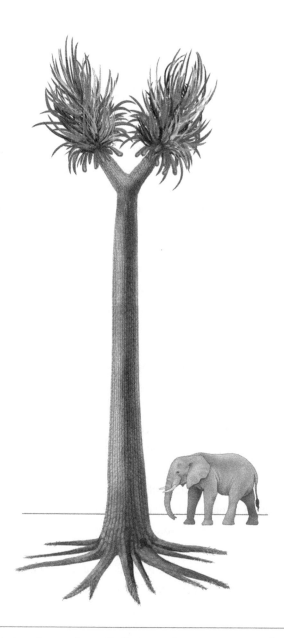

FIGURE 3.9 The lycophyte scale tree *Sigillaria* was about 20 to 25 meters (66 to 82 feet) tall, here compared to a four-meter- (13-foot-) tall male African elephant.

Source: Illustration by Mary Persis Williams.

lepidodendrids, averaging about 20 to 25 meters (66 to 82 feet) in height (Taylor and Taylor 1993), with crowns that did not branch as profusely as the lepidodendrids (compare figs. 3.3 and 3.9).

All in all, 87 percent of the Moscovian trees species and 33 percent of the ground cover and vine species went extinct in the Kasimovian. In addition, the proportion of trees to ground cover and vine species also changed radically in the floral changeover. In the Moscovian, the ratio of trees to ground cover and vines was 30 to 18; in the Kasimovian, that ratio changed to 18 to 25, an almost complete reversal of the floral structure of the rainforest (Pfefferkorn et al. 2008). Why were there so many more vines in the Kasimovian rainforests? Michael Krings, a paleobotanist at the Bavarian State Museum of Paleontology, and his colleagues argue that the evolution and proliferation of new climbing plant species in the Kasimovian was driven by a radical change in the canopy structure of the tropical rainforests during the Late Carboniferous (Krings et al. 2003). The Bashkirian and Moscovian tropical rainforests were dominated by the peculiar arborescent lycophytes, polelike scale trees that did not possess large leafy crowns even during their reproductive phase, when they grew multiple drooping branches at the tops of their trunks. Moreover, their canopy branches were covered by small, microphyll-type leaves that would cast relatively little shadow on the ground below the tree. The combination of these two structural characteristics of the lycophyte trees resulted in a rainforest with a very open canopy, with a great deal of sunlight reaching the ground surface below the trees.

That canopy structure of tropical rainforests changed in the Kasimovian with the extinction of most of the lycophyte trees and their replacement with euphyllophyte tree ferns. These trees possessed larger, macrophyll-type leaves that would catch more sunlight in the canopy and hence extensively shade the ground below the tree. In addition, the canopy of the marattialean filicophyte tree ferns existed throughout the life of the tree, not just during a limited reproductive phase as with the lycophyte trees, so the shading of the ground below was a more continuous phenomenon of the new-style rainforest canopies. For the first time in geologic time, terrestrial forests evolved relatively closed canopies much more similar to those seen in our modern forests today (DiMichele and Hook 1992). For understory plant species, the evolution of closed canopies was a disaster. Sunlight became a much more limited resource, and thus selective pressure existed for the evolution of plants that could climb trees and reach the canopy level of the forest where sunlight was available. The proliferation of vines in the Kasimovian rainforests was the result (Krings et al. 2003).

What could have caused the great rainforests of the Earth to shrink by 53 percent in area and, furthermore, caused the demise of the towering, polelike lepidodendrid scale trees? One hypothesis to explain the Kasimovian crisis in the great rainforests invokes a climatic kill mechanism: the interpulse-drying model (table 3.3). The great ice cap that had covered much of central and western Gondwana in the Serpukhovian and Bashkirian retreated in the Moscovian. In eastern Gondwana, present-day eastern Australia, the late Bashkirian C4-glacial-pulse continental ice sheets began to wane from the middle to the end of the Moscovian (table 3.2, fig. 2.5) (Fielding, Frank, Birgenheier et al. 2008), and in the Northern Hemisphere the ice melted about 310 million years ago. The Earth was entering a major interpulse period (end of C4 to beginning of P1 glacial pulses; table 3.2), a period of warmer and drier climates that would last for some nine million years in the Kasimovian and Gzhelian.

Hermann Pfefferkorn and his colleagues argue that the demise of the great lepidodendrid scale-tree forests was triggered by the onset of these warmer and drier interpulse conditions: "The transition from the middle to late Pennsylvanian witnessed a major floristic change within the wetland biome. . . . This involved a wholesale change from wetlands dominated by

TABLE 3.3 Summary of the kill-mechanism models that have been proposed for collapse of the great rainforest ecosystems in the early Kasimovian.

I. Climatic Kill Mechanism

   A. Interpulse-drying model

   B. Continental-positioning model

   C. Refugia-constriction model

II. Tectonic Kill Mechanism

   A. Continental-uplift model

III. Hyperoxic Kill Mechanism

   A. Carbon-depletion model

IV. Fungal Kill Mechanism

   A. Tree-blight model

Note: See text for model sources.

lepidodendrid lycopsids to dominance by marattialean tree ferns. . . . These patterns appear to be a response to a strong pulse of global warming at or near the Westphalian-Stephanian boundary" (Pfefferkorn et al. 2008, 309). The Westphalian and Stephanian are older geologic time divisions of the Late Carboniferous that were widely used in Europe before the acceptance of the modern epochs and ages (table 1.2). The Westphalian spanned the interval of time from the middle Bashkirian through most of the Moscovian, and the Stephanian spanned the interval of time from the latest Moscovian through the middle Gzhelian (Carroll 2009, fig. 4.1).[8] The paleobotanists Christopher Cleal and Barry Thomas (2005) date the collapse of the great rainforests to the Moscovian-Kasimovian boundary. Their data indicate that the great rainforests reached an all-time maximum extent of 2,395,000 square kilometers (924,470 square miles) in the late Moscovian and shrank to only 1,131,000 square kilometers (436,566 square miles) in the early Kasimovian (table 3.2).

Pfefferkorn and colleagues argue that it was the interpulse drying-out of many of the formerly extensive tropical mires that killed the lepidodendrid scale trees. They point out that not all of the peculiar lycophyte trees perished—the sigillarian scale trees survived (fig. 3.9). Most of the lycophyte trees required standing water for fertilization and germination of their spores, and thus they "preferred habitats in which water tables were high, and only one group—the sigillarians—possessed ecological tolerances that allowed them to colonize better-drained soils (DiMichele and Phillips, 1996). It was the sigillarians that survived the climatic perturbation and continued into the Stephanian [= Kasimovian], where they co-occur with pteridosperms [= non-core seed plants] in mineral substrate settings [= not in peat-accumulating wetlands]" (Pfefferkorn et al. 2008, 313).

The lepidodendrid scale trees were ecologically replaced by marattialean tree ferns, and although tree ferns also reproduce by spores, they do not require standing-water conditions for germination and fertilization. Rather, tree fern spores "require moist conditions for germination and sufficient rainfall during a season when gametes can be produced and fertilization occurs. Tree ferns do not require standing-water conditions for this to happen. Hence, the reproductive strategy of this group involving high fecundity and airborne distribution of spores allowed them to colonize soils developed under a climate characterized by more seasonal rainfall" (Pfefferkorn et al. 2008, 313). The high fecundity of ferns is due to their production of small spores in very large numbers. Ecologically, high fecundity is a characteristic of *weeds*—they grow rapidly, produce large numbers of offspring, and die quickly. This type of life-history strategy is very useful to modern-day "disaster species," enabling them to rapidly colonize areas where the normal vegetation has been destroyed, such as in a forest fire. And in the Moscovian-Kasimovian extinction, the Moscovian-style rainforests were destroyed forever.

Finally, the paleobotanical effect of the return of continental glaciers in the Permian phase of the Late Paleozoic Ice Age (to be considered in detail in chapter 5) gives further support to the interpulse-drying climatic kill model. Massive ice sheets once again formed across Gondwana in the Asselian and Sakmarian Ages in the early Permian, and extensive rainforests once again formed as well. These rainforests were not as geographically extensive as the Carboniferous rainforests and were dominated by tree ferns rather than the peculiar lycophyte trees of the Carboniferous, but it is notable that these new rainforests formed at the same time as the return to colder, wetter climates in the Permian world.

Only in the far east, on the large islands of the Sino-Korean continental block (present-day Korea and North China) and the Yangtze continental block (present-day South China), did the lycophyte-dominated rainforests flourish in the Asselian Age (fig. 3.7). The equatorial regions to the west, covered by lush lycophyte-dominated tropical mires in the late Moscovian of the Carboniferous (fig. 3.7), were now populated by plants that had once been found mostly in highland regions or in savannah environments in the lowlands (Cleal and Thomas 2005). As discussed above, the post-Kasimovian-crisis forests of marattialean tree ferns, gigantopterid seed-fern trees, and cordaitean conifers (table 3.1) were adapted to drier climates and did not produce peat bogs that would become coal deposits.

In contrast to the interpulse-drying model is the alternative continental-positioning climatic kill model (table 3.3): First, the gradual tectonic assembly of all the continents of the Earth into the "all continent" or "world continent" Pangaea[9] resulted in the Earth's landmasses being more or less symmetrically arranged north and south of the equator. Second, it is proposed that the presence of equivalent areas of land straddling the equator caused monsoonal atmospheric circulation to develop, shifting rainfall away from the tropics along the equatorial region of Pangaea. Third, with less rainfall, the wetland habitats of the equatorial regions were disrupted and eventually dried out, leading to the extinction of the equatorial rainforests (Phillips and Peppers 1984; DiMichele et al. 1985). This model does not necessarily negate the interpulse-drying model; it could be argued that both climatic-drying mechanisms operated simultaneously at the Moscovian-Kasimovian boundary to trigger the demise of the hydrophyllic lepidodendrid scale trees.

The continental-positioning model has been questioned by Cleal and Thomas, who point out that "most of the terranes that eventually gave rise to Pangaea were already assembled by the early Bashkirian Age" (Cleal and Thomas 2005, 21) and that, rather than declining and shrinking, tropical rainforests began to flourish and spread during the Bashkirian (table 3.2). As an alternative to either the interpulse-drying or continental-positioning climatic kill mechanism, they propose a tectonic kill mechanism: "It has been suggested that the changes in the geographical distribution of the palaeotropical coal forests were mainly driven by climatic change, but our analysis instead suggests that the underlying controlling-factor was plate tectonics" (Cleal and Thomas 2005, 21). They argue that the onset of the tectonic uplift of the Variscan mountain chain in the middle Moscovian in the European region of Laurussia lowered the water table throughout Europe. Lowering the water table disrupted and drained many of the wetland habitats, and uplifted mountainous regions shed more eroded sediment into lowland habitats and thus also disrupted and filled in low-lying wetland regions. Tectonic uplifting is proposed to have spread westward across the Laurussian supercontinent, disrupting and draining wetland habitats in the North American region by the end of the Moscovian. Likewise, in central Asia to the east, Cleal and Thomas propose that "similar effects of tectonics can be seen in the central palaeotropical coal forests, where they disappeared following the so-called 'Hercynian' tectonic activity resulting from the suturing of the Kazakhstan and Angara plates" (Cleal and Thomas 2005, 22). They argue that only China saw declining tectonic activity in the Moscovian, and note that in China the rainforests persisted into the Kasimovian.

In rebuttal of the tectonic kill mechanism, Pfefferkorn and colleagues argue that "the rate of change of tectonics and rate of vegetation change at the Westphalian-Stephanian are simply not congruent. Tectonic uplift requires durations of time on the order of millions of years," but "[t]he floral change at the Westphalian-Stephanian boundary . . . is nearly instantaneous, implying a widespread and fast-acting causative agent" (Pfefferkorn et al. 2008, 312). They argue that only global climatic change would affect the entire planet simultaneously, and the rapidity of global climatic change from glacial to interglacial phases is well known to us: only 11,000 years ago, huge glaciers covered northern Europe and North America, yet now, in an interglacial period, northern continental glaciers remain only in Greenland. And the rainforests did indeed persist into the Kasimovian Age in China, but their floral composition changed radically: the peculiar lepidodendrid scale trees vanished and were replaced by marattialean tree ferns. If the lack of major tectonic uplift and the lack of lowered water tables in China during the Kasimovian permitted the rainforests to persist, then why did the lepidodendrid scale trees not persist in China as well? Even in these remaining wetland regions, major floral changes occurred, a phenomenon more compatible with global climate change as a triggering mechanism.

Finally, a totally different climatic interpretation for the Carboniferous tropical peat mires has been proposed by Howard Falcon-Lang, a paleobotanist at the University of London, who argues that during Late Carboniferous glacial episodes, tropical climates were cool and seasonally *dry*, not wet, and in some cases even semiarid. He further argues that interglacial episodes were warm and *humid*, not dry, and it was during the *interglacial* episodes that the lepidodendrid-dominated tropical mires formed (Falcon-Lang 2004). Falcon-Lang then teamed up with the Smithsonian Institution paleobotanist William DiMichele to propose a new potential climatic kill mechanism: the refugia-constriction model (table 3.3) (Falcon-Lang and DiMichele 2010). The paper in which this model is proposed is particularly interesting in that the two authors state clearly in the introduction that they disagree on the paleoclimatic reconstruction for the Late Carboniferous coal deposits: Falcon-Lang argues that the coal strata represent transgressive sea-level to highstand sea-level deposits, formed during the glacial-to-interglacial interval when the ice sheets were melting on Gondwana and global sea level was rising to its eventual maximum highstand. In contrast, DiMichele argues that the coal strata represent early to late lowstand sea-level deposits, formed during glacial episodes when ice sheets were spreading to their maxima on Gondwana and global sea level was falling to its eventual minimum lowstand (Falcon-Land and DiMichele 2010).[10]

Despite their diametrically opposing views on paleoclimate conditions during the formation of the coal strata, Falcon-Lang and DiMichele agree that glacial periods were cool and *dry* with globally low sea level and that interglacial periods were warm and *humid* with globally high sea level. They agree that the lepidodendrid-dominated rainforests were most geographically widespread during interglacial humid periods, and most geographically restricted during glacial dry periods. They note that during the numerous smaller, short-term glacial-interglacial rhythms, or cyclothems (Heckel 2008), that occurred in the Late Carboniferous, "coal forests dominated during humid interglacial phases, but were replaced by seasonally dry vegetation during glacial phases. After each glacial event, coal forests reassembled with largely the same species composition," in that less than 10 percent species turnover occurred in each disappearance-reappearance event in the rainforests (Falcon-Land and DiMichele 2010, 611). They then conclude: "There is only one credible solution to this paradox. Coal forests contracted into geographically restricted refugia during glacial phases and surviving taxa then repopulated tropical lowlands when climate became more humid at the start of the next interglacial" (Falcon-Lang and DiMichele 2010, 613). In this model, they propose that the obligate-wetland lepidodendrid scale trees were confined to wetland refugia within river valleys during glacial periods: "While we cannot demonstrate unequivocally that dispersed refugia survived in equatorial valleys during glacial periods, it remains a viable hypothesis that is amenable to further testing. It is worth emphasizing that incised valley systems acted as long-term refugia for wetland plant taxa during periods of dry climate at other times in geological history (Demko et al., 1998)" and that "valley drainages have been proposed as refugia for tropical rainforest species during Pleistocene glacial phases (Meave and Kellman, 1994)" (Falcon-Lang and DiMichele 2010, 614).

How, then, is the Kasimovian crisis in the lepidodendrid-dominated rainforests to be explained in this different climatic kill model—particularly in that the crisis occurred during the major climatic transition from the end of the C4 glacial phase in eastern Gondwana to the beginning of the nine-million-year-long C4-to-P1 interpulse phase in the Kasimovian and Gzhelian (table 3.2)? Surely the lepidodendrid-dominated rainforests should have experienced a resurgence in the reestablishment of humid conditions in this major interpulse interval, if indeed humid conditions did occur (and not dry conditions, as argued in the interpulse-drying and continental-positioning climatic kill models). Instead, the obligate-wetland lepidodendrid scale trees abruptly disappeared and were replaced by dry-tolerant tree ferns during the C4-to-P1 interpulse interval (table 3.2).

Falcon-Lang and DiMichele argue that the intensity of climatic swings in the short-term glacial-interglacial cyclothem rhythms increased steadily throughout the Moscovian C4 glacial interval; that the "effect of increasing glacial intensity through the late Middle Pennsylvanian [= late Moscovian] would have been to progressively reduce the size and connectivity of tropical refugia from one glacial phase to the next"; and that "[t]his trend culminated in maximum intensity at the Middle-Late Pennsylvanian [= Moscovian-Kasimovian] boundary, marked by an extreme marine regression (Heckel, 1991) . . . and valley incision (Easterday, 2004)" (Falcon-Lang and DiMichele 2010, 614). That is, they argue that the C4 glacial interval ended with an extreme ice-buildup pulse that drastically lowered global sea levels (Heckel 1991), and hence river-erosion base levels, triggering erosive downcutting and valley incision in river valleys. On a global scale, with falling sea level, coastal river valleys became progressively deeper and narrower—and the obligate-wetland lycophyte species were confined to refugia in these same river valleys. They then conclude that "the abrupt step change from coal forests dominated by lycopsids to those dominated by tree ferns around the Middle-Upper Pennsylvanian [= Moscovian-Kasimovian] boundary ultimately resulted from hyperconstriction of equatorial refugia during an extreme glacial phase. These changes presumably reduced the already-diminished lycopsids to unsustainably low numbers that eliminated some species and severely reduced the ability of survivors to recolonize environmentally favorable areas following deglaciation at the start of the Late Pennsylvanian [= Kasimovian]" (Falcon-Lang and DiMichele 2010, 615).

In contrast to either climatic or tectonic kill mechanisms, both of which invoke nonbiological physical factors, the University of Washington paleontologist Peter Ward argues that the great rainforests may have been responsible for their own demise! The rainforests are known to have extracted an enormous amount of carbon dioxide from the atmosphere—hence the worldwide distribution of coal strata of Carboniferous age and the very name "Carboniferous" for this interval of geologic time—but did the great rainforests draw down the carbon dioxide content of the atmosphere *too far*? In essence, did the rainforests run out of carbon?

If carbon-dioxide partial pressures in the atmosphere are too low, and oxygen partial pressures are too high, then the presence of oxygen can interfere with the photosynthetic process that most land plants use to produce sugars. In the middle Cenozoic, about 30 to 50 million years ago, a critical threshold was reached in the evolution of the Earth's atmosphere. This event was caused partly by the activity of land plants themselves, as they have been extracting carbon dioxide from the atmosphere and releasing free oxygen back into the atmosphere for the past 425 million years, since the rapid radiation of land plants in the Late Silurian. Another factor was the formation of the gigantic Himalayan mountain chain, produced by the plate-tectonic-driven collision of the Indian subcontinent with Asia, and subsequent weathering processes that have removed a substantial amount of carbon dioxide from the Earth's atmosphere. In any event, plants began to experience difficulty in fixing carbon using their usual photosynthetic pathway.

The majority of land-plant lineages use an ancient photosynthetic pathway known as C_3, which goes all the way back to the evolution of the first cyanobacteria (Sage 1999). To deal with the problem of too little carbon dioxide, and too much oxygen, in the atmosphere, the flowering plants (angiosperms) evolved a more efficient photosynthetic pathway, known as C_4, in the middle Cenozoic. Interestingly, no less than 34 different lineages of angiosperms have convergently, independently, evolved the C_4 photosynthetic pathway (McGhee 2011).

The catch is that C_4 photosynthesis has only been evolved by the angiosperms, and the angiosperms only evolved in the Early Jurassic—they were not present in the Carboniferous world (table 3.1). In fact, none of the living plant lineages that were present in the Carboniferous world have evolved C_4 photosynthesis today, and there is no evidence that any of the Carboniferous members of these lineages ever evolved C_4 photosynthesis (Sage 1999). C_4 photosynthesis appears to be developmentally constrained to the angiosperm plant lineage; that is, it seems that only the flowering plants have the developmental flexibility to evolve the biochemical and cellular changes necessary for C_4 photosynthesis (McGhee 2011).

Atmospheric models indicate that oxygen contents reached very high levels in the Late Carboniferous (fig. 3.8), and this same period corresponds to a carbon-dioxide minimum in the atmosphere (Berner 2006). The Earth's atmosphere became hyperoxic, and carbon depleted, during the Late Carboniferous to Middle Permian interval of geologic time. These observations led Peter Ward to argue: "As far as is known, C_4 plants did not exist in the Permian and thus the drop in carbon dioxide greatly affected plant life, as shown by the presence of a plant extinction at the time of minimum carbon dioxide levels, at about 305 [= 307, beginning of the Kasimovian in table 3.2] million years ago . . . about two-thirds of all plant species known from coal seams going extinct . . . most authors blame this extinction on a drying of the many coal swamps . . . it seems as likely that the extinction was at least partially caused by the carbon dioxide minimum" (Ward 2006, 137). That is, the extinction of the peculiar lepidodendrid scale trees may have been triggered by their own excessive photosynthetic activity in producing a hyperoxic kill mechanism—the carbon-depletion model for plant extinction (table 3.3).

The hyperoxic kill mechanism is one that would affect the entire planet immediately once the critical threshold ratio of carbon dioxide to oxygen was reached. Thus the carbon depletion model is consistent with the nearly instantaneous extinction of the lepidodendrid scale trees that occurred at the Moscovian-Kasimovian boundary. Experiments by the University of Sheffield botanist David Beerling and his colleagues show that when two modern plant species—*Hedera helix* (English ivy) and *Betula pubescens* (birch), both of which are angiosperms—were grown in laboratory conditions with an atmospheric composition of 35 percent oxygen, their leaf photosynthetic rates were 29 percent lower than those of control plants grown in the present Earth's atmosphere with 21 percent oxygen. Model calculations then suggest that the global net primary productivity of land plants may have decreased as much as 18.7 percent—a decrease of 6,300 million tonnes (6,943 million tons) of carbon fixed per year—if atmospheric concentrations of oxygen reached 35 percent in the Carboniferous (Beerling et al. 1998). Could this proposed oxygen-induced suppression of photosynthetic efficiency have killed the lepidodendrid scale trees?

Interestingly, the density of stomatal pores in the leaves of the plants increased in the higher-oxygen-grown plants. Beerling and colleagues point out that one possible explanation for this effect might be the plants' attempt to increase the diffusion rate of carbon dioxide into the leaf under high-oxygen conditions, and thus to acquire more of the atmospherically depleted carbon for photosynthesis. They further note that Carboniferous leaf fossils typically have much higher densities of stomatal openings than do leaves of plants from other geologic time periods, and that this observation would make sense if the Carboniferous plants had grown in an atmosphere with much higher oxygen levels than that of the present Earth's atmosphere (Beerling et al. 1998).

However, Christopher Cleal and colleagues have reported an observed *decrease in stomatal density* in fossil leaves of the medullosan seed fern *Neuropteris ovata* during the late Moscovian, *before* the abrupt disappearance of the lepidodendrid scale trees at the Moscovian-Kasimovian boundary (Cleal et al. 1999). Given the experimental results of Beerling and colleagues, the decrease in density of stomatal pores in plant leaves could be taken as evidence of a *decrease in oxygen* content in the Earth's atmosphere before the extinction of the lepidodendrid scale trees, not an increase. That is, if the lepidodendrid scale trees were suffering from the effects of high atmospheric concentrations of oxygen in the Moscovian, then they should have begun to recover with a decrease in oxygen content in the atmosphere in the late Moscovian, not suddenly gone extinct instead.

Further difficulties for the carbon-depletion model of land-plant extinction arise from uncertainties in the actual amount of carbon dioxide present in the Late Carboniferous atmosphere. Subsequent computer modeling suggests that the roughly 20 percent drop in terrestrial primary productivity predicted in plants living in an atmosphere of 35 percent oxygen would have occurred only if the carbon-dioxide content of that atmosphere was 0.03 percent or less (similar to that of the atmosphere of the pre-industrial-age Earth). If the carbon-dioxide content of the atmosphere is greater than 0.04 percent, then photosynthetic efficiency actually *increases* in this more carbon-dioxide-rich atmosphere, offsetting the cost of photorespiration caused by the 35 percent oxygen content of the atmosphere, and results in an increase in carbon fixation, not a decrease. Thus David Beerling and Robert Berner conclude in this later study that the "Permo-Carboniferous rise in pO_2 [= the oxygen content of the atmosphere] was unlikely to have exerted catastrophic effects on ecosystem productivity" (Beerling and Berner 2000, 12428).

Another possible biological kill mechanism for the lepidodendrid scale trees is tree blight—that the lepidodendrid trees were attacked by a clade-specific fungus that specialized in infesting these forms of lycophyte scale trees (table 3.3). I have often wondered if the sudden disappearance of the lycophyte trees of the family Lepidodendraceae at the end of the Moscovian was equivalent to the sudden death of huge numbers of American chestnut trees, *Castanea dentata*, in North American forests in 1904. In that year, the invasive fungus *Endothia parasitica* was introduced inadvertently into North America, and within 40 years this parasitic fungus had wiped out the magnificent tree species that had once been so widespread in North American forests. The American chestnut trees used to stand 30 meters (100 feet) tall, with trunks 1.2 meters (four feet) in diameter, but now they are gone (Little 1980).

Forty years is but a blink of the eye in the geologic record, and it is easy to visualize the evolution and rapid spread of a fungal species that specialized in attacking trees of the Lepidodendraceae in the late Moscovian, a scenario I call the "tree-blight model" (table 3.3). Even if it took several hundred years for the fungal blight to spread globally in the great rainforests, that interval of time would appear as instantaneous in the geologic record. Tree blight would explain the sudden absence of the lepidodendrid scale trees in the rainforests that still existed in the Kasimovian. Surely some of the Moscovian lepidodendrid scale trees must have survived the shrinking of the areal extent of the rainforests and lived on in reduced numbers in their reduced swampy habitats in the Kasimovian—but this was not so, they all vanished, and tree blight would explain that empirical observation. The survival of lycophyte trees of the family Sigillariaceae into the Kasimovian could perhaps have been equivalent to the survival of the American-chestnut-related Ozark chestnut, *Castanea ozarkensis*. It is attacked by the same fungal parasite that caused tree blight in the American chestnut, but the Ozark species is more resistant to the fungal infestation and is not threatened with extinction (Little 1980).

Finally, it may well be that the extinction of the lepidodendrid scale trees and the dramatic shrinkage in the geographic extent of the great rainforests in the early Kasimovian were not triggered by any single kill mechanism. That is, the catastrophe may have been the product of several kill mechanisms operating simultaneously: water stress in land plants triggered by a global climatic shift from a cold, wet glacial phase to a warm, dry interpulse phase; major tectonic uplifting and wetland draining in regions of Europe, North America, and Asia; all possibly exacerbated by photosynthesis stress in land plants caused by carbon depletion and hyperoxia in the Earth's atmosphere; and possible tree blight in the Lepidodendraceae. For whatever reason or reasons, the great Serpukhovian-Moscovian lepidodendrid-dominated rainforests were destroyed forever.

In the Permian, the ice would return for the second major phase of the Late Paleozoic Ice Age, and the rainforests would begin to recover and to spread once again (table 3.2). We will continue to examine the evolution of the great rainforests of the Late Paleozoic Ice Age in chapter 5, but first let us explore the biological consequences of the Earth's hyperoxic Carboniferous atmosphere for animals—both land animals and marine animals—in chapter 4.

4

Giants in the Earth . . .

There were giants in the Earth in those days . . .
—Genesis 6:4 (King James version)

THE STRANGE CARBONIFEROUS ANIMALS

The writer of the book of Genesis states that there were giants in the Earth in the early days following the creation. In the chronology outlined in Genesis, those giants existed about 6,000 years ago. In actual fact, the first giant land animals on the Earth existed in the Carboniferous, some 345 million years ago. Both on the land and in the seas, animals began appearing in the Carboniferous that were much larger than any living modern-day counterparts—in some cases, as much as ten to 12 times larger! In this chapter we will examine the empirical evidence for the evolution of animal gigantism in the Carboniferous and then explore the possible causes of that gigantism.

Giant Arthropods

The Carboniferous giant arthropods (table 4.1) began to appear in the Visean Age, not long after the end of the Tournaisian Gap. One notable example is the fossil of a giant scorpion, *Pulmonoscorpius kirktonensis*, discovered in East Kirkton in Scotland, that was

700 millimeters (28 inches) long. Here in North America, different living species of scorpions range in length from 37 to 127 millimeters (1.5 to five inches), and the "giant desert hairy scorpion" of the Southwest, *Hadrurus arizonensis*, reaches a length of 140 millimeters (5.5 inches) (Milne and Milne 1980). In comparison with *Pulmonoscorpius kirktonensis*, we can see that our modern "giant" scorpion is

TABLE 4.1 Terrestrial arthropod lineages of the Late Paleozoic Ice Age.

ARTHROPODA (jointed appendages)
– CHELICERIFORMES
– – Chelicerata
– – – MEROSTOMATA (sea scorpions and kin)
– – – – – *Water scorpion giants*
– – – ARACHNIDA (scorpions and spiders)
– – – – – *Scorpion, trigonotarbid, spider giants*
– MANDIBULATA
– – MYRIAPODA (multi-legged)
– – – Diplopoda (millipedes)
– – – – *Arthropleurid giants*
– – – Chilopoda (centipedes)
– – PANCRUSTACEA
– – – Hexapoda (six-legged)
– – – – Entognatha
– – – – – *Dipluran giants*
– – – – INSECTA (insects)
– – – – – basal insects
– – – – – – *Silverfish giants*
– – – – – PTERYGOTA (winged insects, conquerors of the land)
– – – – – – Ephemeroptera (mayflies)
– – – – – – – – *Mayfly giants*
– – – – – – Metapterygota
– – – – – – – Odonatoptera (dragonflies, damselflies)
– – – – – – – – Meganisoptera
– – – – – – – – – – *Griffenfly giants*
– – – – – – – Palaeodictyopterida
– – – – – – – – *Palaeodictyopteran sap-sucker giants*
– – – – – – NEOPTERA (folding-wing insects)
– – – – – – – – *Cockroach giants*

Source: Phylogenetic classification modified from Grimaldi and Engel (2005) and Lecointre and Le Guyader (2006).
Note: Taxa containing giants are in italics; major clades are in capitals.

FIGURE 4.1 Some Late Paleozoic scorpions, such as the Early Carboniferous *Pulmonoscorpius kirktonensis,* were as long as a midsize dog.
Source: Illustration by Mary Persis Williams.

no giant at all—the Carboniferous scorpion was fully five times larger! Imagine hiking in the Arizona desert today and encountering a *real* giant scorpion, one that is as long as a midsize dog (fig. 4.1). How could a scorpion have achieved such a gigantic size? Yet it was only a harbinger of what was to come.

Giant arthropods really started becoming numerous on the Earth in the Moscovian Age of the Late Carboniferous. Close relatives of the scorpions, spiders and their ancient kin began to achieve gigantic sizes. The extinct trigonotarbids were morphologically very similar to modern spiders, but they did not possess silk glands to spin webs. They started out small in the Devonian, from two to 15 millimeters (0.08 to 0.6 inch) long, not unlike modern spiders. In the Late Carboniferous, however, trigonotarbids evolved that were 50 millimeters (two inches) long—more than three times larger than the largest Devonian trigonotarbids (Shear and Kukalová-Peck 1990). A gigantic spider, *Megarachne servinei* from Argentina, had a leg span of almost 500 millimeters (20 inches) (Shear and Kukalová-Peck 1990; Lane 2002).[1] For those who are arachnophobic, just imagine encountering a spider over a foot and a half long!

Other giant Moscovian arthropods include the 2.5-meter- (eight-foot-) long millipede species *Arthropleura armata* and *A. mammata,* animals that were 150 millimeters (six inches) wide and weighed up to ten kilograms (22 pounds)

(Kraus and Brauckmann 2003).[2] Most millipede species today are tiny, slender animals about 50 millimeters (two inches) long; our largest species today, the African "giant black millipede" *Archispirostreptus gigas*, can reach 385 millimeters (15 inches) in length and 20 millimeters (0.8 inch) in width. Once again, we see that our modern "giant" arthropod species is not a giant—the Carboniferous millipede species were six and a half times larger! In fact, the Carboniferous millipedes were fully as long as a modern alligator (fig. 4.2).

The hexapods and insects also began to evolve species with unprecedented body sizes in the Late Carboniferous (table 4.1). The forcipate diplurans (Entognatha; table 4.1) are usually very small cryptic insects, hiding under logs and leaves on the ground, and are generally four to six millimeters (about one-sixth to one-fourth of an inch) long (Milne and Milne 1980). The gigantic forcipate dipluran species *Testajapyx thomasi*, found in Moscovian-aged Mazon Creek strata in Illinois, was 58 millimeters (2.3 inches) long, including a 48-millimeter body and ten-millimeter antennae (Kukalová-Peck 1987). On average, the Late Carboniferous *Testajapyx thomasi* was 12 times larger than living forcipate diplurans! In the same Mazon Creek strata are found gigantic silverfish insects (basal insects; table 4.1), the species *Ramsdelepidion*

FIGURE 4.2 Some Late Paleozoic millipedes, such as the Late Carboniferous *Arthropleura armata*, were as long as a modern alligator.

Source: Illustration by Mary Persis Williams.

schusteri, that are 60 millimeters (2.4 inches) long. Modern silverfish species can often become pests, living in moist regions of the house, such as under sinks or around water pipes, and emerging at night to feed on book bindings or clothes—but their silvery, flattish bodies are usually small, around nine to 13 millimeters (one-third to one-half inch) long (Milne and Milne 1980). On average, the Late Carboniferous silverfish were more than five times larger—imagine finding a silverfish the length of your little finger munching on your favorite book!

The ground-dwelling millipede and insect giants were not alone. Giant insects flew in the skies of the Late Carboniferous and Early Permian (table 4.1). The most primitive of the living flying insects are the Ephemeroptera, the mayflies (table 4.1). Modern mayflies live for only one day, hence their scientific name "ephemeral wings." The mayfly itself is merely the mobile, reproductive phase of the species—after mating and laying their eggs, they die. The Carboniferous mayflies, however, possessed functioning biting mouthparts and clearly fed as adults, not just as aquatic nymphs. And they were gigantic: the Moscovian Age mayfly species *Bojophlebia prokopi* had a wingspan of 450 millimeters (18 inches) (Kukalová-Peck 1985). In contrast, the largest living mayflies in North America, the golden mayflies, can have a wingspan up to 60 millimeters (2.4 inches)—the Carboniferous mayfly *Bojophlebia prokopi* was over seven times larger. In fact, the wingspan of *Bojophlebia prokopi* was as wide as that of a modern midsize bird, such as a blue jay, rather than an insect.

Astonishingly, even larger flying insects existed in the Late Carboniferous and Early Permian. The most impressive of these were the meganisopterans, the griffenflies (table 4.1), which were flying predators similar in appearance to modern-day dragonflies—but much bigger. Giant griffenflies, such as the Chinese griffenfly *Shenzhousia qilianshanensis*, which had a wingspan of 499 millimeters (20 inches), began to appear in the Bashkirian.[3] The Moscovian griffenfly *Bohemiatupus elegans* in Europe was larger, with a wingspan of 552 millimeters (22 inches), and the Kasimovian-Gzhelian griffenfly *Meganeura monyi* was even larger, with a wingspan of 628 millimeters (two feet) (Clapham and Kerr 2012). In the Early Permian, some griffenflies were larger still, the largest being the Asselian-Artinskian *Meganeuropsis permiana* with a wingspan up to 720 millimeters (28 inches) (Grimaldi and Engel 2005; Clapham and Kerr 2012). Giant griffenflies persisted into the beginning of the Middle

Permian with the presence of *Arctotypus* sp., with a wingspan of 489 millimeters (19 inches) (Clapham and Kerr 2012).

The largest dragonfly in North America—again, like the "giant desert hairy scorpion," found in the Southwest—is the Walsingham's Darner, *Anax walsinghami*, with a wingspan of 150 millimeters (six inches) (Milne and Milne 1980). The largest living dragonfly, the Australian species *Petalura ingentissima*, is slightly larger, with a wingspan of 168 millimeters (6.6 inches) (Dorrington 2015). Similar to our comparison of modern and Carboniferous scorpions, which were five times larger, the wingspan of the Carboniferous and Permian griffenflies was almost five times that of our largest living dragonflies. In fact, these griffenflies had a wingspan more similar to that of a large modern bird, such as a magpie or a gull (fig. 4.3), than a living dragonfly. Instead of being dive-bombed by hungry gulls at the beach, eager to snatch a French fry from your hand, imagine encountering similar-sized dragonflies!

Not only were there giant flying predatory insects in the Late Carboniferous, but the flying herbivorous insects were gigantic as well. These include the Moscovian palaeodictyopteran species *Mazothairos enormis* (table 4.1) from Mazon Creek strata in Illinois, which had a wingspan of up to 600 millimeters (two feet) and thus was aptly given the species name "enormous." These ancient insects did not eat energy-rich animal protein or fat like a griffenfly; they survived on tree sap. Yet these Carboniferous sap-suckers still achieved gigantic sizes (Grimaldi and Engel 2005). Our largest living sap-sucking flying insect, the Malaysian

FIGURE 4.3 Some Late Paleozoic dragonfly-like insects, such as the Early Permian *Meganeuropsis permiana*, had wingspans as wide as a modern seagull.

Source: Illustration by Mary Persis Williams.

cicada species *Pomponia imperatoria*, has a wingspan of 200 millimeters (eight inches)—the Carboniferous flying sap-suckers were three times larger.

Not all of the giant Carboniferous insects were larger than their modern relatives. For example, the modern living cockroach *Blaberus giganteus* of Central America and northern South America is fully 90 millimeters (3.5 inches) long, a size not exceeded by the large Carboniferous roachoids, or cockroach-like insects. However, and in summary, many of the Late Carboniferous and Early Permian arthropod species had body sizes that were five to seven times larger than their modern-day counterparts, and an overall size range from three to as much as 12 times larger.

What type of an Earth could have produced such huge arthropods? How could such gigantic scorpions, millipedes, spiders, silverfish, mayflies, griffenflies, and cicada-like sap-suckers have existed? Gigantism convergently evolved in multiple phylogenetic lineages of arthropods in the Carboniferous: it appeared independently in the two clades of the cheliceriform and the mandibulate arthropods, independently in the myriapod and pancrustacean clades within the mandibulates, and independently within multiple separate insect clades (table 4.1). To make matters even more mysterious, there were other giant animals on the Earth in this interval of geologic time, as we will see in the next section.

Giant Vertebrates

The giant Carboniferous arthropods were not alone—many of the Late Carboniferous tetrapod vertebrates were also gigantic (table 4.2), at least in comparison to their Early Carboniferous ancestors. The majority of the tetrapod species of the Early Carboniferous were small animals, usually only about 300 millimeters (one foot) in total length. In contrast, many of the newly evolved tetrapods of the Late Carboniferous were ten times larger: they were often three meters (3,000 millimeters, or ten feet) long. And, as with the insects, the attainment of large body sizes in the Late Carboniferous tetrapods was independent of diet: both carnivores and herbivores were big animals.

Vertebrates began the invasion of land in earnest in the Visean Age of the Early Carboniferous, following the catastrophe of the end-Devonian tetrapod

TABLE 4.2 Terrestrial vertebrate lineages of the Late Paleozoic Ice Age.

TETRAPODA (limbed vertebrates)

– basal tetrapods

– Neotetrapoda

– – BATRACHOMORPHA (ancestors of living amphibians)

– – – basal batrachomorphs ("temnospondyls")

– – – – *Eryopid giants*

– – REPTILIOMORPHA (ancestors of amniote tetrapods)

– – – basal reptiliomorphs ("anthracosaurs")

– – – Seymouriamorpha

– – – *Diadectomorph giants*

– – – AMNIOTA (amniote tetrapods, conquerors of the land)

– – – – SYNAPSIDA (ancient ancestors of mammals)

– – – – – basal synapsids ("pelycosaurs")

– – – – – – *Caseid, ophiacodont, edaphosaur, sphenacodont giants*

– – – – – THERAPSIDA (closer ancestors of mammals)

– – – – – – Biarmosuchia (basal therapsids)

– – – – – – *Dinocephalian giants*

– – – – – – Dicynodontia

– – – – – – Gorgonopsia

– – – – – – Therocephalia

– – – – – – – CYNODONTIA (very close ancestors of mammals)

– – – – REPTILIA (ancestors of living reptiles)

– – – – – PARAREPTILIA

– – – – – – *Pareiasaur giants*

– – – – – EUREPTILIA

– – – – – – *Captorhinid giants*

– – – – – – DIAPSIDA (ancestors of living lepidosaurs, archosaurs, and dinosaurs—the birds)

Source: Phylogenetic classificaton modified from Benton (2015).

Note: Taxa containing giants are in italics; older paraphyletic tetrapod group names are in quotation marks; major clades are in capitals.

extinctions and the long Tournaisian Gap (table 1.7) (McGhee 2013). The first of the batrachomorphs, the ancestors of modern-day amphibians, and of the reptiliomorphs, the ancestors of modern-day amniote animals like mammals, reptiles, and birds, appear in the fossil record at this time (table 4.2).[4] The basal batrachomorphs are represented in the Visean by a single species, *Balanerpeton*

FIGURE 4.4 Some Late Paleozoic salamander-like amphibians, such as the Early Permian
Eryops megacephalus, were two meters (6.6 feet) long.
Source: Illustration by Mary Persis Williams.

woodi, from East Kirkton in Scotland. Individuals of this species were small,
averaging about 200 millimeters (eight inches) long, and looked much like salamanders (Steyer 2012). Later, in the Early Permian, the batrachomorphs would
evolve the giant *Eryops megacephalus*—its species name means "very large
headed"—which, at 2,000 millimeters (6.6 feet) long, was ten times bigger than
Balanerpeton woodi. Eryops megacephalus looked nothing like a harmless sala-
mander—it had a massive body with thick legs (fig. 4.4), and its large head had
a mouth full of sharp teeth including extra fang teeth located in the palate in the
roof of its mouth—it was a deadly predator (Steyer 2012).

The reptiliomorph clade is represented by several different species in the
Visean Age, of which *Silvanerpeton miripedes, Eldeceeon rolfei,* and *Westlothiana
lizziae* are found in the East Kirkton strata along with the batrachomorph
species *Balanerpeton woodi.* These reptiliomorphs were also small, with body
lengths, from snout to pelvis, of only about 200 millimeters (eight inches). The
tail vertebrae are missing from the fossils, so the precise length of their tails is
unknown, but the total length of these early reptiliomorphs was probably only
around 300 millimeters (one foot). Later, in the Late Carboniferous, the diadectomorph reptiliomorphs would evolve the giant *Diadectes maximus*—its species

FIGURE 4.5 Some Late Paleozoic plant-eating reptile-like animals, such as the Late Carboniferous *Diadectes maximus*, were three meters (ten feet) long.

Source: Illustration by Mary Persis Williams.

name means "the greatest"—that, as with the batrachomorph lineage, was ten times larger than the early reptiliomorphs. Rather than 300 millimeters (one foot), *Diadectes maximus* was fully 3,000 millimeters (ten feet) long (fig. 4.5). *Diadectes maximus* is also very interesting in that it is the oldest known fossil tetrapod species to have evolved herbivory—it was a plant-eater, rather than a carnivore (Steyer 2012).

It is from the early reptiliomorphs that the two amniote clades evolved—the synapsids, ancestors of modern-day mammals, and the reptilians, ancestors of modern-day reptiles and birds (table 4.2). Within our own phylogenetic lineage, the synapsids, giant predatory ophiacodonts began to appear in the Late Carboniferous and Early Permian: *Ophiacodon uniformis* was 2.5 meters (eight feet) long. In this same interval of time, giant edaphosaurs and sphenacodonts began to appear: *Edaphosaurus cruciger* was 3.5 meters (11.5 feet) long, and *Dimetrodon grandis*—its species name means "great size"—was three meters (ten feet) long (fig. 4.6). The ophiacodonts and sphenacodonts were large carnivores, but the edaphosaurs were herbivorous; thus, as in the arthropod lineages, the evolution of giant body sizes was not linked to diet. In the Middle Permian, the herbivorous caseid synapsids (table 4.2) would evolve species of *Cotylorhynchus* that were three meters (ten feet) long, and the dinocephalian therapsids would produce the giant *Moschops* herbivores that were fully five meters (16.4 feet) long, the largest land vertebrates in the Paleozoic (Benton 2015). The dinocephalians—the group name means "terrible headed"—were bizarrely shaped

FIGURE 4.6 Some of our Late Paleozoic synapsid cousins, such as the carnivore *Dimetrodon grandis*, were sail-finned giants more than three meters (ten feet) long.

Source: Illustration by Mary Persis Williams.

with massive, barrel-chested bodies but quite short, stocky legs (fig. 4.7). Given their massively built skulls, it has been suggested that these animals butted their heads together in ritual combat during the mating season, much like modern-day goats and sheep (Benton 2015).

Curiously, the other amniote clade, the reptilians, apparently did not evolve giant individuals until the Late Permian. The parareptiles (table 4.2) produced

FIGURE 4.7 Some of our Late Paleozoic therapsid cousins, such as the Middle Permian *Moschops* herbivores, were the largest land vertebrates in the Paleozoic, twice the size of a modern cow.

Source: Illustration by Mary Persis Williams.

the pareiasaurs with massive, elephant-like legs; muscular humps on their backs associated with thick neck muscles; and odd knobby, hornlike protrusions on their skulls. The African pareiasaur *Bunostegos akokanensis* was three meters (ten feet) long (Steyer 2012). Also in the Late Permian, the African captorhinid reptiles (table 4.2) produced giants like *Moradisaurus grandis*—again, the species name means "great size"—which were two meters (6.6 feet) long (Steyer 2012).

Giant Marine Invertebrates

Giant animals also appeared in marine and coastal ecosystems in the Carboniferous. Not only did the giant 700-millimeter- (28-inch-) long *Pulmonoscorpius kirktonensis* scorpions appear on land, but some of their water-scorpion cousins (table 4.1) were over three times as large! In coastal estuaries and rivers in the Devonian and Early Carboniferous, one could encounter water scorpions up to 2.5 meters (eight feet) long, such as the giant *Jaekelopterus rhenaniae* (fig. 4.8) (Braddy et al. 2007; Steyer 2012, 67; Palmer et al. 2012, 122).

Out in the open oceans of the world, the largest brachiopod shellfish known in Earth history appeared—*Gigantoproductus giganteus*, an animal aptly named the "gigantic giant *Productus*" (see table 4.3 for the phylogenetic position of

FIGURE 4.8 Some Late Paleozoic water scorpions, such as the Devonian *Jaekelopterus rhenaniae*, were over 2.5 meters (eight feet) long.

Source: Illustration by Mary Persis Williams.

TABLE 4.3 Marine invertebrate lineages of the Late Paleozoic Ice Age.

EUKARYA (life with eukaryote cells)
– Bikonta (biflagellate single cells)
– – Chromoalveolata ("brown algae" and kin)
– – – Dinophyta (dinoflagellates)
– – Rhizaria
– – – Foraminifera
– – – – *Fusulinid giants*
– Unikonta (uniflagellate single cells)
– – Choanozoa
– – – METAZOA (multicellular animals)
– – – – Cnidaria (diploblastic metazoans)
– – – – – *Coral giants*
– – – – BILATERIA (triploblastic metazoans with bilateral symmetry)
– – – – – PROTOSTOMIA (bilaterians with protostomous development)
– – – – – – Lophotrochozoa
– – – – – – – Lophophorata
– – – – – – – – Bryozoa
– – – – – – – – – *Bryozoan giants*
– – – – – – – – Brachiopoda
– – – – – – – – – *Brachiopod giants*
– – – – – – – Eutrochozoa
– – – – – – – – Mollusca
– – – – – – – – – *Ammonoid giants*
– – – – – – – – – *Bivalve giants*
– – – – – – Ecdysozoa (see table 4.1 for arthropod giants)
– – – – – DEUTEROSTOMIA (bilaterians with deuterostomous development)
– – – – – – Echinodermata
– – – – – – – *Crinoid giants*
– – – – – – – *Echinoid giants*
– – – – – – Chordata (see table 4.2 for vertebrate giants)

Source: Phylogenetic classification modified from Lecointre and Le Guyader (2006).
Note: Taxa containing giants are in italics; major clades are in capitals.

brachiopods). The shell of the Early Carboniferous *Gigantoproductus giganteus* was 300 millimeters (one foot) wide (fig. 4.9) (Newell 1949; Clarkson 1998; Qiao and Shen 2015). In contrast, the shell of an average productid brachiopod in the Late Devonian seas was only about the size of a quarter,[5] fully 15 times smaller than *Gigantoproductus giganteus* (fig. 4.9).

FIGURE 4.9 Some Late Paleozoic marine shellfish, such as the Early Carboniferous brachiopod *Gigantoproductus giganteus*, had shells that were 300 millimeters (one foot) wide. In contrast, average productid brachiopod shells in the Late Devonian were about the size of a quarter (lower right).

Source: Modified and redrawn from Brunton, Lazarev, and Grant (2000).

Fossils of productid brachiopods are so numerous in Carboniferous marine strata that the Carboniferous seas were often referred to as "*Productus* seas" in older historical geology textbooks (Tasch 1973, 297), and many of the Carboniferous species were giants. And gigantism did not just occur in the ecological niche of the productid brachiopods, which lived mostly buried and hidden in the sediment. The fully epifaunal brachiopods—such as species of *Punctospirifer*— which lived exposed on the seafloor (and thus more vulnerable to predator attack) exhibited fivefold increases in size in the Late Carboniferous to Middle Permian time span. This size-increase trend is so notable that Norman Newell, a paleontologist at the American Museum of Natural History, commented: "The trend for increasing size during the late Paleozoic is well shown by a large number of brachiopod genera. Most of these begin in the early Pennsylvanian or late Mississippian and show progressive increase in linear dimensions throughout the Pennsylvanian and Permian" (Newell 1949, 110–111).

The bryozoans—filter-feeding lophophorates (table 4.3) like the brachiopod shellfish, but colonial rather than solitary—also experienced size increases during the Late Carboniferous to Middle Permian time span. Specifically, the diameters of the zooecial openings in the colony skeletons—a rough measure of the size of the actual tiny zooid individuals that lived in the numerous branches of the colony—increased by 100 percent from the Kasimovian to the Roadian (Newell 1949, 109). Gigantism was not confined to the clade of the lophophorates: in the filter-feeding echinoderm sea lily (crinoid, table 4.3) genus *Calceolispongia*, the volume of the basal plate of the calyx skeleton of the animal increased from two cubic millimeters (0.08 cubic inches) to 800 cubic millimeters (2.6 cubic feet) during the Artinskian Age in the Early Permian. If changes in the volume of the basal plate of the calyx skeleton are taken as a rough proportional measure of biomass-volume changes of the total animal, then the biomass volume in these crinoid species increased by a factor of 400 during the Early Permian (Newell 1949, 112). Some mobile echinoderm groups also experienced more than a tripling in size: species of the sea urchin (echinoid, table 4.3) order Melonechinoida increased in size from a diameter of 45 millimeters (1.8 inches) in the genus *Palaeechinus* to 155 millimeters (six inches) in the genus *Melonechinus* in the span of time from the Tournaisian to Visean Ages in the Early Carboniferous (Newell 1949, 111–112).

Predatory marine animals also experienced gigantism in the Carboniferous. The sessile rugose coral species, such as *Zaphrentis*, typically had horn- or cone-like skeletons about 15 millimeters (0.6 inch) high in the Devonian, whereas the Devonian-Carboniferous *Canina* species had skeletons 100 millimeters (four inches) high and *Siphonophyllia* species had even larger skeletons at 750 millimeters (30 inches) high—a fiftyfold increase in height from the Devonian *Zaphrentis* (Newell 1949, 108). The swimming, squidlike ammonoid predators generally had phragmocones—the section of the shell containing the main biomass of the animal—with a maximum diameter of about 75 millimeters (three inches) in the Devonian and Early Carboniferous, rarely ranging up to a diameter of 150 millimeters (six inches). In contrast, in the Permian, ammonoid species with phragmocone diameters of 150 millimeters were common in several different families of these predators, and species of *Medliocottia* and *Cyclolobus* possessed phragmocones with diameters of 250 millimeters (almost ten inches)—over three times larger than those in the Devonian (Newell 1949, 114).

Probably the most astonishing examples of gigantism that occurred in marine organisms during the Carboniferous and Permian are organisms that were not even multicellular animals, but single-celled fusilinid foraminifera (table 4.3). These single-celled organisms grew tiny, lenticular coiled shells about 0.5 millimeters (0.02 inch) in axial diameter—about the size of the period at the end of this sentence—in the Serpukhovian Age of the Early Carboniferous. They then progressively evolved numerous species and genera whose axial lengths increased to the "acme in microscopic giants (*Parafusulina kingorum* Dunbar and Skinner)" (Newell 1949, 106) in the Early to Middle Permian, a species that had an axial length of 60 millimeters (2.4 inches)—an increase in size by a factor of 120 from the Early Carboniferous. This organism—a single cell—was almost three times the width of a quarter (fig. 4.10).

A complicating factor in fusulinid gigantism is that these foraminifera harbored symbiotic photosynthetic algae in their tissues—that is, they harbored an extra source of food—and this trait may have contributed to their ability to become giants. This trait is mirrored by another astonishing example of invertebrate gigantism—the living *Tridacna gigas* bivalve molluscs with shells over one meter (3.3 feet) in diameter. The Permian gigantic bivalves (table 4.3) of the

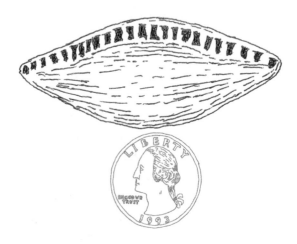

FIGURE 4.10 Some Late Paleozoic single-celled organisms, like the Early Permian marine foraminifera *Parafusulina kingorum*, were 60 millimeters (2.4 inches) long and almost three times the width of a quarter. In contrast, average foraminifers in the Early Carboniferous were about the size of the period at the end of this sentence.

Source: Modified and redrawn from Hoskins, Inners, and Harper (1983).

family Alatoconchidae (Aljinovic et al. 2008; Isozaki and Aljinovic 2009) are also believed to have harbored photosynthetic algae as symbionts in their tissues. We will reconsider both the fusulinids and the alatoconchid bivalves in more detail in the ecosystem evolution and animal gigantism section later in the chapter.

THE CARBONIFEROUS OXYGEN PULSE AND ANIMAL GIGANTISM

The trigger for the evolution of animal gigantism in the Carboniferous had to have been some globally pervasive change in the Earth's environment because that gigantism appeared independently, convergently, in the separate phylogenetic lineages of both the arthropods (table 4.1) and the vertebrates (table 4.2), and it occurred in the same interval of geologic time for both of these separate phylogenetic lineages. Arthropods and vertebrates are radically different types of animals—arthropods are protostomous animals and vertebrates are deuterostomous—and you have to go back in time all the way to the Neoproterozoic to find a common ancestor.[6] That is, the lineages of the protostomous and deuterostomous animals diverged back in the Neoproterozoic, meaning that the arthropods and vertebrates are separated by a vast chasm of more than 660 million years of independent evolution. Yet both of these independent lineages began to evolve giant animals in the Carboniferous—why?

Arthropods and Oxygen

The breathing system used by the insect species offers the best clue to solving the mystery of Carboniferous animal gigantism. Insects breathe with a series of small tubes, the tracheae, that extend from pore-like openings in their exoskeletons down into their body tissues. The insect depends entirely upon diffusion of oxygen in from these tubules (and reverse diffusion of carbon dioxide out of the tubules) to keep their tissues alive. This system of breathing is strongly size limiting. Larger and larger insects have larger and larger volumes of internal tissue, tissues that depend upon the surface areas of the tiny tracheae for gas exchange. With increasing size, the volume of any object increases as a cubic function of linear dimension, whereas surface area increases only as a square function. This constraint—the area-volume effect—means that insects are constrained to sizes small enough to have the proper ratio of internal tissue volume to surface area of tracheal tubes (Harrison et al. 2010).

The area-volume problem is even more acute for flying insects—flying is a highly energetic activity that requires a lot of oxygen. And larger, heavier insects require even more energy—and oxygen—to fly than smaller insects. Thus how could a dragonfly-like insect with a wingspan of almost three-quarters of a meter (fig. 4.3) have existed in the Late Paleozoic? It should have suffocated under its own mass, its tracheae unable to aerate its large volume of internal tissues. But what if the atmosphere had *more oxygen in it* than it presently does? An atmosphere with a higher partial pressure of oxygen—more oxygen molecules concentrated in a given volume of air—could potentially lift the constraint of the area-volume effect to higher ratios of tissue volume to tracheal surface area, thus making possible the existence of much larger insects (Dudley 1998; Harrison et al. 2010).

The mystery of giant-arthropod breathing deepens when it is revealed that the gigantic *Arthropleura* millipedes (fig. 4.2) had no tracheae at all! How could they breathe without tracheae? The zoologist Otto Kraus, of the University of Hamburg, and the paleontologist Carsten Brauckmann, of the Technical University of Clausthal, commented on this mystery: "Despite the fact that many well-preserved fossils are known and described, not even a trace of spiracles (stigmata) and tracheae has ever been found. . . . The maximum length of the largest lepidopteran larvae is approximately 15 cm. However, *Arthropleura* specimens exceeded this dimension up to more than ten times. Provided that large arthropleurids had really maintained tracheae, they should have been extremely stable and strong tubes. They could hardly be overlooked" (Kraus and Brauckmann 2003, 46–47).

One possibility is that the giant arthropleurid millipedes accomplished gas exchange with the atmosphere directly across the surface area of their bodies, and indeed Kraus and Brauckmann note: "In contrast to other very large terrestrial arthropods . . . arthropleurids were apparently *not* equipped with a heavily sclerotized or even mineralized exoskeleton" (Kraus and Brauckmann 2003, 44–45; original emphasis). As anyone who has stepped on a large beetle has experienced, an audible crunch is produced as the beetle's exoskeleton is crushed. The crunch sound is produced because the beetle has a very rigid, stiff exoskeleton whose cuticular tissues contain a great deal of chitin (its tissues are "heavily sclerotized"). Gas exchange across the rigid body surface area of a beetle is minimal; it accomplishes gas exchange through its tracheae instead. The exoskeleton of the giant Carboniferous millipedes was much softer, more limber and flexible, hence more permeable to gas exchange with the atmosphere. However, even given this very unusual exoskeleton in the arthropleurid millipedes, Kraus and Brauckmann argue that "gas exchange simply via the body surface can certainly be excluded" (2003, 46). Instead, they argue that the giant millipedes must have been semiaquatic, living mostly in water, and may have breathed through a series of plates ("K plates") along the bottom surface of their body, plates that perhaps may have functioned somewhat like the gills in other aquatic arthropods.

Preserved fossil trackways made by giant arthropleurid millipedes on land are well known and have even been given their own ichnotaxon designation, or trace-fossil name—*Diplichnites*. Thus clearly the giant Carboniferous millipedes spent a lot of time on land. Instead of being semiaquatic and spending most of their time in water, what if the giant millipedes spent most of their time on land (like normal millipedes) and did indeed successfully breathe directly across their body surface areas because there was *more oxygen in the atmosphere* than in our modern world? The slowly crawling giant millipedes would not have required as much oxygen as the highly energetic flying griffenflies, and gas exchange across their body surface area may have been so efficient in a world with a hyperoxic atmosphere that they did not even require tracheae in order to breathe—hence these structures became vestigial and were later lost in the evolutionary lineage of the arthropleurid millipedes.

It is difficult to see how such large insects could have existed without having an atmosphere that contained much more oxygen than is present in the modern atmosphere of the Earth. And indeed, it is exactly in this period of time—the Late Carboniferous through the Middle Permian—that the empirical models of the ancient atmosphere of the Earth, which we considered in detail in chapter 3, predict that oxygen concentration levels in the atmosphere began to exceed 30 percent and eventually reached a high of around 35 percent (fig. 3.8). Imagine living in a world that has two-thirds more oxygen in the atmosphere than in the air that we currently are breathing.

The problem is, those same empirical models give differing results on exactly when oxygen began to exceed 30 percent in the atmosphere and when the maximum oxygen content of the atmosphere was reached (fig. 3.8). Table 4.4 shows the known fossil occurrences of some of the giant arthropod groups alongside the oxygen-content levels in the atmosphere predicted by the Rock-Abundance model and the Geocarbsulf model. Note that the known stratigraphic distribution of giant arthropod fossils matches the oxygen predictions of the Rock-Abundance model much better than those of the Geocarbsulf model: giant scorpions appear in the Visean Age, and the Rock-Abundance model predicts that the Earth's atmosphere contained 25.8 percent oxygen in the Visean—4.8 percentage points more oxygen than is present in the atmosphere of the Earth today.

TABLE 4.4 Known fossil occurrences of giant terrestrial arthropods in the Carboniferous and Permian compared with the oxygen content in the Earth's atmosphere.

Geologic Age	Atmospheric O$_2$		Giant Arthropod Fossil Occurrences
	(1)	(2)	
Permian			
Changhsingian	24.1%	23.6%	
Wuchiapingian	26.5%	29.0%	
Capitanian	28.3%	30.0%	
Wordian	30.0%	31.0%	← Griffenflies
Roadian	30.6%	30.9%	← Griffenflies
Kungarian	33.0%	30.5%	← Griffenflies
Artinskian	35.0%	30.0%	← Griffenflies
Sakmarian	34.8%	29.3%	← Griffenflies
Asselian	34.5%	28.5%	← Griffenflies, millipedes
Carboniferous			
Gzhelian	34.2%	27.9%	← Griffenflies, millipedes
Kasimovian	33.5%	26.5%	← Griffenflies, millipedes
Moscovian	32.8%	25.8%	← Griffenflies, millipedes, spiders, silverfish
Bashkirian	32.0%	25.0%	← Griffenflies
Serpukhovian	30.0%	23.0%	
Visean	25.8%	18.8%	← Scorpions
Tournaisian	23.0%	17.0%	

Source: Column (1) is the Rock-Abundance model (Berner et al. 2003); column (2) is the Geocarbsulf model (Berner 2006). Oxygen data are extrapolated from figure 3.8 using the timescale of Gradstein et al. (2012).

Note: Atmospheric oxygen contents of 25% or higher are underlined; those of 30% or higher are in bold.

Giant flying griffenflies appear in the fossil record in the Bashkirian Age, when the Rock-Abundance model predicts that the oxygen content of the atmosphere had risen to 32 percent, 11 percentage points more oxygen than our present-day atmosphere of 21 percent. Giant millipedes appear in the fossil record in the Moscovian Age, when the Rock-Abundance model predicts that the oxygen content of the atmosphere of the Earth had risen even further, to 32.8 percent.

In contrast, the Geocarbsulf model (table 4.4) predicts that the atmosphere of the Earth contained only 18.8 percent oxygen in the Visean Age, 1.2 percentage points *lower* than that of the present Earth's atmosphere! How could a scorpion 700 millimeters (28 inches) long have breathed in an atmosphere that contained less oxygen than our world today, with its much smaller scorpions? The Geocarbsulf model predicts that atmospheric oxygen had risen to only 25 percent in the Bashkirian—just 4 percentage points higher than our present world— yet at this same time giant, energetically flying griffenflies appear in the fossil record. How could a dragonfly-like insect with a wingspan of 720 millimeters (28 inches) have breathed in an atmosphere containing only 4 percentage points more oxygen than in our world, where our large dragonflies have a wingspan of only 150 millimeters (six inches). This mismatch between the Geocarbsulf data and the fossil record data of giant arthropods (table 4.4) in the Carboniferous supports Scott and Glasspool's argument that we previously considered in chapter 3: "Collectively, these data suggest levels of O_2 modeled for this interval rising from 17 percent to 23.5 percent [in the Geocarbsulf model] are inappropriate and instead favor prior, higher levels modeled at ≈23–31.5 percent [in the Rock-Abundance model], values further supported by the occurrence of very large arthropods at this time" (Scott and Glasspool 2006, 10863). Indeed, the Geocarbsulf model predicts that atmospheric oxygen on the Earth did not reach the 30 percent level until the Artinskian Age of the Early Permian (table 4.4), some 33 *million years after* the appearance of the first griffenflies!

We can also contrast the differing predictions of the Rock-Abundance and Geocarbsulf models by reversing our argument: if the appearance of giant arthropods in the fossil record is linked to high oxygen levels in the atmosphere, then the *disappearance* of giant arthropods in the fossil record may be linked to lower oxygen levels. The giant griffenflies declined in abundance during the Capitanian in the Middle Permian, were rare in the Late Permian, and did not survive the end-Permian mass extinction (Nel et al. 2009; Clapham and Kerr 2012). In the Rock-Abundance model, oxygen levels in the atmosphere had dropped to 28.3 percent in the Capitanian—almost as low as the 25.3 percent level predicted for the Visean, a time during which no griffenflies were present on the Earth (table 4.4). Indeed, the griffenflies had already begun to experience difficulties back in the late Early Permian, when one of the two subfamilies of meganeurid griffenflies went extinct; the paleoentomologist André Nel of the National Museum of Natural History in Paris and his colleagues note that the "Meganeurinae apparently did not survive after the Early Permian as the Middle to Late Permian griffenflies are all Tupinae" (Nel et al. 2009, 115). In the Rock-Abundance model, oxygen levels in the Early Permian rose from 34.5 percent to a maximum of 35 percent, fell slightly to 33 percent and then dropped still further to 30.6 percent in the Roadian Age at the beginning of the Middle Permian—and the Meganeurinae became extinct.

Although only meganeurid griffenflies of the subfamily Tupinae survived in the Middle Permian, Nel and colleagues note that the "diversity of griffenflies was very high during the Middle Permian, while the more advanced groups of Protozygoptera and Triadophlebiomorpha were also flourishing" (Nel et al. 2009, 115). The Protozygoptera and Triadophlebiomorpha were smaller dragonfly-like insects, and they did not decline in abundance in the Late Paleozoic, nor did they go extinct in the end-Permian mass extinction. Thus, the giant griffenflies went extinct unlike the coexisting "contemporaneous Protozyogpera or Triadophebiomorpha that were still flourishing during the Triassic" (Nel et al. 2009, 115). Yet in the Geocarbsulf model, oxygen levels in the atmosphere persisted as high as 30 percent in the Capitanian at the end of the Middle Permian and as high as 29 percent at the beginning of the Late Permian in the Wuchiapingian (table 4.4), a pattern that does not match the demise of the giant griffenflies during this same interval of time. In contrast, the Rock-Abundance model predicts a much lower oxygen level of 28.3 percent in the Captianian and 26.5 percent in the Wuchiapingian, which is more in accord with the demise of the giant griffenflies.

Vertebrates and Oxygen

In contrast to the Visean Age giant scorpions and Bashkirian Age giant griffenflies (table 4.4), the first of the giant vertebrates began to appear a little later, in the Moscovian Age of the Late Carboniferous (table 4.5). Also in contrast to

TABLE 4.5 Known fossil occurrences of giant terrestrial vertebrates in the Carboniferous and Permian compared with the oxygen content in the Earth's atmosphere.

Geologic Age	Atmospheric O$_2$		Giant Vertebrate Fossil Occurrences
	(1)	(2)	
Permian			
Changhsingian	24.1%	23.6%	← Pareiasaurs, captorhinids
Wuchiapingian	26.5%	29.0%	← Pareiasaurs, captorhinids
Capitanian	28.3%	30.0%	← Dinocephalians, caseids
Wordian	30.0%	31.0%	← Dinocephalians, caseids
Roadian	30.6%	30.9%	← Dinocephalians, caseids
Kungarian	33.0%	30.5%	← Batrachomorphs, edaphosaurs, sphenacodonts
Artinskian	35.0%	30.0%	← Batrachomorphs, edaphosaurs, sphenacodonts
Sakmarian	34.8%	29.3%	← Batrachomorphs, edaphosaurs, sphenacodonts
Asselian	34.5%	28.5%	← Batrachomorphs, edaphosaurs, sphenacodonts
Carboniferous			
Gzhelian	34.2%	27.9%	← Diadectomorphs, ophiacodonts, edaphosaurs
Kasimovian	33.5%	26.5%	← Diadectomorphs, ophiacodonts
Moscovian	32.8%	25.8%	← Diadectomorphs, ophiacodonts
Bashkirian	32.0%	25.0%	
Serpukhovian	30.0%	23.0%	
Visean	25.8%	18.8%	
Tournaisian	23.0%	17.0%	

Source: Column (1) is the Rock-Abundance model (Berner et al. 2003); column (2) is the Geocarbsulf model (Berner 2006). Oxygen data are extrapolated from figure 3.8 using the timescale of Gradstein et al. (2012).

Note: Atmospheric oxygen contents of 25% or higher are underlined; those of 30% or higher are also in bold.

the giant arthropods, numerous vertebrate giants persisted into the late Middle Permian and Late Permian; examples include the Capitanian five-meter-long (16.4-foot-long) dinocephalian therapsid *Moschops capenis* (fig. 4.7), the Late Permian three-meter-long (ten-foot-long) pareiasaur parareptile *Bunostegos akokanensis,* and the two-meter-long (6.6-foot-long) captorhinid reptile *Moradisaurus grandis.*

The effect of a hyperoxic atmosphere on vertebrates is more subtle than for arthropods with their tracheal breathing system (Graham et al. 1995, 1997). Increased body size is clearly one possible effect, and, not only in the Carboniferous, increased levels of oxygen in the atmosphere in the early Cenozoic have been linked to the evolution of very large mammalian vertebrates during this same interval of time (Falkowski et al. 2005). It is surely no coincidence that very large vertebrates coexisted with very large arthropods in the Carboniferous and Permian landscapes (tables 4.4 and 4.5).

Michael McKinney, a paleontologist at the University of Tennessee, has conducted an extensive analysis of body-mass increases in the ophiacodont, edaphosaur, and sphenacodont (fig. 4.6) synapsid clades (table 4.2) during the Early Permian time span (McKinney 1990). During the Asselian and early Sakmarian Ages, the largest sphenacodont in his sample had a body mass of about 35 kilograms (77 pounds), the largest ophiacodont had a body mass of about 25 kilograms (55 pounds), and the largest edaphosaur had a body mass of about 15 kilograms (33 pounds). By the late Sakmarian Age, the body mass of the largest edaphosaur in his sample had increased to around 90 kilograms (198 pounds), and by the Kungurian Age, at the end of the Early Permian, the body mass of the largest edaphosaur was 330 kilograms (728 pounds). The body mass of the largest ophiacodont increased to 120 kilograms (265 pounds) by the Artinskian Age, and still further to 230 kilograms (507 pounds) by the Kungurian. Finally, the body mass of the largest sphenacodont increased to 120 kilograms (265 pounds) by the Artinskian and to 250 kilograms (551 pounds) by the Kungurian (McKinney 1990).

Thus during the 26-million-year span of the Early Permian, the body masses of edaphosaurs increased by a factor of 22, of ophiacodonts by a factor of 9.2, and of sphenacodonts by a factor of 7.1. These are incredible size increases— an edaphosaur at the end of the Early Permian was fully 22 times larger than one at the beginning of the Early Permian! And, in the Rock-Abundance model (table 4.5), oxygen levels in the Earth's atmosphere were well above 30 percent for this entire span of time. Note again the mismatch between the Geocarb-sulf model predictions and the actual distribution of giant vertebrates in time: in the Geocarbsulf model, oxygen did not even reach the 30 percent level in the atmosphere until the middle of the Early Permian, in the Artinskin Age (table 4.5), long after the dramatic size increases in the synapsid vertebrates were well under way.

Giant batrachomorph amphibians, such as the two-meter-long (6.6-foot-long) *Eryops megacephalus* (fig. 4.4) also existed during the Early Permian. It is possible that these giant salamander-like amphibians augmented their lung breathing by also breathing directly across their skin surface area, much as the present-day lungless plethodontid salamanders do. The plethodontids are small, only 100 to 200 millimeters (four to eight inches) long, but in our present-day 21 percent oxygen atmosphere they have totally lost their lungs and live solely by gas exchange across the surface area of their naked skin. In an atmosphere containing over 34 percent oxygen (Rock-Abundance model, table 4.5), such an augmentative breathing mechanism may have enabled the Early Permian amphibians to attain their giant size despite their very large body volume relative to skin surface area.

Marine Invertebrates and Oxygen

In contrast to the temporal pattern of gigantism seen on land in the vertebrates (table 4.5), the temporal distribution of giant marine invertebrates (table 4.6) matches more closely the pattern seen in the terrestrial invertebrates, the arthropods (table 4.4). Giant invertebrates first appear in the oceans in the Visean Age of the Early Carboniferous, as do the giant scorpions on land (fig. 4.1), and giant marine invertebrates are generally not present in the Late Permian, just as giant arthropods are not. The terrestrial arthropods are all protostomous animals, however, whereas in the oceans gigantism occurred in the clades of both the protostomes (brachiopods, fig. 4.9; bryozoans) and the deuterostomes (echinoids, crinoids), in animals that are not bilaterians (corals), and even in organisms that are not animals—the single-celled foraminifera (fusulinids, fig. 4.10).

Most marine invertebrates depend upon diffusion-mediated respiration; thus, seawater containing high concentrations of oxygen should facilitate the same size-increase effects in marine organisms as in terrestrial ones (Graham et al. 1995). Seas that existed under a hyperoxic atmosphere would have become oxygen rich via diffusion and might also have experienced little in the way of seasonal periods of anoxia that occur in many oceanic regions today. However, it can be demonstrated that at least some of the size increases seen in marine organisms that occurred in the Carboniferous and Permian were also the result of increased nutrient availability made possible by the evolution of photosynthetic symbioses, a topic that will be explored further in the next section of the chapter.

TABLE 4.6 Known fossil occurrences of giant marine invertebrates in the Carboniferous and Permian compared with the oxygen content in the Earth's atmosphere.

Geologic Age	Atmospheric O_2		Giant Invertebrate Fossil Occurrences
	(1)	(2)	
Permian			
Changhsingian	24.1%	23.6%	
Wuchiapingian	26.5%	29.0%	
Capitanian	28.3%	30.0%	← Fusulinids, alatoconchid bivalves
Wordian	30.0%	31.0%	← Brachiopods, fusulinids
Roadian	30.6%	30.9%	← Brachiopods, fusulinids, bryozoans
Kungarian	33.0%	30.5%	← Brachiopods, fusulinids, bryozoans
Artinskian	35.0%	30.0%	← Brachiopods, fusulinids, bryozoans, crinoids
Sakmarian	34.8%	29.3%	← Brachiopods, fusulinids, bryozoans
Asselian	34.5%	28.5%	← Brachiopods, fusulinids, bryozoans
Carboniferous			
Gzhelian	34.2%	27.9%	← Brachiopods, fusulinids, bryozoans
Kasimovian	33.5%	26.5%	← Brachiopods, fusulinids, bryozoans
Moscovian	32.8%	25.8%	← Brachiopods, fusulinids
Bashkirian	32.0%	25.0%	← Brachiopods, fusulinids
Serpukhovian	30.0%	23.0%	← Brachiopods, fusulinids, corals
Visean	25.8%	18.8%	← Brachiopods, echinoids, corals
Tournaisian	23.0%	17.0%	

Source: Column (1) is the Rock-Abundance model (Berner et al. 2003); column (2) is the Geocarbsulf model (Berner 2006). Oxygen data are extrapolated from figure 3.8 using the timescale of Gradstein et al. (2012).

Note: Atmospheric oxygen contents of 25% or higher are underlined; those of 30% or higher are also in bold.

ECOSYSTEM EVOLUTION AND ANIMAL GIGANTISM

Not all of the animal gigantism that appeared in the Carboniferous and Permian may be directly attributable to the hyperoxic atmosphere that existed on the Earth during that period of time. Other factors that have been proposed to explain this animal gigantism include the evolution of interspecies symbioses, of predator-prey interactions, and of competitive-exclusion interactions.

Many modern species of corals harbor photosynthetic algal species, collectively called zooxanthellae,[7] in their tissues in a complex interspecies symbiosis.

In the symbiosis, the corals metabolize the hydrocarbons, or food, synthesized by the algae and the oxygen produced by the algae as a waste product in photosynthesis. In turn, in their metabolism the corals produce phosphate and nitrate wastes and carbon dioxide that they recycle back to the algae, which use them to synthesize more hydrocarbons via photosynthesis. In addition, the corals provide the algae with protection from predation by grazing species of organisms by holding them within their tissues.

Corals are sessile predators and generally survive by eating the prey they capture. However, symbiotic coral species can use the additional energy they receive from their symbionts to produce skeletal calcium carbonate at accelerated rates and hence to grow the giant reef tracts found in shallow water regions of the modern world, such as the Great Barrier Reef in Australia. In addition, some modern species of bivalve molluscs are also able to grow to gigantic sizes by utilizing the extra energy provided by photosynthetic symbionts living within their tissues. Individuals of the giant clam *Tridacna gigas* can have shells 1300 millimeters (4.3 feet) long, compared to an average large clam with a shell about the size of your hand.

Given the occurrence of gigantism in some living coral cnidarians and bivalve molluscs that are symbiotic, it is reasonable to propose that the Carboniferous and Permian giant marine invertebrates may have been able to achieve their large sizes because they possessed symbionts, particularly the rugose corals (tables 4.3, 4.6). It is known that the giant fusulinid foraminifera (tables 4.3, 4.6) that existed in the Carboniferous and Permian contained symbiotic photosynthetic algae in their tissues: many of these giant species evolved specialized skeletal morphologies to contain and protect their symbionts.[8] Similar to the shells of the living giant lophotrochozoan mollusc (table 4.3) *Tridacna gigas*, the giant shells of the Permian bivalves of the family Alatoconchidae (Aljinovic et al. 2008) and the Carboniferous lophotrochozoan brachiopod *Gigantoproductus giganteus* (table 4.3, fig. 4.9) may indicate that algal symbioses were widespread in the clade of the ancient lophotrochozoan animals.

However, no photosymbiotic species are known to have evolved in the deuterostome echinoderm lineages of the crinoids and echinoids (table 4.3), the protostome lophotrochozoan lineage of the colonial bryozoans (table 4.3), or the lophotrochozoan molluscan lineage of the ammonoid predators (table 4.3), yet these invertebrate groups also evolved gigantism in the Carboniferous (table 4.6).

Thus, while some of the gigantism exhibited by the Late Paleozoic marine invertebrates may be linked to the evolution of interspecies symbioses, it is highly unlikely that mobile predators like ammonites would ever have evolved photosymbiosis. In addition, no terrestrial arthropod or vertebrate species has evolved photosymbiosis, so gigantism in land animals cannot be attributed to that mechanism.

On land, defense against predation is often invoked as a selective mechanism that would favor the evolution of large body size. A modern example is the combination of large body size and herd behavior in African elephants, both of which are powerful deterrents to lion predators. An adult elephant is simply too big, relative to the size of a lion, and there are too many elephants together in the herd—lions know that attacking them could lead to serious injury or death to the lion, not the elephants. Thus lions concentrate on attacking small, juvenile elephants, if they can be separated from the herd, or on attacking older, less agile and formidable adults. The same ecological strategy apparently evolved in ancient Mesozoic ecosystems with the evolution of giant body sizes in the sauropod dinosaurs and the evolution of herd behavior as well—traits that likely would have been deterrents to theropod dinosaur predators.[9]

Thus it could be argued that the evolution of giant body sizes in the sap-sucking palaeodictyopteran winged insects in the Carboniferous was a defensive response to the evolution of the giant meganisopteran griffenfly predators (table 4.1). How, then, can we ecologically explain the evolution of gigantism in the griffenflies? The griffenflies had no flying predators to defend themselves against—the first aerial, gliding vertebrate predators evolved in the Late Permian *after* the demise of the giant griffenflies (Nel et al. 2009). In addition, modern-day aerial bird and bat predators preferentially select large flying insects, not small ones, as prey (Clapham and Kerr 2012). If the giant griffenflies were ecological predatory equivalents of birds and bats, then it could be argued that predation was a selective mechanism favoring the evolution of *small* body size, not large, in the palaeodictyopteran prey species.

It has also been suggested that the absence of competition in a stable, open niche was a selective mechanism favoring the evolution of gigantism. In this scenario, the evolution of large body sizes is seen as the consequence of animals' evolving in optimal environmental conditions with little or no competition or predation (Briggs 1985; Harrison et al. 2010; Dorrington 2015). That is, the evolution of gigantism in stable, open niches could be argued to be the result of a form of ecological release, in contrast to the circumstances of species evolving in habitats with numerous competitors and/or predators (Blackburn and Gaston 1994). Clearly competition for limited resources such as food, could be a selective mechanism that would limit size. And, as discussed above, preferential predation on large prey species could be a size-limiting selective mechanism as well. Thus it could be argued that the giant griffenflies evolved as an ecological consequence of the absence of competitors and predators in the Carboniferous skies—the reverse of the defense-against-predation argument for the evolution of gigantism. How, then, can we ecologically explain the evolution of gigantism in the palaeodictyopteran sap-suckers, species that did not evolve in an environment free of predators?

In summary, the evolution of photosymbioses in marine invertebrate organisms clearly provides an alternative to a hyperoxic atmosphere as a physiological pathway to gigantism, although the two phenomena could also operate in concert. It is more difficult to argue that the *simultaneous* evolution of gigantism in numerous species of both terrestrial arthropods and vertebrates in the Carboniferous (tables 4.4 and 4.5) was the result of predator-prey and/or competitive-exclusion ecological interactions within the separate evolutionary lineages of these two very different groups of organisms. It is easier to argue that a common pervasive factor—the presence of a hyperoxic atmosphere in the Carboniferous—was the trigger for the evolution of gigantism in terrestrial arthropods and vertebrates. Further support for this argument comes from the fact that in another period of geologic time in which the Earth had a very oxygen-rich atmosphere—the early Cenozoic—gigantism evolved in mammalian vertebrates (Falkowski et al. 2005).

In this section of the chapter, we have considered physiological and ecological hypotheses for the evolution of animal gigantism as alternatives to the hypothesis that gigantism was triggered by hyperoxia. In the next section, that argumentation will be reversed: we will consider arguments that changes in the oxygen content of the Earth's atmosphere actually were the trigger for major physiological and ecological events in animal evolution.

LATE PALEOZOIC OXYGEN AND ANIMAL
EVOLUTIONARY EVENTS

The hyperoxic atmosphere of the Earth during the Carboniferous and Permian may have been the trigger for far more evolutionary events than just the appearance of animal gigantism. Jeffrey Graham, at the University of California in San Diego, and his colleagues have proposed that "global atmospheric hyperoxia possibly aided the vertebrate invasion of land"; that "the Carboniferous diversification of both the insects and the vertebrates correlates with the rise of atmospheric oxygen"; that "hyperoxic air may have also been critically important in the evolution of the cleidoic egg" in the first amniotes; and that the Permian "diversification of synapsids seems to be partially attributable to the effects of hyperoxia and a denser atmosphere on activity enhancing specializations such as metabolic heat retention," hence leading to the evolution of endothermy (Graham et al. 1995, 119–120).

A comparison of some of the major events in the evolution of vertebrates in the Carboniferous and Permian and the modeled oxygen content in the Earth's atmosphere during this span of time is given in table 4.7. The invasion of land by vertebrates began in the Devonian with the evolution of the first tetrapods from the sarcopterygian fishes, but these early invaders were "aquatic tetrapods," spending most of their time in the rivers and lakes of the terrestrial realm. Only in the Visean Age of the Early Carboniferous, following the Tournaisian Gap (table 1.7), did the invasion of dry land by the vertebrates begin in earnest (McGhee 2013). It is therefore of interest that the Rock-Abundance model predicts a rise in oxygen in the Earth's atmosphere to 25.8 percent in the Visean, 4.8 percentage points higher than in the present-day atmosphere, and that giant scorpions appeared on land and giant invertebrates appeared in the oceans at this same time (tables 4.4 and 4.6).

Graham and colleagues argue that the elevated oxygen content in the Earth's atmosphere during the Visean helped the vertebrates to successfully invade dry land in three ways. First, the hyperoxic atmosphere helped elevate primitive lung effectiveness in oxygen uptake. Second, it helped the early vertebrates to lower their rate of desiccation in breathing dry air by allowing them to take fewer breaths yet still obtain sufficient oxygen. Third, the hyperoxic atmosphere helped boost the metabolic rates of the vertebrates, thus assisting them in

TABLE 4.7 Major events in the evolution of terrestrial vertebrates in the Carboniferous, Permian, and Early Triassic compared with the oxygen content in the Earth's
atmosphere.

Geologic Age	Atmospheric O$_2$		Vertebrate Evolutionary Events
	(1)	(2)	
Triassic			
Olenekian	22.5%	20.0%	← Endothermic therapsids?
Induan	23.3%	21.8%	← Endothermic therapsids?
Permian			
Changhsingian	24.1%	23.6%	← End-Permian mass extinction
Wuchiapingian	26.5%	29.0%	← Therapsid and reptilian diversifications
Capitanian	28.3%	30.0%	← End-Capitanian extinction
Wordian	30.0%	31.0%	
Roadian	30.6%	30.9%	← Therapsid diversification
Kungarian	33.0%	30.5%	← Olson's extinction
Artinskian	35.0%	30.0%	
Sakmarian	34.8%	29.3%	
Asselian	34.5%	28.5%	
Carboniferous			
Gzhelian	34.2%	27.9%	
Kasimovian	33.5%	26.5%	← Tetrapod diversification, ecological innovation
Moscovian	32.8%	25.8%	
Bashkirian	32.0%	25.0%	← Synapsid and reptilian amniotes present
Serpukhovian	30.0%	23.0%	← Origin of amniotes?
Visean	25.8%	18.8%	← Invasion of land; origin of amniotes?
Tournaisian	23.0%	17.0%	

Source: Column (1) is the Rock-Abundance model (Berner et al. 2003); column (2) is the Geocarbsulf model (Berner
2006). Oxygen data extrapolated from figure 3.8 using the timescale of Gradstein et al. (2012).

Note: Atmospheric oxygen contents of 25% or higher are underlined; those of 30% or higher are also in bold.

overcoming the force of gravity in walking on dry land. Dehydration is a problem that land-dwelling animals still face today, as are the metabolic demands of
locomotion while enduring the constant pull of gravity—they point out that,
as water is 1,000 times more dense than air, an early tetrapod that was essentially weightless in water would be 1,000 times heavier on dry land (Graham
et al. 1997). But note that this invasion-assistance hypothesis works only for the
atmospheric-oxygen predictions (table 4.7) of the Rock-Abundance model for

the Visean—the Geocarbsulf model predicts a Visean atmosphere containing *less* oxygen than that of the present-day Earth!

We know from the fossil record that the amniote vertebrates had evolved by the Bashkirian Age at the beginning of the Late Carboniferous and that both clades of the amniotes were present—the reptilian amniote clade is represented by body fossils of the species *Hylonomus lyelli* and the synapsid amniote clade by body fossils of the species *Protoclepsydrops haplous*, both found in the famous Joggins strata with their preserved fossil tree stumps and vertebrate skeletons in Nova Scotia, Canada (Benton 2015). However, as discussed in chapter 2, trace fossil evidence suggests that more derived species of synapsid amniotes than *Protoclepsydrops haplous* were also present in the Bashkirian, not just the basal one known from body fossils. This evidence comes from the trackway ichnospecies *Dimetropus*, found in Bashkirian-aged strata in Germany (Voigt and Ganzelewski 2010). This trackway was made by quite a large animal—its hind feet were around 140 millimeters (5.5 inches) long, and its forefeet around 70 millimeters (2.8 inches) long. The trackway is most similar to those produced by more-derived ophiacodontid, edaphosaurid, or sphenacodontid synapsids, all of which were large animals. In contrast to the large *Dimetropus* footprints, both the known basal synapsid and reptilian amniotes were quite small—the animals were only 200 millimeters (eight inches) or so in length. If the ichnospecies *Dimetropus* was produced by a highly derived synapsid amniote in the Bashkirian, then the split between the synapsid and saurposid amniotes had to have occurred even earlier than the Bashkirian—sometime in the Early, not the Late, Carboniferous.

As discussed in chapter 2, an intriguing partial body fossil suggests that the amniotes may have evolved in the Visean Age, a possibility that we will explore in more detail here. This fossil is of the enigmatic species *Casineria kiddi*, found in a sedimentary nodule that contains most of the body skeleton—but unfortunately the head and tail are missing. The animal was tiny: its back, from the base of its neck to its pelvic girdle, was only 80 millimeters (three inches) long, making it the smallest known Early Carboniferous tetrapod (Carroll 2009).

The vertebrae of *Casineria kiddi* are solidly ossified, and the animal had long, slender, curved ribs. The well-preserved forelimb and forefoot of the animal held the greatest surprise, as described by the Cambridge University paleontologist Jennifer Clack: "The humerus is much more slender than that of any other

Early Carboniferous tetrapod, with an obvious shaft and with the two ends set at different angles to each other (known as torsion). The radius and ulna are also slender, with an olecranon process on the ulna.... These features alone strongly suggest a fully terrestrial animal." In the forefoot, "the manus has five slender digits.... The last phalanx on each digit (the ungual) is noticeably curved, and the whole arrangement suggests a hand capable of grasping. No other Early Carboniferous tetrapod shows digits like this, but they are similar to those found in Late Carboniferous early amniotes" (Clack 2002, 199). The McGill University paleontologist Robert Carroll also observes: "Uniquely, *Casineria* had curved terminal phalanges (unguals) forming claws, as in many early amniotes" (Carroll 2009, 74). In summary, Clack notes: "It has been suggested that the origin of amniotes is connected with an evolutionary step involving small size ... and here is a specimen that accords well with that theory" (Clack 2012, 277). Because the head of the animal is missing, it is impossible to prove that *Casineria kiddi* was indeed the first known amniote; clearly, however, it is the most derived species of tetrapod yet known from the Visean Age.

Graham and colleagues argue that a hyperoxic atmosphere—whether in the Visean, Serpukhovian, or Bashkirian—also assisted in the evolution of the first amniote vertebrates. A key step in the evolution of the amniotes was the evolution of the amniote egg: "With specialized membranes to prevent water loss (amnion), enhance respiratory gas exchange (chorion), and collect waste products (allantois), the amniotic egg (also termed the cleidoic egg), which appeared in the Carboniferous, eliminated amphibian reliance upon aquatic egg laying and larval development.... Accordingly, natural selection leading to the development of the three amniote egg membranes enabled a further increase in egg size. Moreover, the hyperoxic Carboniferous atmosphere would have allowed the development of large amniote eggs by minimizing the ratio of water loss to oxygen uptake" (Graham et al. 1997, 153). Note again that this amniote-evolution-assistance hypothesis works only for the atmospheric-oxygen predictions of the Rock-Abundance model for the Visean Age (25.8 percent oxygen) and not the Geocarbsulf model (table 4.7), which predicts a Visean atmosphere containing less oxygen (18.8 percent) than in our present world (21 percent).

Following the evolution of amniotes (whether in the Visean or Serpukhovian), a major diversification of amniotes occurred in the Moscovian-Kasimovian interval of time. The first fossils of the advanced ophiacodontid synapsids are found in Moscovian strata, and fossils of the first advanced varanopid, edapho-saurid, and sphenacodontid (fig. 4.6) synapsids are found in Kasimovian strata. In the reptilian amniote clade, the first fossils of the advanced diapsid, *Petrola-cosaurus kansensis*, is found in Kasimovian strata (McGhee 2013). Of particular note is the evolution of herbivory in both the batrachomorph amphibian and the synapsid amniote clades in the Late Carboniferous. The evolution of the ability to efficiently use living plants as a food source was a key element in the construction of the terrestrial ecosystem familiar to us today, with its energy flow from living plants to herbivores to carnivores. Within the batrachomorph clade the first fully herbivorous vertebrates, the diadectomorphs, are represented by the Moscovian species *Limnostygis relictus*, and the herbivorous edaphosaurid synapsid species *Ianthasaurus hardestii* and *Xyrospondylus ecordi* appeared in the following Kasimovian Age (McGhee 2013).

Graham and colleagues further argue that "the Carboniferous diversification of both the insects and vertebrates correlates with the rise in atmospheric oxy-gen" (Graham et al. 1995, 119, fig. 2). However, establishing a causal link between increased speciation and evolutionary innovation and the presence of a hyper-oxic atmosphere is not as straightforward as it is for the evolution of large body sizes or the evolutionary transition from aquatic habitats to terrestrial habitats. Why should higher oxygen levels trigger evolutionary innovation, such as the evolution of herbivory, or increased speciation rate? Graham and colleagues simply speculate that "increased oxygen availability may have also fuelled the diversification and ecological radiation of late Palaeozoic groups by acting as a substrate for the evolution of behavioural, physiological and ecological adapta-tions, permitting greater exploitation of aquatic habitats and the newly evolv-ing terrestrial biosphere" (Graham et al. 1995, 120). It is known that both clades of the amniotes were present in the Bashkirian (table 4.7), when oxygen levels reached 32 percent in the atmosphere (in the Rock-Abundance model). Was the increase in oxygen in the atmosphere to 33.5 percent in the Kasimovian really the trigger for amniote diversification and ecological innovation?

In contrast, the University of Bristol paleontologist Sarda Sahney and col-leagues argue that the Carboniferous diversification of tetrapods was triggered by the ecological effects of the Kasimovian crisis in the great rainforests that we considered in chapter 3. They argue that the Kasimovian crisis resulted in the fragmentation of the tropical rainforests into numerous small habitat islands, and demonstrate that a major increase in the number of endemic species of tetrapods occurred from the Moscovian to the Kasimovian. Not only did the evolution of numerous isolated pockets of endemic species occur, but that "rain-forest collapse was also accompanied by acquisition of new feeding strategies (predators, herbivores), consistent with tetrapod adaptation to the effects of habitat fragmentation and resource restriction" (Sahney et al. 2010, 1079), and "our data, which show elevated extinction rates, increased endemism, and eco-logical diversification, apparently represent a classic community-response to habitat fragmentation" (Sahney et al. 2010, 1081). They also point out that the batrachomorph amphibians experienced high extinction rates—losing nine families—and major ecological turnover from the Moscovian to the Kasimov-ian, whereas the amniotes lost no families to extinction, probably because they had an "ecologic advantage in the widespread drylands that developed" in the Kasimovian (Sahney et al. 2010, 1081).

Another animal evolutionary event during the Carboniferous and Permian that may have had no relationship to the presence of a hyperoxic atmosphere on Earth is Olson's extinction[10] in synapsid amniotes, which occurred in the Early to Middle Permian transition (table 4.7). In this vertebrate extinction event, the long-lived and dominant "pelycosaurs" (table 4.2), the non-therapsid synapsids, were eliminated. Of the existing six families of pelycosaurs present before the extinction, three went extinct in the late Artinskian, a fourth perished in the early Kungurian, and the final two families succumbed in the Captanian (Kemp 2006). The exact cause of the demise of the previously highly successful clades of non-therapsid synapsids remains unknown. Oxygen levels in the atmosphere are predicted to have fallen by 2 percent from the Artinskian to the Kungurian in the Rock-Abundance model (table 4.7)—but could that have triggered the extinction of six entire families of pelycosaurs? Olson's extinction will be considered in more detail in chapters 5 and 6, where the possibility that this vertebrate extinction event was triggered by paleoclimatic changes at the end of the Late Paleozoic Ice Age will be explored.

Following a gap in the fossil record, named Olson's Gap (Lucas and Heck-ert 2001; Lucas 2004), a major diversification in therapsid synapsids occurred in the Roadian (table 4.7). The basal therapsid group, the Biarmosuchia, had evolved by the Early Permian and is represented by the species *Tetraceratops insignis*. However, it was only in the Middle Permian that the dinocephalian therapsids evolved—a diverse group both in terms of numbers, over 40 genera, and in terms of ecology, as both carnivorous and herbivorous dinocephalians existed (Benton 2015). The dinocephalians also included some of the largest land animals to exist in the Permian, such as the five-meter-long (16.4-foot-long) dinocephalian herbivore *Moschops capenis* (fig. 4.7). The dinocephalians appear in the fossil record after the demise of the "pelycosaurs" (table 4.2), the non-therapsid synapsids; thus their replacement of these earlier synapsid herbivores and carnivores in the Permian appears to be a case of passive ecological replacement rather than active, competitive ecological replacement.

The dinocephalians themselves were driven to extinction at the end of the Capitanian (table 4.7) and were replaced by the dicynodont and gorgonopsian therapsids (table 4.2). The end-Capitanian extinction will be considered in more detail in chapters 5 and 6, but like Olson's extinction, it appears not to have had any relationship to the oxygen content of the Earth's atmosphere. Oxygen levels fell only 1.7 percent from the Wordian to the Capitanian (in the Rock-Abundance model, table 4.7), even less than the 2 percent drop that preceded Olson's extinction. Instead, the end-Capitanian extinction is thought to have been triggered by the onset of mantle-plume volcanism in China during this interval of time.[11]

The Late Permian dicynodonts were a diverse group of herbivores, over 60 genera, and the gorgonopsians comprised about 35 genera of carnivores (Benton 2015). The dicynodonts and gorgonopsians were the dominant herbivores and carnivores of the Late Permian, and together they constituted about 80 percent of the terrestrial vertebrate diversity (Erwin 1993). Finally, the therocephalian and cynodont therapsids (table 4.2) also appeared in the Late Permian—the cynodonts are particularly of note because the first mammals would evolve in the cynodont clade in the Late Triassic, represented by the species *Adelobasileus cromptoni*.

Although the therapsids dominated Late Permian terrestrial ecosystems, significant evolutionary diversification and innovation also occurred in the clade of the reptilian amniotes. The pareiasaurs and captorhinids appeared, some species of which were large animals (tables 4.2, 4.5). Also, the very first gliding vertebrates evolved in the Early Permian: the diapsid reptile *Coelurosauravus jaekeli*.[12] The first reptiles capable of true powered flight, the pterosaurs, would evolve some 50 million years later in the Triassic (Steyer 2012).

Do any of these vertebrate evolutionary events have any relationship with the amount of oxygen present in the atmosphere in the Permian? Graham and colleagues argue that in the Permian the "synapsids underwent a pronounced diversification, proceeding from the pelycosaurs to the therapsids, a diverse assemblage of herbivores and carnivores. . . . The diversification of synapsids seems to be partially attributable to the effects of hyperoxia and a denser atmosphere on activity enhancing specializations such as metabolic heat retention. The large 'sails' of pelycosaurs such as *Dimetrodon* [fig. 4.6] are thought to have functioned in heat transfer. The capacity to regulate heat gain and loss was an important precursor to endothermy" (Graham et al. 1995, 120). The distinctive "sails," or crests, on the backs of both the carnivorous sphenacodonts and herbivorous edaphosaurs are indeed thought to have had a thermoregulatory function, allowing these large animals to shed heat to the surrounding air when they became too hot or to absorb heat from sunlight when the animals were cold. Thus, although the animals were still ectotherms, they had a sophisticated anatomical and behavioral mechanism for regulating their body temperature. If, in addition, the sphenacodonts (fig. 4.6) and edaphosaurs had elevated metabolic rates—triggered by the hyperoxic atmosphere of the Early Permian—the need for such a thermoregulatory mechanism would have been more acute.

The non-therapsid synapsids were not the only vertebrates evidencing elevated metabolic rates during the Permian. In the Early Permian, two separate lineages of reptiles appear to have convergently evolved the capability to stand up and locomote on their hind limbs only—they evolved bipedalism.[13] The case for *Aphelosaurus lutevensis*, a basal diapsid reptile, is less clear—it may have been mostly arboreal, spending most of its time climbing around in trees and only occasionally descending to the forest floor to locomote on its hind limbs. The case for the anapsid bolosaur *Eudibamus cursoris* is unequivocal—this animal was clearly bipedal, capable not only of walking or running on its hind limbs only but also of jumping, giving it the nickname "kangaroo reptile" among vertebrate paleontologists (Steyer 2012, 134). Bipedal locomotion is metabolically costly, and its appearance in the reptiles gives evidence for the presence of elevated metabolic rates in both the synapsid and reptilian amniote clades during the Early and Middle Permian.

The problem is, the dominant sphenacodonts and edaphosaurs went extinct in the Kungurian and the therapsid diversification in the Roadian occurred only after the disappearance of the non-therapsid synapsids—that is, following Olson's Gap. Thus there was not a continuous diversification of synapsids "proceeding from the pelycosaurs to the therapsids," as argued by Graham and colleagues, and the Roadian therapsid diversification appears to have been a separate evolutionary event. It is not clear how this diversification and ecological innovation, in the evolution of both herbivores and carnivores, in the dinocephalian therapsids could have been driven by atmospheric oxygen. Only the giant size of some of the dinocephalians, like the *Moschops* herbivores (fig. 4.7), can clearly be attributed to the effect of hyperoxia.

Still, in a later study, Graham and colleagues expanded their argument that major evolutionary changes in the synapsid lineage were driven by changes in the oxygen content of the Permian atmosphere, now adding both hyperoxia and *hypoxia* to the equation:

> A significant event in tetrapod ecological physiology was the evolution of endothermy . . . The large sails of *Dimetrodon* (from the Lower Permian) and other sphenacodontids suggest the presence of a complex behavioral repertoire revolving around the capacity to regulate heat transfer. . . . The discovery of turbinate bones in the nasal passages of therapsids indicates the presence of a water-conserving mechanism linked to frequent ventilation and endothermy and correspondingly suggests that the evolution of a "mammalian" metabolic rate has occurred by the Late Permian (Hillenius, 1992; 1994). We suggest a two-part scenario for the evolution of a mammalian-level of metabolism in the hyperoxic Carboniferous-Permian biosphere. First . . . synapsids may have undergone natural selection for a relatively high metabolic rate and also increased their body size (thermal inertia). . . . Increased metabolic expenditures . . . would have been favored by an abundance of environmental oxygen. Second, the presence of these metabolically specialized and hyperoxia adapted organisms in a Permian environment characterized by progressive atmospheric hypoxia could have intensified natural selection on certain lineages for an increased ventilation frequency (hence the appearance of turbinal bones in therapsids) and improved cardiac efficiency for oxygen delivery to the tissues (i.e., separation of systemic and pulmonary circulation) . . . this finding suggests a mechanism through which the Permian oxygen decline could have influenced the evolution of a four chambered heart. (Graham et al. 1997, 158–160)

Graham and colleagues note that "only the crocodiles, among the reptiles, and the birds and mammals have a four-chambered (i.e., completely separated pulmonary and systemic circulations) heart" (Graham et al. 1997, 155), and of these animal groups both the birds and mammals are endothermic. Thus Graham and colleagues argue that the onset of hypoxia in the latest Permian triggered the evolution of the four-chambered heart, and perhaps full endothermy, in the latest Permian therapsids (presumably the cynodonts, the ancestors of the mammals, although this is not explicitly stated in their argument).

The Rock-Abundance model does indeed predict progressive drops in the oxygen content of the Earth's atmosphere during the span of the Late Permian into the Early Triassic—falling to 26.5 percent in the Wuchiapingian (a 1.8 percentage point drop from the Capitanian), to 24.1 percent in the Changhsingian (a 2.4 percentage point drop from the Wuchiapingian), to 23.3 percent in the Early Triassic Induan Age (a 0.8 percentage point drop from the Changhsingian), and finally to 22.5 percent in the Early Triassic Olenekian Age (another 0.8 percentage point drop from the Induan) (table 4.7). A predicted Early Triassic atmosphere containing 22.5 percent oxygen is still richer in oxygen than our present world, and is certainly not hypoxic.

In the Geocarbsulf model, however, atmospheric oxygen is predicted to have fallen precipitously from 29 percent in the early Wuchiapingian to 23.6 percent in the early Changhsingian—a 5.4 percentage point drop in only 5.6 million years—with an additional 3.6 percentage point drop to a hypoxic 20.0 percent by the beginning the Early Triassic Olenekian Age (table 4.7). Yet Graham and colleagues do not attribute the end-Permian mass extinction to the onset of hypoxia in either model, stating that the "rate of oxygen decline was too gradual, however, to have been the primary cause of the end-Permian extinction" (Graham et al. 1995, 117).

The problem with the hypoxia-onset model for triggering the evolution of a four-chambered heart and the evolution of an endothermic metabolic rate is the temporal pattern of evolution seen in the geologic record for both the synapsid and reptilian lineages. As noted by Graham and colleagues, only the crocodilian archosaurs, the avian dinosaurs (birds), and the mammals possess four-chambered hearts today. However, the first known mammal, *Adelobasileus cromptoni*, appeared only in the Late Triassic—long after the Late Permian–Early Triassic hypoxic interval. The first known bird, *Archaeopteryx lithographica*, appeared only in the Late Jurassic, even further removed in time from the Late Permian–Early Triassic hypoxic interval.

If the cynodont therapsids (table 4.2) had evolved a four-chambered heart and an endothermic metabolism in the Late Permian or an Early Triassic hypoxic interval, why was the evolution of the first mammals—animals that unequivocally possess both of these traits—delayed over 15 million years until the Late Triassic? Rather than the synapsids, a better temporal case for the hypoxia-onset model could be made for the reptiles: the crocodilian archosaurs (table 4.2) possess a four-chambered heart (but not an endothermic metabolic rate), and the crocodilian archosaurs did indeed evolve in the Early Triassic.

In summary, we have examined thus far many of the proposed consequences of the evolution of the great Carboniferous rainforests, both for the composition of the Earth's atmosphere and for the evolution of both trees and animals during the Late Paleozoic Ice Age. In the Permian, the Late Paleozoic Ice Age would come to an end—with catastrophic consequences for both plant and animal life. In the next chapter, we will examine the prelude to disaster—the beginning of the end.

5 | The End of the Late Paleozoic Ice Age

A massive expansion of ice occurred at the Pennsylvanian-Permian bound-ary, and glaciation became bipolar at that time. Ice sheets are inferred to have been at their maximum extent during the Asselian and early Sakmarian, after which they decayed rapidly over much of Gondwana.

—Fielding, Frank, and Isbell (2008, 343)

THE EARLY PERMIAN DEEP FREEZE

Nine million years had passed since the end of the C4 continental glaciation in eastern Gondwana (present-day Australia) and the massive extinction in the great rainforests in the Kasimovian Age of the Late Carboniferous (table 3.2). But the Late Paleozoic Ice Age was not finished yet, and the dawn of the Permian saw the return of massive ice sheets across the supercontinent Gondwana in the Southern Hemisphere, both in the west and the east, and by the early Sakmarian Age in the Northern Hemisphere as well (table 5.1). For the second time, the Late Paleozoic Ice Age had become bipolar, as stated in the epigraph of this chapter. Or had it? Once again, the bipolar controversy concerning the Late Paleozoic Ice Age—which we first encountered in the Carboniferous in chapter 2—returns when we enter the Permian.

The Early Permian chill is astounding: the Earth went from the balmy Kasimovian-Gzhelian interpulse greenhouse interval in the Late Carboniferous (table 3.2) to a globally frozen state with new ice caps in the South in the Asselian—and perhaps in the North Pole in the Sakmarian (table 5.1). At the beginning of the Permian, the ice

TABLE 5.1 Climatic events, area of rainforest coverage, and biological events during the Permian phase of the Late Paleozoic Ice Age.

Age	Geologic Time (Ma)	Climatic Events			Extent of Rainforests	Extinction Events
		(1)	(2)	(3)		
(Triassic)	252					
CHANGHSINGIAN	253					← End-Permian Mass Extinction
	254				140,000 km²	
WUCHIAPINGIAN	255				183,000 km²	
	256	ICE			236,000 km²	
	257	ICE			289,000 km²	
	258	ICE			342,000 km²	
	259	ICE			395,000 km²	
CAPITANIAN	260	ICE-P4			373,000 km²	← Capitanian Extinction
	261				350,000 km²	
	262				328,000 km²	
	263	ICE			305,000 km²	
	264	ICE			283,000 km²	
	265	ICE			261,000 km²	
WORDIAN	266	ICE			239,000 km²	
	267	ICE			227,000 km²	
	268	ICE	ICE?		216,000 km²	
ROADIAN	269	ICE	ICE?		194,000 km²	
	270	ICE	ICE?		179,000 km²	
	271	ICE-P3	ICE?		164,000 km²	
	272		ICE?		150,000 km²	
KUNGURIAN	273				127,000 km²	
	274				105,000 km²	← Olson's Extinction
	275				105,000 km²	
	276				105,000 km²	
	277				105,000 km²	
	278				105,000 km²	
	279				105,000 km²	
ARTINSKIAN	280				332,000 km²	
	281				482,000 km²	
	282				633,000 km²	
	283				784,000 km²	

		(1)	(2)	(3)	
	284		ICE		1,011,000 km²
	285		ICE		1,115,000 km²
	286		ICE		1,220,000 km²
	287		ICE		1,325,000 km²
	288		ICE		1,430,000 km²
	289		ICE		1,690,000 km²
	290		ICE		1,640,000 km²
SAKMARIAN	291		ICE	ICE?	1,590,000 km²
	292		ICE-P2	ICE?	1,590,000 km²
	293			ICE?	1,590,000 km²
	294	ICE	ICE	ICE?	1,590,000 km²
	295	ICE	ICE	ICE?	1,590,000 km²
ASSELIAN	296	ICE	ICE		1,590,000 km²
	297	ICE	ICE		1,422,000 km²
	298	ICE	ICE-P1		1,255,000 km²
(Carboniferous)	299				1,087,000 km²
	300				1,095,000 km²

Source: Gondwana data from Isbell et al. (2003), Frank et al. (2008), Fielding, Frank, Birgenheier et al. (2008), and Metcalfe et al. (2015); Siberian data from Ustritsky (1973), Epshteyn (1981b), Chumakov (1994), and Raymond and Metz (2004). Rainforest data from Cleal and Thomas (2005), timescale modified from Gradstein et al. (2012).

Note: Climatic events column (1) is West and Central Gondwana and column (2) is East Gondwana (Australia), both in the Southern Hemisphere; column (3) is Siberia (northeastern Russia) in the Northern Hemisphere. Bold type indicates the existence of continental glaciers; normal type indicates the presence of alpine glaciers, with glaciations in eastern Gondwana designated P1 through P4, from oldest to youngest.

sheets in western and central Gondwana formed their third and final phase of continental glaciation in the Late Paleozoic Ice Age, and in eastern Gondwana the ice sheets formed the first glaciation phase, P1, of the Permian.

The climatic condition of the North Pole is much more uncertain. Early stratigraphic data from the northeastern tip of Russia have been use to argue for the presence of continental ice in this region, which at that time was very close to the North Pole,[1] during the Sakmarian, but other workers have argued that these data were misdated and that they were in fact of Bashkirian-Muscovian age in the Late Carboniferous (Ustritsky 1973; Epshteyn 1981a; Isbell et al. 2016). For this reason, I have placed question marks on the Sakmarian northern hemisphere data in table 5.1.

A brief warming phase then occurred in the middle Sakmarian Age in the Southern Hemisphere, and the continental glaciers in Gondwana retreated (table 5.2). Cooling resumed in the late Sakmarian, triggering the P2 continental glacial phase in eastern Gondwana—but only in eastern Gondwana, as western and central Gondwana remained free of continental glaciers. Still, in eastern Gondwana the Early Permian P1 and P2 continental glacial pulses spanned a total of some 14 million years of ice cover (table 5.1).

But change was in the air. By the end of the Sakmarian, the ice cap in the Northern Hemisphere had melted away—if it existed!—and by the middle to late Artinskian, the P2 continental glaciation phase in eastern Gondwana had come to an end as well (table 5.1). The twilight of the Late Paleozoic Ice Age had arrived.

THE MELTING OF THE ICE AGE

The Earth experienced a 12-million-year interpulse greenhouse period spanning the period of time from the middle of the Artinskian Age to the early Roadian Age.[2] In the early Roadian Age, the planet once again cooled—but to a lesser extent than in any previous cooling event in the Late Paleozoic Ice Age. Two more phases of ice sheets formed in eastern Gondwana, but only in the highlands and mountains (table 5.1). The earlier phase of alpine glaciation, P3, was colder, and it has been argued that the last glaciation in the Northern Hemisphere also occurred at this time (Ustritsky 1973; Epshteyn 1981b; Chumakov 1994; Raymond and Metz 2004), but both events were brief on geologic timescales.

The last claim to continental glaciation in the Northern Hemisphere (table 5.1), like all of the others, has been challenged by John Isbell and colleagues. In this instance, they have actually shown that reportedly glacial, continental sedimentary deposits in northeastern Russia were in fact deepwater marine in nature and were formed by submarine gravity-flow slumps and turbidity currents—at least in three stratigraphic sections in the Okhotsk Basin (Isbell et al. 2016). Still, the very high latitude position of the northeastern tip of Russia in the Late Permian—almost positioned on the North Pole itself[3]— makes it difficult to understand how it could have remained ice free while the Southern Hemisphere was glaciated (table 5.1). Isbell and colleagues note this climatic anomaly and state that "our findings constrain boundary conditions such that future climate modeling can better determine factors that allowed

Gondwana glaciation to occur while inhibiting the development of land-based ice in Northeastern Asia" (Isbell et al. 2016, 297).

In the Southern Hemisphere, the last ice sheets in the Late Paleozoic Ice Age, the alpine glaciation phase P4 in eastern Gondwana, lasted for some five million years before they also melted away (table 5.1). The Late Paleozoic Ice Age had ended.

Why did the Late Paleozoic Ice Age end? One factor that contributed to its end was paleogeographic: the tectonic plate holding the giant continent of Gondwana was finally moving off of the South Pole (fig. 1.4). For over 110 million years, the South Pole—the coldest spot in the Southern Hemisphere—had been located on Gondwana, first in the west and then moving slowly and erratically to the east as the tectonic plate holding Gondwana shifted with time over the South Pole. By the end of the Permian, the South Pole had come to be located on the edge of eastern Gondwana—the southeast edge of present-day eastern Australia (fig. 1.4). The open oceanic waters of Panthalassa[4]—the "all ocean" or "world ocean" surrounding the single world continent Pangaea[5]—were located just offshore to the south, and they would have ameliorated temperature extremes on the neighboring continental margin just as oceanic waters do on continental margins today. As anyone knows who lives on a coastline near an ocean, summers are usually cooler and winters are usually warmer than they are in the continental interior. Cities like Chicago, located near large bodies of freshwater like Lake Michigan, also usually experience a smaller range of temperatures from summer to winter than cities like Tucson, surrounded only by land and rock. This effect occurs because water takes a long time to heat up and, once warm, takes a long time to cool down. The opposite is true of bare soil and rock—they heat up rapidly and become very hot in the summer, then just as rapidly lose heat and become very cold in the winter.

In figure 1.4 we saw that the position of the South Pole moved from the continental interior of Gondwana, present-day Antarctica, to the coastline about 265 million years ago. During the last phase of the Late Paleozoic Ice Age—the entire P4 glacial interval (table 5.1)—the South Pole was located near the southeastern margin of Gondwana and hence near the oceanic waters of Panthalassa. The entire P4 interval was also the least cold of the Permian phases of the Late Paleozoic Ice Age, consisting of alpine glaciers in the highlands of Gondwana, and the continental ice sheets were unipolar, occurring only in the Southern Hemisphere.

Another factor that contributed to the end of the Late Paleozoic Ice Age was global climate change. The Earth became hotter and drier with the passage of time in the Late Permian, and this global climatic change also contributed to the death of the great lycophyte rainforests—a topic that will be examined in detail in the next section of the chapter. The P4 glacial pulse spanned the period from the late Capitanian to the mid Wuchiapingian (Metcalfe et al. 2015), and thus the beginning of the last glacial pulse of the Late Paleozoic Ice Age coincided in time with the Capitanian extinctions (table 5.1) (see Metcalfe et al. 2015, 75, fig. 14). The environmental trigger for the Capitanian biodiversity crisis will be explored in detail in the next chapter, but here it will be noted that massive volcanic eruptions also occurred in South China during the late Capitanian. The geographic area of these eruptions is known today as the Emeishan Large Igneous Province, and we have geochemical evidence that enormous amounts of carbon dioxide and methane—both powerful greenhouse gases—were vented into the Earth's atmosphere in the late Capitanian. With an atmosphere laden with heat-trapping greenhouse gases, the entire Earth had to have become hotter, and not just in eastern Gondwana where the last of the P4 glaciers eventually melted.

THE END OF THE GREAT RAINFORESTS

Not only did massive glaciers return on the Earth at the dawn of the Permian, but the great lycophyte rainforests began to return as well. The extent of the rainforests increased by 168,000 square kilometers (64,848 square miles), or about 15 percent, at the beginning of the Asselian Age—from a low of 1,087,000 square kilometers (419,582 square miles) at the close of the Carboniferous to 1,255,000 square kilometers (484,430 square miles) when the ice sheets returned to Gondwana (table 5.1). The rainforests continued to expand in the Early Permian, reaching a maximum area of 1,690,000 square kilometers (652,340 square miles) in the early Artinskian Age (table 5.1). Thus the maximum extent of the Permian rainforests was comparable to the size of the rainforests present at the Bashkirian/Moscovian boundary in the Late Carboniferous (table 3.2), some 23 million years earlier. However, the maximum size of the Permian rainforests never quite reached the maximum of the Carboniferous rainforests, being 70 percent of the rainforest cover present on the Earth during the Moscovian Age (table 3.2).

As noted in chapter 3, the Permian rainforests were also more geographically restricted than those of the Carboniferous. Only in the far east, on the large islands of the Sino-Korean continental block (present-day Korea and North China) and the Yangtze continental block (present-day South China), did the lycophyte-dominated tropical mires continue to exist (fig. 3.7). The equatorial regions of Pangaea to the west, once covered by the lycophyte-dominated rainforests in the Carboniferous (fig. 3.7), were now populated by plants that had previously been found mostly in highland regions or in savannah environments in the lowlands (Cleal and Thomas 2005).

The Permian lycophyte rainforests began to decline in the Artinskian, coincident with the melting of the P2 continental glaciations (table 5.1). From that point onward, the rainforests were never larger in area than one million square kilometers, and they shrank in size by almost an entire order of magnitude—from 1,011,000 to 105,000 square kilometers (390,346 to 40,530 square miles)—in just five million years following the end of the P2 continental glaciation (table 5.1). Eventually the lycophyte rainforests vanished entirely on the island of the Yangtze continental block (fig. 3.7), and only to the north, on the Sino-Korean block, did the rainforests survive into the Kungurian. By the late Kungurian, the Sino-Korean rainforests had shrunk to just over 100,000 square kilometers (38,600 square miles)—about the size of the modern state of South Korea. The drastic reduction in the expanse of the Permian rainforests in the mid-Artinksian to late Kungurian had a devastating effect on the pelycosaurs, the non-therapsid synapsids; we will examine the demise of these tetrapods in detail in the next section of the chapter.

With the onset of the smaller P3 alpine glaciations in Gondwana, and perhaps the return of ice in the Northern Hemisphere, the rainforests recovered somewhat and expanded in size by almost a factor of four, reaching an extent of 395,000 square kilometers (152,470 square miles) by the dawn of the Wuchiapingian Age (table 5.1). This brief recovery can be attributed largely to the reestablishment and spread of the rainforests on the Yangtze continental block, as the Sino-Korean rainforests to the north remained the same size as they had been in the Kungurian (Cleal and Thomas 2005).

The rainforests began their final decline following the end of the P4 alpine glaciation in eastern Gondwana (table 5.1). By the end of the Wuchiapingian, the rainforests on the Sino-Korean continental block had vanished—only those on the island of the Yangtze block survived (Cleal and Thomas 2005). The paleobotanists Christopher Cleal and Barry Thomas argue that the lycophyte-dominated rainforests shrank during the Permian "as water-stress and increasing wildfires made the habitat unsuitable for the dominant lycophytes (Wang and Chen, 2001.) . . . Towards the end of the Permian, even drier conditions caused the forests to contract further, and increasing numbers are found of Mesophytic [drier adapted] plants. . . . This trend towards drier and hotter conditions may have been in part a consequence of the rain-shadow effect from the rising Northern Border Highlands to the north. However, Enos (1995) favoured the progressive drift north of [North] China, so that by the Changhsingian Age the area was outside of tropical latitudes. This is compatible with the persistence through the end of the Permian of some coal forests in South China, which were still in tropical latitudes" (Cleal and Thomas 2005, 21). That is, Cleal and Thomas attribute the paleoclimatic trend toward "drier and hotter conditions" in rainforest regions toward the end of the Permian to tectonic effects—both mountainous uplift on the Sino-Korean continental block and movement of the block to the north out of the equatorial region of the Earth.

But what if the trend toward hotter and drier conditions in the late Permian was *global*, occurring across the entire planet, and not just in the Sino-Korean region? In the next chapter we will examine the geological evidence that massive volcanic eruptions occurred on the Yangtze continental block during the late Capitanian, along with the geochemical evidence that enormous amounts of powerful greenhouse gases were vented into the Earth's atmosphere beginning in the late Capitanian.

Regardless of the ultimate cause of the late Permian paleoclimatic trend toward hotter and drier conditions, by the mid-Changhsingian Age the Yangtze rainforests had also vanished. The great Late Paleozoic rainforests of the Earth were gone, and their 70-million-year history was at an end (tables 3.2 and 5.1).

THE PRELUDE TO DISASTER FOR PALEOZOIC LIFE

The close of the Early Permian Epoch saw a significant change in the synapsid amniote faunas of the world. Most of the long-lived and numerically dominant pelycosaurs, the non-therapsid synapsids (table 4.2), died out about 274 million years ago in the late Kungurian Age (table 5.1). This event has been named Olson's extinction, and it does not appear to have been triggered by changes in the oxygen content of the Earth's atmosphere, as discussed in chapter 4.

The non-therapsids in the synapsid clade of amniote animals were ecologically replaced by the more derived therapsids—particularly the dinocephalian therapsids (table 4.2)—in the following late Roadian and Wordian Ages of the Middle Permian. However, as discussed in chapter 4, there exists a gap in geologic time spanning most of the Roadian Age—Olson's Gap (table 5.2)—between the last of the sphenacodonts in the late Kungurian and the diversification of the dinocephalians in the late Roadian–Wordian (Lucas 2004; Blieck 2011). Thus, the more derived dinocephalians do not appear to have competitively displaced their older synapsid relatives in a case of active ecological replacement. Rather, the ecological replacement that did occur appears to have been passive—the dinocephalians simply moved into vacant ecospace created by the demise of the non-therapsid synapsids. The dinocephalians flourished in the Middle Permian, producing over 40 genera of both carnivorous and herbivorous animals, and some of the largest land animals to exist in the entire Permian, as discussed in chapter 4.

This standard passive-ecological-replacement model for the non-therapsid versus therapsid synapsids turnover has been questioned by Tom Kemp, a paleontologist at the Oxford University Museum of Natural History. He has proposed a more nuanced model of ecological replacement for these two evolutionary faunas that contains both passive and active phases. First, he argues that the non-therapsid pelycosaurs (table 4.2) were adapted for life in the ever-warm, ever-wet, equatorial zone created by Late Paleozoic Ice Age climatic conditions (fig. 5.1) and that they experienced no selective pressures for lifestyle changes in these stable environments. The areal extent of the great rainforests—and the zone they inhabited—progressively shrank in the late Artinskian–early Roadian interval between the end of the P2 glaciation and the start of the P3 glaciation (table 5.2), and it is in this same interval of time that four of the six families of non-therapsids perished.[6] Thus, the extinction of these non-therapsid families was driven by changing climatic conditions and was independent of any competition by the therapsids, which were to appear in numbers only later in the late Roadian of the Middle Permian (table 5.2).

TABLE 5.2 Climatic events, area of rainforest coverage, and biological events during the transition from the latest Early Permian (Kungurian Age) to the earliest Middle Permian (Roadian Age).

Age	Geologic Time (Ma)	Climatic Events		Extent of Rainforests rainforests:	Evolutionary Events
		(2)	(3)		
WORDIAN	266	ICE		239,000 km²	
	267	ICE		227,000 km²	
	268	ICE	ICE?	216,000 km²	
ROADIAN	269	ICE	ICE?	194,000 km²	← Therapsid Radiation
	270	ICE	ICE?	179,000 km²	← Olson's Gap
	271	ICE-P3	ICE?	164,000 km²	← Olson's Gap
	272		ICE?	150,000 km²	← Olson's Gap
KUNGURIAN	273			127,000 km²	← Olson's Gap
	274			105,000 km²	← Olson's Extinction
	275			105,000 km²	
	276			105,000 km²	
	277			105,000 km²	
	278			105,000 km²	
	279			105,000 km²	
ARTINSKIAN	280			332,000 km²	
	281			482,000 km²	
	282			633,000 km²	
	283			784,000 km²	
	284	ICE		1,011,000 km²	
	285	ICE		1,115,000 km²	
	286	ICE		1,220,000 km²	
	287	ICE		1,325,000 km²	
	288	ICE		1,430,000 km²	
	289	ICE		1,690,000 km²	
	290	ICE		1,640,000 km²	
SAKMARIAN (pars)	291	ICE	ICE?	1,590,000 km²	
	292	ICE-P2	ICE?	1,590,000 km²	

Note: Climatic events column (2) is East Gondwana (Australia) in the Southern Hemisphere, and column (3) is Siberia (northeastern Russia) in the Northern Hemisphere; see table 5.1 for data sources.

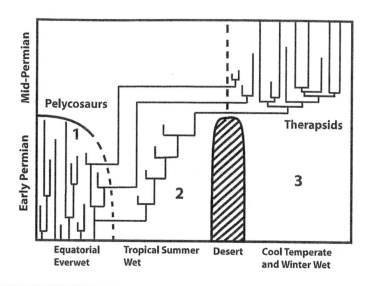

FIGURE 5.1 Kemp's paleoecological model for the evolution of the therapsid synapsids from the non-therapsid pelycosaurs during the Early to Middle Permian time interval; see text for discussion.

Source: From Kemp (2006), reprinted with permission.

Second, Kemp argues that one family of the Early Permian pelycosaurs, the predaceous sphenacodontids (figure 4.6), gave rise to the earliest therapsids—but not in the ever-warm, ever-wet equatorial zone. Instead, this evolutionary event took place in cooler and seasonally arid regions of the Earth—geographic regions that expanded in area during the P2–P3 interpulse greenhouse period. This seasonally arid, savanna-like environment Kemp labeled "tropical summer wet" (fig. 5.1), noting that not only did the first therapsids evolve there but the two remaining families of the non-therapsids, the Caseidae and Varanopidae, also migrated there as the equatorial ever-warm, ever-wet region shrank.

Third, Kemp argues that the early therapsids invaded even harsher environments—the cool-temperate/winter-wet zone—in the Middle Permian (fig. 5.1). Previously these higher-latitude regions were isolated from equatorial regions by the temperate-zone desert bands that formed in the Late Paleozoic Ice Age. As sea level fell with the return of expanding ice sheets of the P3 glaciation (table 5.2), the early therapsids were able to migrate north along the exposed east coastline of Laurussia and enter the higher-latitude regions. Kemp argued that the key to their spectacular success in this harsh region was the suite of morphological traits they had evolved in dealing with the less-harsh conditions of the cool and seasonally arid regions in which they had evolved. These traits included much higher metabolic rates and sustained activity levels, faster growth, and—most important—the ability to regulate their body temperatures to a greater degree than any previous synapsid group (as discussed in chapter 4). The therapsids rapidly diversified in this new environment, producing some nine new clades of both carnivores and herbivores (Kemp 2006).

Finally, Kemp notes that the fossil record shows that the last of the non-therapsids, the caseids and the varanopids, also managed to invade the cool temperate region and coexisted with the newly evolved therapsids for a few more million years (fig. 5.1). He argues that the last caseids and varanopids were eventually competitively displaced by the more-advanced therapsids; thus, the final extinction of the last of the non-therapsid pelycosaurs was the result of active ecological replacement, not passive like the initial extinction of four entire families of pelycosaurs in the Early Permian.

Whatever its ultimate cause, Olson's extinction triggered the loss of some two-thirds of the biodiversity of terrestrial vertebrates and thus was not a trivial event (Sahney and Benton 2008; Blieck 2011). Other clades of land animals—not just the synapsid amniotes—were also affected by contraction of the ever-warm, ever-wet, equatorial zone that occurred in the late Artinskian–early Roadian interval of the Early Permian. Ecological diversity was also lost: post-Olson's-extinction vertebrate communities possessed a lower number of guilds than in any other period of Permian history, being comprised almost entirely of piscivorous carnivores and browsing herbivores (Sahney and Benton 2008).

But what about Olson's Gap—a hiatus in the fossil record that the University of Lille paleontologist Alain Blieck describes as being of global extent and having a duration equivalent to most of Roadian time (Blieck 2011, 207)? This gap in the fossil record occurs in phase 2 of Kemp's ecological replacement model (fig. 5.1)—namely, the period of time in which the first therapsids are modeled as having evolved from— some of the last sphenacodontids, an event that took place in seasonally arid, savanna-like environments only marginal to the usual geographic range of the sphenacodontids. The oldest known therapsid species is the basal biarmosuchian *Tetraceratops insignis* (table 4.2) found in Early

Permian strata in Texas in North America (Benton 2005, 2015; McGhee 2013), yet the fossil appearance of the oldest Middle Permian therapsid species is far away to the east in Russia (Kemp 2006). Olson's Gap may be a preservational artifact; that is, the late Kungurian and early Roadian newly evolved therapsid species may have existed in such small population sizes that they had a very low probability of preservation in the fossil record.

On the other hand, Olson's Gap might be a real ecological-evolutionary phenomenon and not a preservational artifact. The University of Bristol paleontologists Sarda Sahney and Michael Benton argue that "Olson's extinction was a dramatic extinction 'trough' that is a prolonged period of very low diversity after a long and sustained diversity rise and probably the result of prolonged environmental stress" (Sahney and Benton 2008, 761). That is, the low diversity of fossils found in the strata during the period of Olson's Gap did not result from a failure of representation of the actual diversity of species because of small population sizes of those species, and hence a low probability of their being preserved in the fossil record; rather, the low diversity of fossils actually records the low diversity of species present on the Earth during the Olson's Gap period of time.

In contrast to extinction and biodiversity loss on land at the end of the Early Permian, life in the oceans was rebounding from the long period of evolutionary stagnation triggered by the Serpukhovian crisis that we examined in detail in chapter 2. Starting in the early Sakmarian Age (Stanley 2007), species diversity in marine ecosystems steadily increased through the remainder of the Early Permian and through the Middle Permian up to the middle of the Capitanian Age. That is, while on land the glaciers slowly melted away and the great equatorial rainforests progressively shrank to critical minimum levels during the late Early and early Middle Permian (table 5.2), in the oceans normal evolutionary turnover rates returned, diverse ecological specialist species re-evolved, and rising sea levels created new shallow-water habitats on the continents that were invaded by marine life around the Earth.

At the beginning of the Capitanian Age, life was good not only in the oceans but also on land as therapsid species continued to diversify in the cool-temperate zones of the Earth. Then the Earth began to heat up more rapidly than ever seen in any of the interpulse greenhouse periods of the Late Paleozoic Ice Age. In the oceans, dead regions of oxygen-depleted waters formed and marine species began to go extinct around the world. On land, arid desert regions began to expand, driving plant and animal species to extinction. Even in the wet regions of the Earth, plant and animal species mysteriously began to go extinct. The Capitanian crisis had begun.

My colleagues Matthew Clapham, Peter Sheehan, Dave Bottjer, and Mary Droser and I have demonstrated that global ecosystems of the world began to collapse during the Capitanian Age, triggering the fifth-most-severe ecological disruption in the Phanerozoic Eon (table 1.3) (McGhee et al. 2013). In the oceans, the ecological impact of the Capitanian biodiversity crisis at the end of the Late Paleozoic Ice Age was greater than that of the Serpukhovian crisis at the beginning of the "Big Chill" phase of the Late Paleozoic Ice Age (table 2.2), but not as great as that of the Late Devonian biodiversity crisis at the onset of the Late Paleozoic Ice Age (table 1.3)—that is, *if* the Late Devonian crisis marked the onset of the Late Paleozoic Ice Age, a controversy discussed in chapter 1. On land, Sahney and Benton note that "the ecological impact of the Guadalupian [Middle Permian] events is catastrophic; 8 (out of a possible 12) guilds are lost from the Artinskian high of 10 guilds. These are recovered in the last stages of the Permian before being devastated again by the end-Permian event. . . . A dramatic change in diet type also occurs: proportions of piscivores, insectivores, predators and browsers are thrown out of balance during each extinction pulse" (Sahney and Benton 2008, 761).

Then, as if the Capitanian crisis were not enough, Peter Sheehan, Dave Bottjer, Mary Droser, and I have determined that the ecological impact of the end-Permian mass extinction—in both marine and terrestrial ecosystems—was the worst ecological catastrophe in Earth history (McGhee et al. 2004). The magnitude of the ecological disruption triggered by the end-Permian mass extinction was so great that recovery was impossible—the world of the Paleozoic Era came to an end.

What possible catastrophes could have happened on the Earth in the Permian that would have had such a devastating effect that they triggered the fifth-most-severe and first-most-severe ecological crises in Earth history (table 1.3) in the short time span of only seven million years (table 5.1)? To readers who subscribe to the Gaia hypothesis—that the geochemical and biochemical cycles of the Earth are buffered to protect and nurture life—the answer to that question will be shocking, and that horrific answer will be examined in detail in chapter 6.

6 | The End of the Paleozoic World

The end of the Permian period is marked by global warming and the biggest known mass extinction on Earth. The crisis is commonly attributed to the formation of the Siberian Traps Large Igneous Province. . . . Heating of organic-rich shale and petroleum bearing evaporites around sill intrusions led to greenhouse gas and halocarbon generation in sufficient volumes to cause global warming and atmospheric ozone depletion. . . . The gases were released to the end-Permian atmosphere partly through spectacular pipe structures with kilometre-sized craters.

—Svensen et al. (2009, 490)

THE CAPITANIAN CRISIS

The Capitanian Age witnessed the opening salvo in the volcanic assault that would bring an end to the Paleozoic world. Deep beneath South China, a huge plume of hot magma was slowly rising to the Earth's surface. South China in the Capitanian was a large island, separate from present-day North China, and was located in the tropics at the Earth's equator (fig. 6.1). The mantle plume would eventually intersect the bottom of South China, burn its way up through the continental crust, and pour an incredible amount of hot liquid lava across a huge expanse of the Earth's surface. In the process, it would create what we now call the Emeishan "large igneous province," or LIP in the vernacular of the volcanologists.[1]

What is a LIP? A typical LIP is the product of mantle-plume volcanism in which a plume of magma rises from deep in the Earth's mantle and, when it reaches the Earth's surface, erupts almost unimaginable volumes of basaltic lava and injects a huge amount of

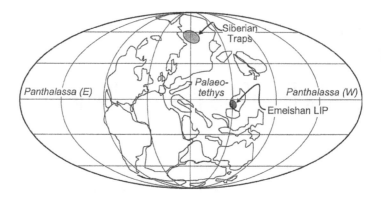

FIGURE 6.1 Paleogeography of the Late Permian world showing the locations of the Emeishan and Siberian large igneous provinces, both of which erupted during the last nine million years of the Permian.

Source: From *Lithos*, volume 79, pp. 475-489, by J. R. Ali, G. M. Thompson, M.-F. Zhou, and X. Song, "Emeishan large igneous province, SW China," copyright © 2005 Elsevier. Reprinted with permission.

hot gases into the atmosphere. A LIP eruption is thus characterized by (1) its large volume, hundreds of thousands to millions of cubic kilometers of molten magma; (2) its large areal extent, with lava covering hundreds of thousands to millions of square kilometers of land or seafloor; (3) its rapid eruption, consisting of numerous years- to decade-long volcanic pulses spread over only one to five million years—a very short period of time on geologic timescales; and (4) the type of lava usually produced in the eruption—basaltic.[2] Basaltic lava is quite liquid and flows for considerable distances before slowly freezing into solid rock, which partially explains why such large areas of land are covered by lava in a "flood basalt" LIP volcanic event.[3] The other explanation for the large size of LIPs is the fantastic amount of lava that pours out onto the Earth's surface through numerous cracks and fissures produced in the Earth's crust by the huge head of the hot mantle plume located beneath it.

Vincent Courtillot, a geophysicist at the University of Paris, argues that the initial head of a mantle plume could exceed 200 kilometers (124 miles) in diameter when it initially forms at the boundary between the molten outer core of the Earth and the lower mantle, located some 2,900 kilometers (1,800 miles) below the surface (fig. 6.2). As it slowly rises through the mantle, he argues, the head of the plume may expand to some 500 kilometers (310 miles) in diameter

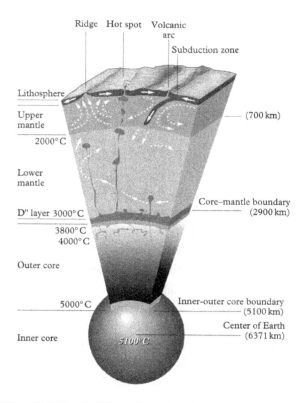

FIGURE 6.2 Cross-section of the interior of the Earth showing a hot-spot LIP region (top center of figure) produced by mantle plumes rising to the Earth's surface from a depth of 2,900 kilometers (1,800 miles).

Source: From *Evolutionary Catastrophes*, by Vincent Courtillot, copyright © 1999 Cambridge University Press. Reprinted with permission.

as it melts through mantle rock and then further expand to some 2,000 kilometers (1,240 miles) in diameter as it impacts and spreads out under the continental crust located above it at the Earth's surface (Courtillot 1999, 108).

A mantle plume can exist for a significant period of geologic time, producing a relatively stable "hot spot" (fig. 6.2) near the surface of the Earth with the gigantic plates of the Earth's lithosphere slowly moving across it. The Hawaiian Islands were produced in this fashion, with each volcanic island forming sequentially as the mantle-plume hot spot burned through the lithospheric plate moving across it. By dating the ages of the Hawaiian Islands and associated submerged seamounts, it can be shown that the mantle plume producing the island chain has been stable in its position for at least 74 million years. The hot spot is presently located beneath the largest island, Hawaii, which is the largest volcano on Earth—although this is not apparent because most of the volcano is submerged with only its tip protruding above the waters of the Pacific Ocean. Measuring from its tip down to the seafloor, the volcano is a ten-kilometer-high (6.2-mile-high) mass of basaltic lava produced by mantle-plume volcanism (Courtillot 1999). The largest known volcanoes produced by mantle-plume hot-spot eruptions are not on Earth, however, but on our sister planet, Mars. The largest known volcano in the solar system is the Martian Olympus Mons, which is a towering 21.3-kilometer-high (13.2-mile-high) pile of basaltic lava flows (fig. 6.3). The Martian volcanoes Ascraeus Mons, Pavonis Mons, Arsia Mons, and Elysium Mons are all taller than ten kilometers (hence taller than Earth's Hawaii volcano) (Croswell 2003), and they were all formed by mantle-plume volcanism. Mars is a smaller planet with a smaller mantle and core, and the movement of the plates of its crust were not as dynamic as those seen on Earth before the core of Mars eventually cooled and the Martian magnetic field collapsed. As a result, most of the volcanoes on Mars were created by mantle plumes, not, as on Earth, by the

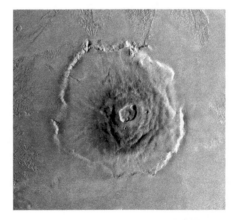

FIGURE 6.3 The Martian volcano Olympus Mons, the largest known volcano in the solar system, a towering 21.3-kilometer-high (13.2-mile-high) pile of basaltic lava flows produced by mantle-plume hot-spot eruptions.

Source: Photograph courtesy of NASA.

melting of plate boundaries in subduction zones (for example, the "ring of fire" distribution of volcanoes surrounding the Pacific Ocean is created by subduction of Pacific Ocean plates). Because plate motion ceased early in Martian history, its plates did not move progressively across mantle-plume hot-spot locations. Almost all of the lava that erupted at a hot spot on Mars accumulated into a single volcano—rather than a chain of volcanoes like the Hawaiian Islands—resulting in the even more gigantic size of the Martian LIP volcanoes.

Eleven LIP volcanic eruptions are known to have occurred on Earth in the past 260 million years (Courtillot and Renne 2003; Saunders 2005; Courtillot et al. 2010). No less than two of these massive eruptions occurred during the short time span of the last nine million years of the Permian—producing the Emeishan and the Siberian LIPs (figure 6.1). Our best radiometric data indicate that the Emeishan LIP began to erupt between 257.6 ± 0.5 and 259.6 ± 0.5 million years ago, near the Capitanian/Wuchiapingian boundary (Shellnutt 2013), whereas biostratigraphic data indicate that the lava eruptions may have begun earlier in the mid-Capitanian (Bond, Wignal, et al. 2010). The Emeishan LIP may have continued to erupt sporadically for about 20 million years (Racki and Wignall 2005; Shellnut 2013), up into the Triassic, but the majority of the volcanic material in the LIP was emplaced in less than 1.5 million years, essentially at the Capitanian/Wuchiapingian boundary (Shellnutt 2013). The Emeishan eruption is estimated to have produced 300,000 to 600,000 cubic kilometers (71,700 to 143,400 cubic miles) of lava, and its main lava outcrop covers about 250,000 square kilometers (96,500 square miles) of land today (Ali et al. 2005; Shellnut 2013). It is difficult to estimate the original size of the Emeishan LIP because of extensive erosion of the original lava flows and major tectonic fragmentation of the South China continental block as it collided with and sutured together with the North China and Indochina blocks in the early Mesozoic, plus the Indo-Eurasian block collisions in the early Cenozoic (Shellnutt 2013). In our modern world, outcrops of the original Emeishan LIP volcanics can be scattered as much as 300 kilometers (186 miles) away from one another across southwest China (fig. 6.4) (Wignall et al. 2009). Thus some estimates put the original land area covered by the Emeishan flood basalts at over two million square kilometers (772,000 square miles) (Racki and Wignall 2005).

A volume of 300,000 to 600,000 cubic kilometers of molten lava is a staggering amount—nothing like the Emeishan LIP eruption has ever occurred within

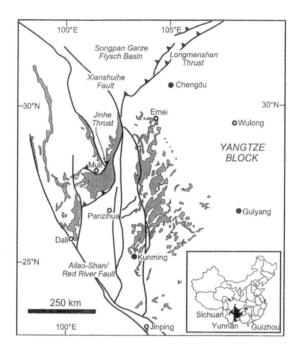

FIGURE 6.4 Geologic map showing the exposed basalt outcrops (shaded regions) of the Emeishan large igneous province in southwest China (inset map in lower right corner of the figure).

Source: From *Lithos*, volume 79, pp. 475–489, by J. R. Ali, G. M. Thompson, M.-F. Zhou, and X. Song, "Emeishan Large Igneous Province, SW China," copyright © 2005 Elsevier. Reprinted with permission.

human history. The largest flood-basalt eruption in recorded history is the Eldgjá eruption in Iceland, which occurred from AD 934 to AD 940 and produced 19.6 cubic kilometers (4.7 cubic miles) of basaltic lava (Thordarson and Self 2003). However, the best-documented LIP eruption in history is the Laki eruption, also in Iceland, which started on June 8, 1783, and continued for eight months until February 7, 1784 (Thordarson et al. 1996; Thordarson and Self 2003). The Laki LIP eruption is the second largest in human history, producing 14.7 cubic kilometers (3.5 cubic miles) of basaltic lava. This lava poured from 140 volcanic vents in fissures that extended in a row some 27 kilometers (17 miles) long, flowing away from the fissures to covered some 580 square kilometers (224 square miles) of land in southeast Iceland (Thordarson and Self 2003; Stone 2004).

The Laki LIP eruption produced a plume of gases that extended some 15 kilometers (9.3 miles) up into the atmosphere. The total eruption ejected approximately 122 million tonnes (134 million tons) of sulfur dioxide (SO_2) into the atmosphere, which reacted with water vapor to produce about 200 million tonnes (220 million tons) of sulfuric acid (H_2SO_4) droplets. About 175 million tonnes (193 million tons) of sulfuric acid precipitated out of the atmosphere as acid rain all across Europe, and the remaining 25 million tonnes (28 million tons) remained in the upper tropopause to lower stratosphere for over a year (Thordarson and Self 2003). The volcanic rifts also emitted huge clouds of chlorine and fluorine gas, which reacted with water vapor to produce about seven million tonnes (7.7 tons) of hydrochloric acid (HCl) and 15 million tonnes (16.5 tons) of hydrofluoric acid (HF) in the atmosphere, of which one million tonnes (1.1 million tons) of hydrofluoric acid fell as acid rain on the island of Iceland alone (Thordarson and Self 2003; Stone 2004).

The effect of the Laki LIP eruption on the human population of Iceland was catastrophic: 10,000 people were killed—about 20 percent of the total population of Iceland at the time (Stone 2004). Over 75 percent of the grazing livestock animals were killed (Schmidt et al. 2011). Rain on the island was so acid that it burned holes in the leaves of trees—most of the trees and shrubs died and did not return for some three to ten years. Cultivated grasses—food for livestock—withered. Many people died of starvation in a famine that lasted for three years (Jackson 1982). Others died from sulfur dioxide and sulfuric acid inhalation (Schmidt et al. 2011). But many people and livestock animals died from chronic fluorine poisoning. The hydrofluoric acid rain produced fluoride salts on grasses eaten by farm animals, poisoning the animals. The animals were eaten by humans, and they became poisoned. Even the water was poisoned by fluorine—drinking water contained as much as 30 times the level of fluorine compounds permitted in modern drinking water. Merely drinking a glass of water would make you sick, yet the islanders had no choice—there was no other source of drinking water (Stone 2004).

The effects of the Laki LIP eruption extended much farther away than just Iceland. All across Europe, sunlight dimmed as the sulfuric aerosol cloud spread east from the Icelandic flood basalts. This volcanic air pollution was called the "dry fog" in English-speaking areas, as it looked like fog but was not damp like fog; in German-speaking areas it was described as the *Höhenrauch*, or smoke in the skies. Acid rain withered summer wheat crops across Europe, and the green leaves on trees in the middle of summer turned yellow and brown and fell to the ground as if it were late autumn. The dry fog persisted for some five to six months, and winter came unusually early.[4] The winter of 1783–1784 was the harshest recorded in 250 years, and lemon crops far to the south in Italy were destroyed by frost. In England alone, some 20,000 people died from weather-related illnesses; in France the death rate was 25 percent higher than normal (Stone 2004). If an equivalent eruption were to occur today, computer simulations predict that some 142,000 people would die in Europe from air-pollution effects alone (Schmidt et al. 2011).

In North America, the winter of 1783–1784 was the most severe in recorded history in the New World. In the fledgling United States of America, Benjamin Franklin proposed that the harsh winter was probably caused by a volcanic eruption that he had heard about in Europe. In Europe itself, the French naturalist Mourgue de Montredon attributed the intense cold directly to the Laki eruption in Iceland (Thordarson and Self 2003),[5] and Neale (2010) considers the environmental effects of the Laki eruption to have helped spark the French Revolution. The volcanic cloud of sulfuric aerosols produced in the Laki LIP eruption eventually covered the entire Northern Hemisphere from a latitude of about 35°N up to the North Pole. The mean surface temperature of the Earth in this region dropped by about −1.3°C (−2.3°F), and the colder-than-normal period lasted from two to three years in the Northern Hemisphere (Thordarson and Self 2003; Schmidt et al. 2012).

The mantle plume beneath the Icelandic hot spot is predicted to produce a Laki-style flood-basalt eruption every 200 to 500 years (Thordarson and Larsen 2007; Schmidt et al. 2012). It is a historical fact that four such large-volume basaltic eruptions have occurred in the past 1,200 years, including the largest in recorded history (Eldgjá) and the second largest (Laki), giving an average eruption frequency of one per 300 years (Thordarson and Self 2003). For a comparison with the Emeishan LIP eruption in the Middle Permian, take the volume of lava produced in the largest known eruption in history, 19.6 cubic kilometers, and for simplicity in arithmetic round it up to 20 cubic kilometers (4.8 cubic miles). Now take the shortest predicted Icelandic LIP-eruption frequency, one per 200 years, and divide 100,000 years by that. In this worst-case scenario for the Icelandic mantle plume—the largest eruption occurring in the shortest predicted period of time—some 500 eruptions, each producing 20 cubic kilometers of lava, would occur in 100,000 years. Thus the total volume of Icelandic hot-spot volcanism would be 10,000 cubic kilometers (2,400 cubic miles) of lava per 100,000 years, or 150,000 cubic kilometers (36,000 cubic miles) of lava per 1.5 million years.[6] In contrast, the Emeishan LIP produced a minimum of 300,000, and perhaps as much as 600,000, cubic kilometers of lava in less than 1.5 million years. Thus the Emeishan mantle-plume lava production was at least twice as large, and perhaps as much as four times as large, as that of the Icelandic hot spot.

The Emeishan LIP eruption was coincident in time with the Capitanian extinction (Zhou et al. 2002; Wignall et al. 2009; Bond, Wignall, et al. 2010). It is thus logical to explore a causal relationship between the two events—that is, to seek the cause of the biodiversity crisis in the environmental effects of this mantle-plume volcanic eruption. But first let us examine some of the biological effects of the extinction. The loss in marine biodiversity that occurred in the Capitanian event was once thought to be the third largest in Phanerozoic history (Sepkoski 1996; Bambach et al. 2004), but it is now known that many of the extinctions that were once thought to have occurred in the Capitanian in fact occurred later, in the Changhsingian mass extinction (Clapham et al. 2009). Instead of a 36 to 47 percent loss of standing biodiversity, the true magnitude of the biodiversity loss that occurred in the Capitanian extinction was more like 25 percent (McGhee et al. 2013). Ecologically, however, the Capitanian extinction was the fifth-most-severe event in the Phanerozoic (table 1.3) (McGhee et al. 2013). Thus the Capitanian extinction is particularly interesting because it is one of the bioevents in Earth history in which the ecological impact of the event—relatively large—was markedly different from the magnitude of the biodiversity loss—relatively small. Why did the Capitanian extinction have such a large ecological impact? The answer may perhaps be found in the cause of the event.

The kill scenario (Retallack and Krull 2006; Retallack et al. 2006; Retallack and Jahren 2008; Bond, Hilton, et al. 2010; Clapham and Payne 2011; Payne and Clapham 2012) goes something like this: Back in the Capitanian Age, the super-plume rising beneath South China first caused the land above to push up into a 1,000-kilometer-wide (620-mile-wide) dome[7] and then began to form huge cracks and fissures in the surface of the Earth, accompanied with numerous earthquakes. Fluid basaltic lava began to pour from the fissures, and explosive craters formed along the fissures where hot gases vented into the atmosphere. The hot gases were primarily sulfur dioxide , carbon dioxide (CO_2), and methane (CH_4). The sulfur dioxide and carbon dioxide gases reacted quickly with water vapor to produce droplets of sulfuric acid and of carbonic acid (H_2CO_3), and oxidation of the methane produced even more carbon dioxide.[8]

In the lower atmosphere of the Earth, clouds of sulfuric and carbonic acid precipitated out as acid rain on the land and in the neighboring ocean waters. In the upper atmosphere, clouds of sulfuric acid droplets injected by volcanic plumes reflected some of the incoming light from the sun and caused the temperature of the Earth below to drop. Each volcanic pulse triggered a cooling pulse in which the Earth remained cold for a period of a decade or more. However, the venting of carbon dioxide and methane from the eruption eventually triggered the reverse—global warming caused by the greenhouse effect of those gases. As the temperature of the atmosphere and the oceans increased, methane frozen in ice in high-latitude permafrost regions on land and buried in submarine sediments in the oceans became increasingly unstable and began to dissolve, bubbling even more methane up into the water and into the atmosphere. Methane is a more potent greenhouse gas than carbon dioxide, and methane releases triggered even steeper increases in the Earth's temperature.

Ocean waters that had become acidified from sulfuric and carbonic acid rain now became hot and stagnant, producing large "dead zone" areas of anoxic water. With each passing year, the global climate became warmer and the dead regions of oxygen-depleted ocean waters grew larger and larger. Marine organisms began to die from acid poisoning (acidosis), carbon-dioxide poisoning (hypercapnia), and asphyxiation (hypoxia). On land, plants began to die from acid-rain poisoning. Terrestrial areas became more arid with the increased heat of the atmosphere; deserts began to expand and spread. Oxidation of methane in the atmosphere caused oxygen levels to fall and carbon-dioxide levels to rise.[9] Land animals died in increasing numbers as a result of climatic stress—from hyperthermia during the summer heat and from hypercapnia in breathing the carbon-dioxide-laced air. Lower atmospheric oxygen levels caused animals to retreat from the highlands of the Earth, and they began to suffer from fluid buildup in and swelling of their lung tissues (pulmonary edema). Herbivorous animals died from starvation as their food, the land plants, died from either acid-rain poisoning or lack of water in newly formed desert regions. Carnivorous animals died from starvation as their food, the herbivores, died. The global eco-system of the Capitanian world began to collapse, and the fifth-most-severe eco-logical disruption in the Phanerozoic Eon was the result (McGhee et al. 2013).

As in our modern oceans, reef organisms were particularly sensitive to the change to higher water temperatures and acid-water chemistry in the Capitanian Age. The tropics of the Capitanian world contained many reefs constructed by large demosponges, ancient rugose and tabulate corals, and reef-building microbes (Weidlich 2002; Kiessling and Simpson 2011). These reefs were destroyed in the Capitanian crisis, and, with the exception of some small biostromes and bryozoan reefs, no significant reef-building occurred until the recovery of reef ecosystems some five to seven million years later in the late Wuchiapingian and Changhsingian (Fan et al. 1990; Flügel and Kiessling 2002; Weidlich 2002). Ecologically, the Capitanian extinction trig-gered a global restructuring of reef ecosystems from complex, multicellular-animal-dominated reef structures to unicellular-microbe-dominated reefs (Flügel and Kiessling 2002).

The reef-building demosponges with their hypercalcified mode of growth (Kiessling and Simpson 2011) were particularly hard hit by the acidification of the world's oceans by the Emeishan eruption. Another group of large, highly calcified organisms that perished in the extinction were the giant unicells of the photosymbiotic fusulinid foraminifera that we considered in chapter 4. Acid waters impeded the growth of the calcium-carbonate-rich skeletons of the large-bodied and morphologically complex foraminiferal species that had specialized skeletal structures for nurturing and protecting their photosymbi-ots.[10] All of the large-bodied foraminifera were driven to extinction, leaving as survivors only the smaller and morphologically simpler species.[11] Large-bodied bivalves with thick calcareous shells that also hosted photosymbionts in their tissues also went extinct.[12] The simultaneous extinction of the hypercalcified, large-bodied, photosymbiont-hosting members of what the University of Tokyo Yukio Isozaki and University of Zagreb Dunja Aljinović paleontologists describe as the "Tropical Trio"—the "waagenophyllid corals, verbeekinid foraminifers, and alatoconchid bivalves" (Isozaki and Aljinović 2009)—is a diagnostic fea-ture of the effect of the Capitanian crisis.

The Capitanian extinction also had a major ecological impact on the ammonoids, which were actively swimming predators in the Capitanian seas.

The typical Paleozoic ammonoid faunas were completely replaced by ammonoid faunas that were to become the typical Mesozoic forms—some eight million years before the start of the Mesozoic (Leonova 2009)! Ammonoids of the order Ceratitida first appeared at the Early/Middle Permian boundary, but by the Capitanian only five genera of the order Ceratitida existed, representing 18 percent of the ammonoid diversity. Following the Capitanian crisis, the diversity of ceratite genera jumped to 55 percent in Wuchiapingian and further increased to 74 percent in the Changhsingian (fig. 6.5) (Leonova 2009). The shift in abundance was just as or even more abrupt than the shift in diversity that occurred between the Capitanian and Wuchiapingian, coincident with the Capitanian extinction. These ammonoid faunal changes are reflected in an ecological shift in marine ecosystem structure from bottom-swimming (nekto-benthic) ammonoid faunas to open-ocean (pelagic) ammonoid faunas, an ecological shift to a more Mesozoic-style ecosystem structure that we will consider in more detail later in this chapter.

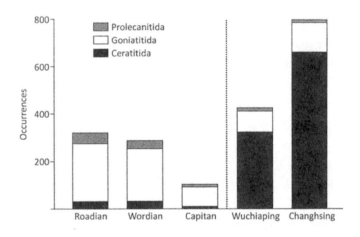

FIGURE 6.5 Occurrence counts for the three late Paleozoic ammonoid orders from data in the Paleobiology Database (www.paleodb.org). An occurrence is the record of a species from a single fossil collection, and is a proxy for abundance. The Capitanian biodiversity crisis is marked by the vertical dotted line. Geologic timescale abbreviations: Capitan, Capitanian; Wuchiaping, Wuchiapingian; Changhsing, Changhsingian.

Source: From McGhee et al. (2013).

The Capitanian extinction may have been more severe ecologically than it was taxonomically because of the environmentally selective nature of the event. For most animal groups, the physiological stresses induced by the Emeishan LIP eruption may not have been sufficiently intense to drive major taxonomic extinctions. In contrast, taxa such as hypercalcified sponges, corals, and larger foraminifera suffered because they had poor physiological buffering and would have been particularly susceptible to the effects of ocean warming and acidification (Clapham and Payne 2011; McGhee et al. 2013). Their elimination triggered major ecological restructuring, particularly in reef ecosystems. Other typical Paleozoic benthic organisms with less developed respiratory systems and low metabolic rates—such as the bryozoan and brachiopod lophophorates—suffered differentially from hypercapnia and hypoxia, as opposed to the more energetic gastropod and bivalve molluscs that were to become dominant in the Mesozoic (Knoll et al. 1996; Retallack and Krull 2006).

The cause of the shift in dominance structure in ammonoid pelagic ecosystems, from the typical Paleozoic mollusc orders of the Goniatitida and Prolecanitida to the more typical Mesozoic order of the Ceratitida (fig. 6.5), is less clear because the biology of these extinct groups is uncertain. It is of note, however, that the ceratite ammonoids also survived the end-Permian mass extinction—also associated with a LIP eruption, as we will see in the next section of the chapter—whereas the goniatites and prolecanitids did not (McGhee et al. 2013).

Finally, the selective extinction of the large-bodied photosymbiont-bearing foraminifera and bivalves during the Capitanian may not have been driven entirely by acidosis. Another possibility is that all photosynthetic organisms—including plants on land—suffered radiation poisoning during the Capitanian extinction (Ross 1972; Bond, Wignall, et al. 2010). In this alternative kill scenario, chlorine and fluorine gases emitted from the Emeishan LIP eruption seriously damaged the ozone layer of the Earth's atmosphere, permitting the flux of high-energy ultraviolet radiation to greatly increase at the surface of the Earth. As yet, it has not been demonstrated that enough chlorine or fluorine gas was released during the Emeishan eruption to disrupt the ozone layer, but we will see below that the radiation-poisoning kill scenario does indeed appear to apply to the end-Permian mass extinction.

The reader may have noticed a major difference in the scenario outlined above for the environmental effects of the Emeishan LIP eruption and those environmental effects actually recorded in the 1783–1784 Laki LIP eruption that we considered previously. The opening of huge fissures in the Earth, the outpouring of flood basalts, the clouds of hot sulfur-dioxide gas, the formation of sulfuric acid in the atmosphere, the effect of acid rain on land plants and animals, and short-term global cooling all occurred in the historical Laki eruption and are also proposed to have occurred in the Emeishan eruption. What is different is the major presence of *two additional gases* in the Emeishan eruption—carbon dioxide and methane—and the environmental effects of the long-term global warming produced by these greenhouse gases in the atmosphere of the Capitanian world. The Laki LIP eruption produced in insignificant amount of carbon dioxide compared to the amount of sulfur dioxide it released into the atmosphere (Self et al. 2005). And the Laki LIP is not unusual in this respect—volcanic-generated carbon-dioxide release from the typical basaltic magmas that produce LIPs usually is not large in comparison to the amount of carbon dioxide already present in the atmosphere (Self et al. 2005; Ganino and Arndt 2009).

Why was the Emeishan eruption so different from Laki—what evidence do we have for the additional production of huge amounts of carbon dioxide or methane in the Emeishan eruption some 360 million years ago? There are two lines of evidence: (1) the composition of the sedimentary rocks encountered by the molten magma of the Emeishan eruption on its rise to the Earth's surface and (2) the composition of the types of carbon released into the Earth's atmosphere and ocean waters during the time of the eruption.

First, when the rising Emeishan super-plume eventually intersected the bottom of South China and began to melt its way up through the continental crust, it encountered thick layers of sedimentary rock that were present above it in the Sichuan Basin—thick layers of reef limestone (rocks rich in calcium carbonate; $CaCO_3$), dolomite (rocks rich in calcium magnesium carbonate; $CaMg(CO_3)_2$), and, in some areas, shales containing petroleum deposits (Ganino and Arndt 2009). Clément Ganino and Nicholas Arndt, geologists at the Université Joseph Fourier de Grenoble, estimate that the Emeishan mantle-plume melting and heating—cooking—of the Sichuan Basin limestone and dolomite strata released between 41.3 and 97.6 *trillion* tonnes (45.5 to 107.6 trillion tons) of carbon dioxide into the Earth's atmosphere. This is in addition to the much smaller 11.3 trillion tonnes (12.5 trillion tons) of carbon dioxide emitted directly from the LIP magma itself; thus the total amount of carbon dioxide released into the Earth's atmosphere during the Emeishan eruption was between 52.6 and 108.9 trillion tonnes (58 to 120 trillion tons).[13]

In contrast, an Emeishan-sized flood-basalt eruption that occurs where the magma does not encounter any carbonate strata in the surrounding country rock is estimated to inject some 4.2 trillion tonnes (4.6 trillion tons) of sulfur dioxide into the Earth's atmosphere in the eruption plume and an additional 1.2 trillion tonnes (1.3 trillion tons) from the surface area of the lava flows, bringing the total amount of sulfur-dioxide gas emitted to some 5.4 trillion tonnes (6 trillion tons).[14] Thus, in the actual Emeishan eruption, the estimated amount of sulfur dioxide released into the atmosphere—5.4 trillion tonnes— is 9.7 to 20.2 times *smaller* than the amount of carbon dioxide emitted into the atmosphere—52.6 to 108.9 trillion tonnes. The emission of millions of tonnes of sulfur dioxide in the 1783–1784 Laki LIP eruption chilled the Northern Hemisphere of the Earth for two to three years (Thordarson and Self 2003; Schmidt et al. 2012), and the emission of *billions* of tonnes of sulfur dioxide with each major flood-basalt flow in the Emeishan eruption is predicted to have triggered global cooling pulses that persisted for a decade or longer (Self et al. 2005). Still, the climatic effect of injecting 52.6 to 108.9 trillion tonnes of the greenhouse gas carbon dioxide into the atmosphere would eventually produce global warming of a magnitude that would swamp the cooling effects of periodic sulfur-dioxide emissions. The planet would heat up and remain hot for hundreds of thousands of years, or for as long as the carbon-dioxide content of the atmosphere remained high (Wignall 2001; Bond and Wignall 2014). In the end analysis, the production of sulfuric acid from the 5.4 trillion tonnes of sulfur dioxide produced in the Emeishan LIP eruption would have had a much larger killing effect by acidifying the world's oceans and blighting the land areas with acid rain than by periodic global cooling.

Second, the analysis of carbon-isotope ratios not only confirms that an enormous amount of carbon was added to the environment during the time of the Emeishan eruptions, but also raises intriguing questions as to the source of all of the carbon. In short, too much of the lighter isotope of carbon, carbon-12, was injected into the Capitanian atmosphere than could possibly have been produced by a basaltic volcanic eruption alone. How was this conclusion reached? To explain, we know that the carbon atom has two stable isotopes, carbon-12 and carbon-13. Carbon-12 has six neutrons and six protons in the nucleus of the atom, and the heavier isotope carbon-13 has seven neutrons and six protons in its nucleus (carbon has a still heavier isotope, carbon-14, but it is unstable and radioactively decays to nitrogen-14). We also know that living organisms prefer the lighter isotope of carbon in their growth—plants preferentially extract carbon-12 from the atmosphere in their photosynthetic construction of hydrocarbons like sugar,[15] and marine animals preferentially extract carbon-12 from seawater in the construction of their calcium-carbonate skeletons.[16]

Ratio proportions of the two isotopes in geologic strata are customarily reported by an index, $\delta^{13}C$.[17] Positive values of the index $\delta^{13}C$ indicate an enrichment of the strata sample in the heavier isotope carbon-13 resulting from the depletion of the lighter isotope, carbon-12. Positive values of $\delta^{13}C$ are commonly produced by the activity of organisms in marine environments, such as algal blooms, as those organisms preferentially remove carbon-12 from the environment and leave the heavier isotope behind. Negative values of the index $\delta^{13}C$ indicate the opposite—an enrichment of the strata sample in the lighter isotope carbon-12. Negative values of $\delta^{13}C$ are commonly produced by the weathering of organic-rich sediments—that is, sediments enriched in carbon-12.

A major negative carbon isotope anomaly has been discovered to be associated with the onset of Emeishan LIP volcanism and with the Capitanian extinction (Retallack and Krull 2006; Retallack et al. 2006; Retallack and Jahren 2008; Wignall et al. 2009; Bond, Wignall, et al. 2010). The magnitude of the Capitanian negative carbon isotope anomaly is so great that it could not possibly have been produced by volcanic outgassing of carbon dioxide by the magma intrusions and flood-basalt lavas of the Emeishan LIP eruption. Gregory Retallack and Hope Jahren, geologists at the University of Oregon, further argue that the anomaly is too large even to have been produced by the metamorphism and melting of the Sichuan Basin carbonate reef strata that the Emeishan magmas burned through, producing the massive release of carbon dioxide into the atmosphere described by Ganino and Arndt that we examined earlier. A source of even lower carbon-isotope ratios is needed, and Retallack and Jahren argue that that source could only have been methane (Retallack and Jahren 2008; see also Berner 2002).

Where could the methane have come from? Retallack and Jahren point out three possible scenarios for the origin of the methane: (1) deep-ocean water overturn, (2) melting of methane-rich ice deposits,[18] and (3) magma intrusion into carbonaceous sediments—sediments containing coal, petroleum, or natural gas (Retallack and Jahren 2008). The first scenario requires the stratification of large areas of the Panthalasan ocean of the Capitanian world (fig. 6.1). Enormous amounts of methane could accumulate and be trapped in stagnant, oxygen-poor bottom waters contained below an upper-ocean oxygenated water-layer cap. Catastrophic overturn of these ocean waters—perhaps triggered by global warming—would release the deepwater methane into the Earth's atmosphere. Retallack and Jahren argue it is highly unlikely that the huge, globally continuous Panthalasan ocean could have been stratified to such an extent and, moreover, that deepwater overturn would oxygenate the bottom waters, whereas the actual strata and the fossil record indicate that marine anoxia was widespread during the Capitanian extinction.

The second scenario is based on the fact that in our modern world enormous reservoirs of methane exist frozen in the ice of permafrost regions on land and within the pores of sediment located in cold-water, high-pressure regions of the seafloor. If the global climate were to warm substantially, the methane-rich ice located in the high-latitude permafrost regions of the Earth would begin to melt and release the methane gas into the atmosphere. Likewise, with warming ocean waters, the gas hydrates in cold-water submarine sediments would become unstable, dissociate, and release methane into the ocean and subsequently into the Earth's atmosphere. This scenario is very real and of major concern at present, as the warming of our modern world is already triggering the release of methane from permafrost regions of the planet. Why is this of concern? Carbon dioxide is a potent greenhouse gas, but methane is much worse! The Earth could become much hotter much faster with higher levels of methane in the atmosphere.

However, several lines of evidence exist that the negative carbon isotope anomaly seen in the Capitanian could not have been generated by methane outbursts alone. First, the overall anomaly actually consists of several negative isotopic excursions within the span of time of 10,000 years. Methane release from ice-deposit destabilization and melting should occur as a single catastrophic outburst, as it apparently did in the end-Paleocene hothouse world (Dickens et al. 1995). Following such a catastrophic pulse, it is argued, it would take a "recharge time" of at least 200,000 years for the Earth to accumulate enough methane for another catastrophic outburst (Dickens 2001). In contrast, several such bursts would have had to have happened in less than 10,000 years if the negative carbon isotopes excursions seen in the Captanian were due to methane-rich ice destabilization alone (Retallack and Jahren 2008). Another line of argument is that the overall negative carbon isotope anomaly was produced slowly and gradually, not suddenly and catastrophically. In South China, values of the index $\delta^{13}C$ made a major shift in the negative direction at the onset of Emeishan volcanism and the onset of Capitanian extinctions. However, $\delta^{13}C$ values continued to shift in the negative direction in strata deposited after the main pulse of the extinction, and only began to return to pre-extinction levels some two conodont zones above the extinction interval. The University of Leeds geologist David Bond and his colleagues argue that the negative shift in $\delta^{13}C$ values seen in the Chinese sections was gradual, not abrupt, and that it occurred at a decline rate of one to two per mille per twenty meters of limestone (Bond, Wignall, et al. 2010). Converting this stratigraphic-thickness decline rate into one of measured time is problematic. If—and that is a big if—one assumes that the carbonate strata accumulated at a rate of about ten centimeters (four inches) per thousand years, then the Capitanian negative $\delta^{13}C$ shift occurred at a rate of 0.01 to 0.02 per mille per thousand years. In contrast, negative carbon isotope anomalies attributed to the catastrophic release of methane from methane-rich ice destabilization in Jurassic strata show negative $\delta^{13}C$ shifts that occurred at a rate of 1.5 per mille per thousand years—roughly two orders of magnitude faster than the negative shift seen in the Chinese Capitanian strata (Kemp et al. 2005). Given the much slower release of carbon-12 into the atmosphere that occurred in the Capitanian, Bond and colleagues argue against a methane-rich ice source for the carbon-12 (Bond, Wignall, et al. 2010). If, on the other hand, one assumes an equal duration of time per conodont zone—another big if, as that implies a constant, clocklike, rate of evolution in the conodont species—the negative $\delta^{13}C$ shift occurred rapidly in the time interval from the latest *Jinogondolella altudaensis* to the earliest *Jinogondolella prexuanhanensis* conodont zones in South China.

Given these complications with the methane-rich ice-melting scenario, Retallack and Jahren argue that the third scenario—magma intrusion into carbonaceous sediments—was the source of the methane necessary to have produced such large negative carbon isotope anomalies in the Captanian (fig. 6.6). They point out that large diabase dikes (subterranean injection features containing basaltic magma; see fig. 6.6) from the Emeishan LIP, some as much as 200 meters (656 feet) wide, penetrated some 100 meters (328 feet) of Carboniferous coal deposits to the southwest of the main magma mass, and that some of the coal seams in these deposits are up to 12 meters (39 feet) thick. To the northwest of the Emeishan basalts, the Sichuan Basin also contains shale and limestone strata that contain natural gas and petroleum deposits; Ganino

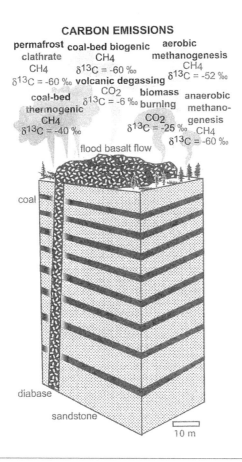

FIGURE 6.6 Model showing the carbon-emission results of intruding hot basaltic lava through strata containing layers of coal: five separate sources of methane (CH_4) generation are shown as well as their characteristic negative carbon-isotope signatures ($\delta^{13}C$).

and Arndt (2009) argue that these strata may have been metamorphosed by hot magma at depth and may also have been a source of methane release into the atmosphere. Methane release by contact metamorphism of surrounding carbon-rich sediments by the subterranean injection of Emeishan hot magmas would be a longer-term, more gradual phenomenon that the sudden explosive release of methane into the Earth's atmosphere resulting from methane-rich ice destabilization.

In summary, the debate concerning the exact sources of all of the carbon-12 injected into the atmosphere during the Emeishan LIP eruption continues to be lively, but there is no dispute that massive amounts of carbon dioxide and methane were released—unlike in the historic Laki LIP eruption, in which no carbonate or carbonaceous strata were metamorphosed by the mantle-plume magmas. The Laki LIP eruption was deadly enough, but to produce a really horrendous degradation of the environment of the entire Earth, a LIP also needs to encounter additional sources of carbon during its eruption.

The size and the environmental effects of the Emeishan LIP eruption were outside anything we have encountered in human history, and the eruption triggered the fifth most ecologically severe biodiversity crisis (table 1.3) in the past 600 million years of Earth history, since the evolution of animals in the Neoproterozoic (table 1.1). Yet the 300,000 to 600,000 cubic kilometers of lava produced by the Emeishan mantle-plume eruption was tiny, miniscule in comparison to what was to come at the end of the Permian. And to make matters much worse, the end-Permian mantle super-plume encountered a huge additional source of carbon, a source that was to ignite like a petrochemical bomb.

THE END-PERMIAN MASS EXTINCTION

We have now arrived at the catastrophic end-Permian mass extinction, which nearly ended animal life on the planet Earth about 252 million years ago (Benton 2003; Shen et al. 2011; Erwin 2015). All forms of life, both terrestrial and marine, suffered in that global catastrophe. If it had been only a little more severe, it would have erased the previous 350 million years of animal evolution, leaving only the simplest animals like jellyfish and sponges as survivors. As it was, it triggered a restructuring of the ecosystems of the entire planet, both on the land

and in the seas (McGhee et al. 2004), and is recorded in the fossil record as the largest loss of biodiversity ever seen in geologic time.

The series of events that led to the greatest catastrophe in Earth history began innocently enough: organic-rich shales began to be deposited in the marine waters occupying the Tunguska Basin in the eastern Siberian region of what is now Russia. Perhaps this region of the Earth has a propensity for catastrophe, for it is the same region in which the Tunguska asteroid exploded in the Earth's atmosphere on June 30, 1908, with a force of some ten to 15 megatons of TNT, flattening trees in the surrounding forest over an area of 2,150 square kilometers (830 square miles) (Farinella et al. 2001). Fortunately, no large loss of life is known to have occurred in that event—but that was not to be case with the end-Permian mass extinction.

The strata of the Tunguska Basin contain the oldest petroleum-bearing deposits in the world, formed by the maturation of organic material in Tonian- and Cryogenian-aged source rocks, Neoproterozoic strata that are 1,000 to 635 million years old (see table 1.1 for the geologic timescale) (Svensen et al. 2009; Polozov et al. 2016). These strata range in thickness from one to eight kilometers (0.6 to five miles) and are overlain by thick deposits of carbonate and evaporite rocks (rocks formed by chemical precipitation in the evaporation of seawater) deposited in the Ediacaran Period of the latest Neoproterozoic and the Cambrian Period of the earliest Phanerozoic. During this period of time, the marine waters in the Tunguska basin shallowed and began to evaporate, leaving behind vast deposits of sea salt ($NaCl$) and anhydrite ($CaSO_4$). At least five different episodes of shallowing and evaporation occurred in the Tunguska basin during the Cambrian, leading to the deposition of up to 2.5-kilometer-thick (1.5-mile-thick) sequences of salt- and anhydrite-rich strata. The largest of these is the Early Cambrian Usolye salt basin, which has an average thickness of 200 meters (656 feet) of salt spread over an area of some two million square kilometers (772,000 square miles) (Svensen et al. 2009; Polozov et al. 2016).

The petroleum-, salt-, and anhydrite-rich strata deposited in the late Neoproterozoic and early Cambrian in the Tunguska basin were then buried successively by carbonates, sandstones, and eventually terrestrial coal deposits as the basin infilled from the Ordovician to the Permian. Under this cover of younger sediments, the oldest petroleum-rich deposits in the world sat quietly like a ticking time bomb. In the Late Permian, that petrochemical bomb would be triggered.

Deep in the Earth beneath the Tunguska basin in Siberia, the second salvo in the volcanic assault that would bring an end to the Paleozoic world was fired—a mantle plume so huge that it has been designated as the "Siberian super-plume"—was rising toward the Earth's surface. The Siberian super-plume volcanic eruptions would produce the largest known continental LIP in Earth history (Reichow et al. 2009). The total volume of magma produced in the Siberian eruption—in extrusive lava flows and intrusive dikes and sills—is estimated to be at least five million cubic kilometers (1,195,000 cubic miles), with some estimates ranging as high as 16 million cubic kilometers (3,824,000 cubic miles) (Dobretsov and Vernikovsky 2001; Racki and Wignall 2005; Payne and Clapham 2012). Even the lower estimate of five million cubic kilometers is an almost unimaginable volume: try to visualize a black cube of basaltic rock that towers 171 kilometers (106 miles) into the sky, is 171 kilometers wide, and is 171 kilometers long. The Earth's oceans are only about six kilometers (four miles) deep, on average, so if you were to place such a cube in the middle of the ocean, it would still tower 165 kilometers (102 miles) into the sky, reaching all the way up into the ionosphere of the Earth's atmosphere. If you were to drive a car at a constant 100 kilometers per hour (62 miles per hour) alongside the base of this cube, it would take you almost two hours to get from one corner of the cube to the next. The minimum estimate for the Siberian LIP volume—five million cubic kilometers—is 8.3 to 16.7 times larger than the 300,000- to 600,000-cubic-kilometer volume estimate for the Emeishan LIP; that is, at the very least, the eruption of the Siberian LIP was equivalent to the eruption of 8.3, and possibly as many as 16.7, Emeishan LIP events taking place all at once.

The Siberian LIP eruption covered five million square kilometers (1,930,000 square miles) of land with flood-basalt lavas. That is a land area larger than one-half of the 48 contiguous states of the United States of America—imagine 62 percent of the U.S. map covered with molten lava— not once, but multiple times: the flood-basalt pile of the Siberian LIP is composed of layer after layer of lava flows. The present-day outcrop of the Emeishan LIP lavas covers some 250,000 square kilometers—the Siberian LIP area is 20 times larger.

What, then, are the expected environmental consequences, in comparison to the Emeishan LIP, of the vastly larger Siberian LIP eruption? Given the huge magma volume and flood-basalt coverage of the eruption alone, one might extrapolate and expect that the environmental effects of the Siberian eruption might be 8.3 to 20 times more severe than those of the smaller Emeishan LIP eruption. But that calculation omits the consequences that were to follow when the rising Siberian super-plume ignited the petrochemical bomb buried in the strata of the Tunguska basin.

Using thermomechanical modeling, the German Georesearch Center (GFZ) petrologist Stephen Sobolev and colleagues argue that the Siberian super-plume arrived below the lithosphere in the northern border of the Siberian Shield about 253 million years ago (Sobolev et al. 2011). It was extremely hot, with a temperature of around 1,600°C (2,900°F), and had a plume-head diameter of around 800 kilometers (497 miles). It began to melt the lower part of the lithosphere at a depth of between 130 and 180 kilometers (81 and 112 miles), and the plume head spread out to a width of 1,200 kilometers (745 miles) just below the lithosphere. Massive intrusion of molten magma via sills and dykes into the lithosphere then began to occur. The super-plume continued to rise and began to break through the lithosphere and into crustal rocks in only 100,000 to 200,000 years. The top of the super-plume was now located at a depth of about 50 kilometers (31 miles), and it began to melt the continental crust. This model predicts that between six and eight million cubic kilometers (1.4 to 1.9 million cubic miles) of molten magma was then intruded into crustal rocks (Sobolev et al. 2011), a prediction that is in accord with the empirical estimates of the volume of the Siberian LIP (Dobretsov and Vernikovsky 2001; Racki and Wignall 2005; Payne and Clapham 2012).

The huge volume of molten magma intruded into crustal rocks through subterranean sills and dykes now encountered the petroleum-, salt-, and anhydrite-rich strata buried between one and three kilometers (0.6 to five miles) below the surface of the Earth in the Tunguska Basin. Beginning around 252 million years ago, the contact of hot magma with the organic- and evaporite-rich strata buried in the basin resulted in the generation of huge amounts of gas at high pressure, subterranean gas that exploded and produced gigantic blast pipes that reached all the way to the surface of the Earth.

About 250 blast pipes filled with magnetite-rich breccia are located in the southern part of the Tunguska Basin in the region characterized by thick deposits of Cambrian salts at a depth of some 2,000 meters (6,560 feet) below the surface of the Earth. Several of these pipes contain iron-ore concentrations of commercial value and are currently being mined for magnetite (Fe_3O_4). In the northwest region of the Tunguska Basin, outside the region underlain by the Cambrian evaporites, are more than 500 additional blast pipes filled with basalt (Svensen et al. 2009; Polozov et al. 2016).

A geologic cross section of one of these explosive pipes, the Scholokhovskoie pipe, is shown in figure 6.7. Because of the commercial value of the magnetite ores contained in the pipe, the strata both within the pipe and surrounding it have been extensively drilled. A massive, 200-meter-thick (656-foot-thick) intrusive-magma sill of dolerite (labeled "Upper Sill" in fig. 6.7)[19] is present in the 400 meters (1,310 feet) of Late Cambrian sandstones and siltstones of the Verkholensk Suite strata (topographic elevations of 100 to 400 meters in fig. 6.7). At greater depths, some 800 to 900 meters (2,620 to 2,950 feet) below

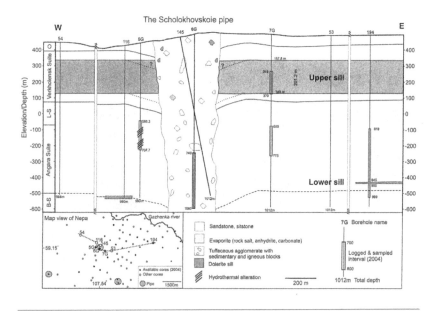

FIGURE 6.7 Geologic cross section of the upper 1,000-meter (3281-foot) thickness of the strata containing the Scholokhovskoie blast pipe, obtained by subsurface data from numerous drill holes into the strata (inset map, lower left corner of the figure). Strata abbreviations in the left margin of the figure: O, Ordovician; L-S, Litvinstsev Suite; B-S, Angara and Bulay Suites.

Source: From *Earth and Planetary Science Letters*, volume 277, pp. 490–500, by H. Svensen, S. Planke, A. G. Polozov, N. Schmidbauer, F. Corfu, Y. Y. Podladchikov, and B. Jamtveit, "Siberian Gas Venting and the End-Permian Environmental Crisis," copyright © 2009 Elsevier. Reprinted with permission.

the surface of the Earth, thinner dolerite sills are located in the thick Early Cambrian salt and anhydrite evaporite strata of the Angara and Bulay Suite strata (lower sills at topographic depths of −400 to −550 meters in fig. 6.7). The blast pipe itself contains large breccia blocks of the Cambrian evaporite strata that have been explosively ejected up the pipe and, back at the end of the Permian, out the mouth of the surface crater onto the surrounding countryside. Also contained within the pipe are breccia blocks from the dolerite sills, proving that the explosions that produced the blast pipe were contemporaneous with the intrusion of the magma intrusions that produced the sills. These magmatic breccia blocks are rich in glass, produced by rapid cooling of the hot intruded magma as it was exposed to air within the pipe.

Note that the Scholokhovskoie pipe at the surface of the Earth (fig. 6.7) has a diameter of some 430 meters (1,410 feet) and narrows to a diameter of 300 meters (984 feet) at a depth of 1,000 meters (3,280 feet) below the surface—a geometry produced by the upward and outward force of the explosions that formed the pipe. Other pipes in the region have surface craters that are some 1.6 kilometers wide (one mile wide), an exit-blast diameter that is a mute geologic witness to the spectacular magnitude of the explosions that occurred in this region of the Earth 253 million years ago.

All in all, over 6,400 explosive pipes exist in the Tunguska Basin, and drill-hole geologic samples from many of them have been extensively studied by the University of Oslo geologist Henrik Svensen and his colleagues (2009). The geophysical model they propose for the formation of the Siberian LIP blast pipes is shown in figure 6.8. In the first frame of the sequence (from left to right), the intrusion of a horizontal sill of hot volcanic magma, marked with "V" symbols in figure 6.8, occurs at depth in the Proterozoic strata underlying the petroleum-rich Cambrian evaporite strata (petroleum pools are black ovals marked with white "P" letters in fig. 6.8). In the second frame, a larger sill of hot magma is shown intruding into the Cambrian strata, the thermal volatilization of the petroleum deposits within the strata begins to occur, and the resultant gas expands, shown by the stippled-bounded region and the upward-pointing arrows. In the third frame, the gas explodes at depth and produces the upward blast force (upward-pointing arrows) that creates the pipe, carrying large blocks of the sill rock and surrounding Cambrian strata upward and out of the blast crater at the surface of the Earth. Additional contact-metamorphism and gas

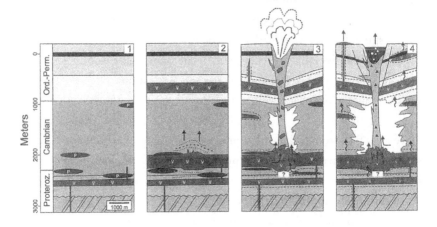

FIGURE 6.8 Model of the sequential events (from left to right) that led to the formation of the Siberian LIP blast pipes; see text for discussion.

Source: From *Earth and Planetary Science Letters*, volume 277, pp. 490–500, by H. Svensen, S. Planke, A. G. Polozov, N. Schmidbauer, F. Corfu, Y. Y. Podladchikov, and B. Jamtveit, "Siberian Gas Venting and the End-Permian Environmental Crisis," copyright © 2009 Elsevier. Reprinted with permission.

production is shown surrounding the degassing sill at shallower depths within the Ordovician strata (upper sill marked with "V" symbols in the upper part of the figure). Finally, in the fourth frame, continued degassing (wavy upward-pointing arrows) is shown from the petroleum-rich Cambrian evaporite strata and the magma layers, and is shown continuing to occur in the pipe itself—illustrated by the bubbles in the waters of the lake now shown occupying the blast crater. In addition, intrusive sills are now shown to come into contact with shallow-buried coal deposits (the uppermost black layer in the top of the figure), and contact metamorphism of the coal produces even more methane degassing, shown by the wavy upward-pointing arrow (Svensen et al. 2009).

Simply trying to imagine the eruption of one of these pipes is staggering. Suddenly a vast expanse of land—more than 1.6 kilometers (one mile) across, the size of the downtown area of a small city—exploded into the air with a tremendous blast followed by the ground-shaking roar and hiss of a gigantic column of superheated gases and steam jetting upward into the sky. Huge chunks of country rock and glowing-hot magma blobs were sent rocketing across the heavens, arcing away to fall back to the Earth at incredible distances. Thick black plumes of smoke began to billow up from the blast crater, produced by the subterranean burning of coal beds by the hot magma sills. Coal fly ash from these billowing smoke clouds was transported as far as 20,000 kilometers (12,400 miles) from Siberia, around the top of the world, to settle in Canadian High Arctic sediments (Grasby et al. 2011).

If that scenario is difficult to envision, imagine it happening 6,400 times! Svensen and colleagues point out that the eruption pattern of the Siberian LIP blast pipes is unknown. Obviously, if one pipe exploded per year, then the entire process took 6,400 years, but it is much more likely that pipe formation was clustered in time—probably starting out with the occasional blast of a single or a few pipes erupting as the initial magma intrusions reached the buried Cambrian petroleum-rich evaporite strata, followed by a period of time with numerous pipes exploding simultaneously across the landscape during the phase of maximum heating and magma intrusion from the head of the mantle super-plume located at depth, and ending with a tapering off of pipe formation as the subterranean magma masses cooled. In that scenario, the formation of the Siberian LIP blast pipes took place in a time interval much shorter than 6,400 years.

Svensen and colleagues have conducted extensive heating experiments with geologic samples of the Siberian LIP strata, obtained through numerous drill holes produced by prospectors searching for petroleum, magnetite, and potassium salt deposits. They have concluded that the Siberian LIP blast pipes vented between 39 and 114 trillion tonnes (43 to 125 trillion tons) of carbon dioxide, plus an additional 20 trillion tonnes (22 trillion tons) of carbon dioxide degassed from the lava. In addition, the contact metamorphism of coal and other organic-rich strata by the Siberian mantle super-plume magmas is estimated to have released between 14.3 and 41.9 trillion tonnes (16 to 46 trillion tons) of methane. The coal fly ash produced by the burning of the Siberian LIP coal beds also contained high levels of the toxic metals chromium, arsenic, mercury, and lead (Grasby et al. 2011, 2017).

The carbon dioxide and methane vented from the burning of petroleum- and organic-rich rocks produced a huge injection of the light carbon-12 isotope into the atmosphere (Svensen et al. 2009). Analyses of the resultant negative $\delta^{13}C$ excursion found in end-Permian strata have led the Stanford University geologist Jonathan Payne and colleagues to estimate an even higher volume of vented carbon-dioxide gas from the Siberian LIP—some 100 to 160 trillion tonnes (110 to 176 tons)—than that estimated by Svensen and colleagues (Payne et al. 2010).

In addition to organic-rich rocks, the Siberian LIP magmas encountered a type of rock not present in the Emeishan LIP area—the Cambrian evaporite strata, thick deposits of rock salt and anhydrite. The burning of rock salt, or halite, released sodium metal and chlorine gas, both of which are poisonous. The Siberian LIP chlorine gas interacted with water vapor to produce rock-salt-derived hydrochloric acid, in addition to the amount produced by the magma itself. Even more serious, however, is that the vented chlorine gas combined with the huge amounts of methane also released by the Siberian LIP to produce a very noxious gas—methyl chloride (CH_3Cl). Svensen and colleagues estimate that between 5.2 and 15.3 trillion tonnes (5.7 to 16.9 trillion tons) of methyl-chloride gas were released into the Earth's atmosphere by the Siberian LIP. Finally, the presence of bromine in the Cambrian evaporite strata is estimated to have produced another noxious halocarbon gas in bromine's combination with methane—methyl bromide (CH_3Br). Although smaller than the huge volume of methyl chloride produced by the Siberian LIP, Svensen and colleagues estimate that between 87 and 255 billion tonnes (96 to 281 billion tons) of methyl bromide were emitted into the Earth's atmosphere at the end of the Permian (Svensen et al. 2009). The consequences of the formation of these two halocarbon gases by the Siberian LIP were dire indeed, as we will see shortly.

Finally, it is estimated that the Emeishan LIP eruption produced about 5.4 trillion tonnes of sulfur dioxide, as discussed previously. Since the Siberian LIP flood-basalt eruption was 8.3 to 20 times larger than the Emeishan, one might conclude that the Siberian LIP lavas would have produced 8.3 to 20 times more sulfur dioxide. But that calculation does not consider the effect of the thermal metamorphism of the other evaporite rock type present in the Cambrian evaporite strata—thick deposits of anhydrite. Anhydrite is a calcium sulfate, and burning anhydrite produces sulfur dioxide gas in addition to the amount produced by the magma itself; thus, the potential amount of sulfuric acid produced by the Siberian LIP eruption was even greater than would be anticipated based on the size of the flood basalts alone.

Can the Siberian LIP scenario get any worse? The answer is yes. The thermo-mechanical modeling of the Siberian super-plume by Sobolev and colleagues suggests that about 10 to 20 percent of the magma volume produced did not come from the deep mantle alone but rather from the melting and recycling of oceanic crustal rock located beneath the continental crust of Siberia—oceanic crust that had been subducted beneath the continental crust of Asia in earlier plate-tectonic events. If so, the magma produced by the super-plume would be much more gaseous and carbon rich than the typical basaltic-type lavas seen in LIP eruptions like those of Laki, Iceland. Sobolev and colleagues estimate that the degassing of the modeled super-plume magmas could have produced about 175 trillion tonnes (193 trillion tons) of carbon dioxide—an amount greater than the estimated 39 to 114 trillion tonnes of carbon dioxide vented through the Siberian blast pipes from the subsurface thermal metamorphism of the Tunguska Basin petroleum-rich strata! In addition, a surprising amount of hydrochloric acid is also modeled to have occurred solely from the degassing super-plume itself—some 18 trillion tonnes (20 trillion tons) (Sobolev et al. 2011).

In this scenario, Sobolev and colleagues suggest that "CO_2 from the plume alone may have triggered the main extinction event . . . degassing of the plume, rather than thermogenic gases from sediments, triggered the biotic crises" (Sobolev et al. 2011, 314–315). However, we know that the thick deposits of petroleum-, coal-, and evaporite-rich strata in the Tunguska Basin were thermally metamorphosed, and the explosive result of that metamorphism is evidenced by the 6,400 blast pipes present in the basin today. In addition, we know that the burning of the Tunguska Basin coal deposits produced huge clouds of carbon-rich ash, ash in such quantities that it was transported aloft some 20,000 kilometers away to Canada (Grasby et al. 2011). Thus, rather than an *alternative source* of carbon dioxide, the Siberian super-plume model of Sobolev and colleagues may provide an *additional source* of trillions of tonnes of carbon dioxide to add to the trillions of tonnes produced by thermogenic degassing from the Tunguska Basin petroleum-rich strata—making an originally catastrophic environmental scenario much worse. In this combination scenario, catastrophic degassing from the deeper super-plume itself began shortly before the main flood-basalt eruptions, followed by additional catastrophic degassing from the shallower Tunguska Basin petroleum-rich evaporite strata during the flood-basalt eruptions.

THE END-PERMIAN KILL MECHANISMS

The predicted environmental consequences of the injection of teratonnages of both methane and carbon dioxide into the Earth's atmosphere by the huge Siberian LIP eruption are the same as they were for the earlier and smaller Emeishan eruption, just very much worse (table 6.1). Methane and carbon dioxide are both greenhouse gases—methane much more so than carbon dioxide—and global heating of the planet is expected with high concentrations of these two gases in the atmosphere. Methane is particularly bad in that it retains heat itself and, when oxidized, also produces carbon dioxide, a gas that continues to retain heat in the atmosphere.

Kill Mechanism 1: Heat Death

The nearly unbelievable seriousness of the magnitude of the heating of the Earth that occurred at the end of the Permian and into the Early Triassic can be sensed in the wording of the titles (usually pretty staid and descriptive) of numerous scientific papers in research journals that have revealed the enormity of the event—titles such as "Hot Acidic Late Permian Seas Stifle Life in Record Time" (Georgiev et al. 2011), "Lethally Hot Temperatures During the Early

TABLE 6.1 Pollutants produced by the Siberian LIP eruption and their global environmental impacts.

Pollutant/Pollutant Product	Environmental Impact
Sulfur dioxide/sulfuric acid	Acidification of oceans, acid rain on land
Carbon dioxide/carbonic acid	Acidification of oceans, acid rain on land
Chlorine/hydrochloric acid	Acidification of oceans, acid rain on land
Atmospheric sulfur dioxide	Short-term global cooling
Atmospheric carbon dioxide	Long-term global warming
Atmospheric methane	Long-term global warming, global oxygen depletion
Chlorine/methyl chloride	Atmospheric ozone destruction
Bromine/methyl bromide	Atmospheric ozone destruction
Coal fly ash	Oceanic euxinia and anoxia, metal toxicity

Triassic Greenhouse" (Sun et al. 2012), and "Post-Apocalyptic Greenhouse Climate Revealed by Earliest Triassic Paleosols" (Retallack 1999). The language within many of these scientific papers is just as startling—for example, from a study using the ratios of rhenium (Re) and osmium (Os) isotopes in calculating seawater variation in temperature and acidity, "If temperature was indeed the controlling factor, the extreme ^{187}Re /^{188}Os ratios in the Upper Permian time require global warming at levels unparalleled in the geologic record. . . . Such ratios are unknown from any other period in Earth history" (Georgiev et al. 2011, 397, 399).

The standard technique for calculating past temperatures is to measure the ratios of oxygen isotopes ($\delta^{18}O$) preserved in the skeletal elements of organisms alive during the study interval, usually the shells of brachiopods or the dentary elements of conodonts (McGhee 2013, 196–199). Using the $\delta^{18}O$ index values found in the dentary elements of conodonts, the China University of Geosciences geobiologist Yadong Sun and colleagues have reconstructed seawater temperatures in the late Changhsingian to early Middle Triassic. Their research has revealed that seawater temperatures rose rapidly from 21°C (70°F) to 36°C (97°F) across the Permian/Triassic boundary. This rapid rise in temperature was interrupted by a cool pulse, when seawater temperatures dropped back to a still-quite-warm 32°C (90°F)—because of renewed volcanic eruption of sulfur dioxide?—before rising to the highest temperatures seen in their study: seawater temperatures of 38°C (100°F) and sea-surface temperatures of 40°C (104°F) in the mid-Olenekian Age of the Early Triassic.[20]

Seawater temperatures above 35°C (95°F) are lethal for most marine organisms, triggering hyperthermia (table 6.2). Yet the study by Sun and colleagues show that seawater temperatures remained above 35°C for almost the entire Early Triassic, a span of 5.1 million years, before finally falling back to the very warm range of 32°C to 34°C (90°F to 93°F) at the onset of the Middle Triassic Epoch. At present, sea-surface temperatures in the equatorial zone of the Earth—the hottest part of the Earth—range between 25°C and 30°C (77°F to 86°F). High seawater temperatures also lead to lower oxygen solubility and thus facilitate the development of marine anoxia, a fact we will return to when we consider the hypoxic kill mechanism in detail (table 6.2).

On land, photorespiration starts to seriously interfere with plant photosynthesis at temperatures above 35°C (95°F), and few terrestrial plants can survive

TABLE 6.2 Kill mechanisms triggered by the Siberian-LIP-induced global environmental changes listed in table 6.1.

Kill Mechanism	Organisms Adversely Affected
Hyperthermia (heat death)	Large animals (both marine and land), energetic animals with higher metabolic rates, marine sessile benthic organisms, land plants
Acidosis (acid poisoning)	Marine uni- and mullticellular organisms with calcified skeletons, marine sessile benthic organisms, land plants
Hypercapnia (CO_2 poisoning)	Large animals (both marine and land), marine uni- and multicellular organisms with calcified skeletons, marine animals with poorly buffered respiratory physiologies, marine sessile benthic organisms
Hypoxia (suffocation)	Large animals (both marine and land), energetic animals with higher metabolic rates, marine sessile benthic organisms
Radiation poisoning	Land plants, land animals that cannot burrow to escape surface radiation

temperatures higher than 40°C. At temperatures of 45°C (113°F), land animals begin to succumb to hyperthermia (Sun et al. 2012). The Université Claude Bernard geochemist Kévin Rey and colleagues have analyzed the $\delta^{18}O$ values found in the bones and teeth of terrestrial vertebrates in South Africa and have reported that at the end of the Permian "an intense and fast warming of +16°C [+29°F] occurred and kept increasing during the Olenekian" Age of the Early Triassic (Rey et al. 2016, 384). All in all, the end-Permian hot pulse in air temperatures on land "lasted 6 Ma [million years] before temperatures decreased" in the Anisian Age of the Middle Triassic (Rey et al. 2016, 384).

In the "Hot Earth" of the latest Permian and Early Triassic, the equatorial tropics were lethal zones both on land and in the sea—in stark contrast to our modern world, where the equatorial tropics harbor the highest diversity of life on the planet. The equatorial marine zone of the Hot Earth was full of "gaps"—the "marine fish gap," the "marine reptile gap"—marked by the absence of life-forms driven out of the tropics to higher latitudes of the planet where temperatures were survivable (Sun et al. 2012). Individuals of marine species that managed to survive in the Hot Earth equatorial seas were all abnormally small and stunted—a phenomenon known as the Early Triassic "Lilliput effect" (Twitchett 2007), which Sun and colleagues (2012) argue to be the result of the low thermal tolerance range of organisms with large body sizes. The equatorial land areas of the Hot Earth saw the "tetrapod gap," the "coal gap," and the "conifer gap"—the absence of tetrapods, the absence of peat swamps, and the absence of conifers (Retallack et al. 1996). Instead of the highly derived conifer seed plants, the flora of the equatorial Hot Earth were dominated by peculiar primitive spore-reproducing lycophytes, relatives of the ancient lycophytes that dominated the coal swamps of the Carboniferous world that we considered in chapter 3. Only in the marginally cooler Middle Triassic Earth would the tetrapods and conifers once again return to the equatorial regions, and peat swamps once again form (Sun et al. 2012).

Kill Mechanism 2: Carbon Dioxide Poisoning

Carbonic acid formed from both carbon dioxide and methane acidified the oceans and produced acid rain on land. To the acid effect of trillions of tonnes of both gases was added the trillions of tonnes of sulfuric acid and hydrochloric acid produced both from the mantle super-plume itself and from the thermal metamorphism of the Tunguska Basin petroleum-rich evaporite strata (table 6.1). To make matters even worse, the ocean chemistry of the Paleozoic world was not like that of our present world. Jonathan Payne and his colleagues point out that the Earth's oceans today are "buffered against acidification by extensive, fine-grained, unlithified carbonate sediments on the deep-sea floor, which could relatively rapidly dissolve to counter acidification. By contrast, the Late Permian deep sea contained no such carbonate buffer because Permian oceans lacked abundant pelagic carbonate producers such as coccolithophorids and planktonic foraminifers. Consequently, any buffering against acidification via dissolution of carbonate sediments could only have occurred more slowly in the less extensive, coarse-grained, mostly lithified, shallow-marine carbonate platform sediments or via chemical weathering of silicate and carbonate rocks on land" (Payne et al. 2010, 8546; see also Ridgwell 2005) and "In the absence of a large reservoir of fine-grained, unlithified deep-sea carbonate sediment, whole-ocean acidification could likely last for tens of thousands of years (Archer et al., 1997), and few refugia would exist—survival would depend primarily on long-term physiological tolerance of altered conditions" (Payne and Clapham 2012; Clarkson et al. 2015).

Biological evidence corroborates the hypothesis that the acidification of the world's oceans during the end-Permian mass extinction was much worse than it was during the earlier Capitanian crisis. During the eruption of the Emeishan LIP, highly calcified organisms like massive corals, large-bodied fusulinid foraminifera, and giant alatoconchid bivalves suffered high extinction losses relative to organisms with lightly calcified skeletons. However, during the eruption of the Siberian LIP, *all organisms with calcareous skeletons* suffered high extinction losses. In fact, having a skeleton composed of anything other than calcium carbonate was highly beneficial in surviving the end-Permian mass extinction (Clapham and Payne 2011). This differential survival pattern is particularly striking within clades of closely related organisms: many species of calcareous foraminifera went extinct, whereas those foraminifera that formed their skeletons by agglutinating siliceous sedimentary grains survived; many species of calcareous brachiopods went extinct, whereas the inarticulated brachiopods with phosphatic shells survived; many species of calcareous sponges went extinct, whereas siliceous hexactinellid sponges survived; species of calcified corals and algae went extinct, whereas unskeletonized corals (naked cnidarians, like modern sea anemones) and unskeletonized algae survived; and so on (Knoll et al. 2007; Clapham and Payne 2011; Kiessling and Simpson 2011; Garbelli et al. 2017).

A surprising contrast to the deadly effect of acid-rich seawater on calcareous organisms is the discovery of peculiar primary deposits of calcium carbonate on the seafloor in some areas, deposits of odd fan-shaped crystal structures that can be some ten centimeters (four inches) in diameter. These calcium-carbonate fans and crusts are thought to have been produced in the aftermath of an oceanic acidification event, following the death of most benthic organisms with calcareous skeletons, and the upwelling of alkaline deep waters supersaturated with calcium carbonate—but no organisms to take up the calcium carbonate for their skeletal constructions—resulted in these primary sedimentary carbonate deposits (Payne and Clapham 2012). Finally, on land, the deadly effects of acid rain, such as those that were seen in Europe with the historic Laki LIP eruption, were added to the already deadly effects of hyperthermia in killing off land plants.

Added to the deadly effects of sulfuric, hydrochloric, and carbonic acids in marine waters is the related effect of high concentrations of simple carbon dioxide in ocean waters—hypercapnia, or carbon-dioxide poisoning (table 6.2). Hypercapnia preferentially kills many of the same organisms as acidosis, but with a few differences. Animals with poorly buffered respiratory physiologies are least able to deal with the deadly effects of carbon-dioxide poisoning; these animals include the sponges, corals, brachiopods, bryozoans, and most echinoderms. In contrast, energetic animals with well-buffered physiologies are better able to deal with carbon-dioxide stress; these animals include the molluscs, arthropods, chordates, and infaunal organisms in general, which are adapted to living within submarine sediments where oxygen partial pressures are low and carbon-dioxide partial pressures are high (Knoll et al. 1996, 2007; Clapham and Payne 2011). The differential survival rates of these two groups of organisms provide an important key to understanding the radical ecological reorganization that ended the Paleozoic world—an ending we will consider in detail later in this chapter.

Terrestrial "infaunal" animals—that is, burrowers that live underground—also are better able to survive high carbon-dioxide levels in atmospheric gases, for the same reason as with the infaunal marine animals. Animals with more advanced respiratory structures, such as muscular diaphragms for energetic breathing, lungs with large surface areas contained in barrel-chested rib cages, and nasal turbinals are also better able to survive high-carbon-dioxide, low-oxygen conditions (Retallack and Krull 2006; Knoll et al. 2007). Thus the worldwide survival and proliferation of species of the small-bodied, barrel-chested, burrowing tetrapod *Lystrosaurus* in the Early Triassic—at high latitudes, far away from the lethally hot tropics (Sun et al. 2012)—may be attributed to the combined selective effects of high temperatures, high carbon-dioxide levels, and low oxygen levels in terrestrial environments following the end-Permian mass extinction (fig. 6.9).

Kill Mechanism 3: Suffocation

Just as the kill mechanisms of acidosis and hypercapnia adversely affect many of the same organisms, the effects of the kill mechanism of hypoxia—oxygen-deprivation stress all the way to suffocation death—are similar as well (table 6.2). The Siberian LIP eruptions triggered a major drop in the oxygen content of

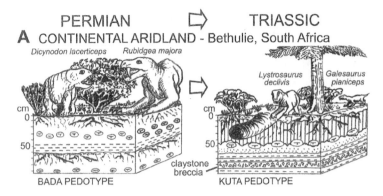

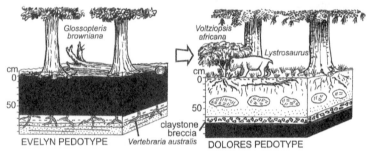

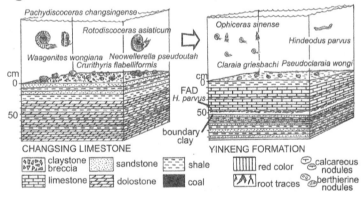

FIGURE 6.9 Summary reconstructions of biological conditions before and after the end-Permian mass extinction: (a) the replacement of Permian large land animals (*Dicynodon lacerticeps* and *Rubidgea majora*) by Triassic small land animals (*Galesaurus planiceps*) and burrowers (*Lystrosaurus declivis*) in dry environments; (b) the replacement of Permian wetland thick peat deposits (black layer) and trees with aquatically adapted deep roots (*Glossopteris browniana*) by Triassic dry conditions; (c) the replacement of Permian large marine animals by Triassic small "Lilliput" marine animals.

Source: From *Geological Society of America Special Paper* 399, pp. 249–268, by G. J. Retallack and E. S. Krull, "Carbon Isotopic Evidence for Terminal-Permian Methane Outbursts and Their Role in Extinctions of Animals, Plants, Coral Reefs, and Peat Swamps," copyright © 2006 Geological Society of America. Reprinted with permission.

the Earth's atmosphere in a cascade of linked environmental effects. First, the burning—the oxidation—of the vast deposits of organic material contained in the Tunguska petroleum-rich strata and coal beds would necessarily consume a tremendous amount of atmospheric oxygen.[21] As we previously considered, one of the very negative consequences of the burning of those subterranean hydrocarbon-rich strata was the production of trillions of tonnes of methane. Second, the oxidation of that vast amount of methane also sucked oxygen out of the air—two oxygen molecules per single methane molecule—leading to a further global decline in the oxygen content of the oceans and of the atmosphere.[22]

The atmospheric modeling of Robert Berner (fig. 3.8) indicates that a decline in oxygen content of the Earth's atmosphere began in the Middle Permian, continued through the Late Permian and Early Triassic, and reached a minimum level of 15.5 percent at the beginning of the Middle Triassic, in the Geocarbsulf model (in the Anisian Age; see table 6.3), or a minimum level of 16.0 percent in the late Middle Triassic (the Ladinian Age; see table 6.3) in the Rock-Abundance model. In the models, the predicted atmospheric-oxygen values are calculated at ten-million-year intervals and thus correspond only roughly to the expected oxygen-depleting effects of the Emeishan and Siberian LIP eruptions, whose catastrophic phases were of much shorter duration than ten million years (table 6.3). Note, however, that the minimum values predicted for the Triassic—an atmospheric oxygen content of 15.5–16.0 percent—are still above the 13 percent level below which wildfires do not take place and charcoal is not formed.

Abundant geologic evidence exists for widespread marine anoxia during and after the end-Permian mass extinction; indeed, this may have been the most geographically widespread anoxic event in the world's oceans for the entire Phanerozoic—a "superanoxic event" (Wignall and Twitchett 2002). Black-shale deposits are typical sedimentary signals of oxygen depletion in marine waters, and black-shale deposits are widespread in shallow marine environments during the end-Permian mass extinction. Thinly laminated sediments rich in pyrite, an iron sulfide (FeS_2), are also widespread and suggest not only anoxic conditions but the presence of deadly hydrogen sulfide (H_2S). Other geochemical signatures of anoxia and of the presence of hydrogen sulfide, such as shifts in the relative isotopic abundances of nitrogen and uranium and the presence of isorenieratane, have also been widely discovered in Permian/Triassic boundary strata (Payne and Clapham 2012).

TABLE 6.3 Major LIP volcanic events in the Permian, Triassic, and earliest Jurassic compared with the oxygen content in the Earth's atmosphere.

Geologic Age	Atmospheric O_2		
	(1)	(2)	
Hettangian	17.0%	14.5%	
Rhaetian	17.0%	18.0%	← Central Atlantic LIP eruptions
Norian	16.0%	16.0%	← Central Atlantic LIP eruptions
Carnian	17.0%	15.5%	
Ladinian	17.6%	16.0%	
Anisian	20.9%	18.7%	
Olenekian	22.5%	20.0%	
Induan	23.3%	21.8%	← Smaller Siberian LIP eruptions continue
Changhsingian	24.1%	23.6%	← Catastrophic Siberian LIP eruptions
Wuchiapingian	26.5%	29.0%	
Capitanian	28.3%	30.0%	← Emeishan LIP eruptions
Wordian	30.0%	31.0%	
Roadian	30.6%	30.9%	
Kungarian	33.0%	30.5%	
Artinskian	35.0%	30.0%	
Sakmarian	34.8%	29.3%	
Asselian	34.5%	28.5%	

Source: Column (1) is the Rock-Abundance model (Berner et al., 2003) and column (2) is the Geocarbsulf model (Berner, 2006). Oxygen data extrapolated from figure 3.8 using the timescale of Gradstein et al. (2012).

Note: Atmospheric oxygen contents of 25% or higher have been underlined; those of 30% or higher are also in bold.

In the oceans, anoxia would have been triggered not only by Siberian-LIP-induced low atmospheric pressure of oxygen and the oxidation of Siberian-LIP-produced methane but also by the high seawater temperatures of the Hot Earth, which substantially lowered oxygen solubility. In addition, nutrients (chiefly iron) from the coal fly ash produced by the volcanic burning of subterranean coal deposits in the Tunguska Basin may have triggered vast algal blooms and eutrophication of shallow marine waters (table 6.1), leading to further oxygen depletion in bottom waters (Grasby et al. 2011).

Still, Jonathan Payne and the University of California paleontologist Matthew Clapham argue that the hypoxic kill mechanism played a lesser role in the total biological devastation of the end-Permian mass extinction, stressing the primary kill mechanisms of hyperthermia, acidosis, and hypercapnia (table 6.2). They point out that surficial and shallow-water anoxic and hydrogen-sulfide-rich conditions "could not have been both temporally persistent and geographically widespread unless atmospheric oxygen levels were far lower than the values indicated by model reconstructions (e.g., Berner 2006)" (Payne and Clapham 2012, 102) (see table 6.3).

For land animals, the effects of hypoxia and hypercapnia are essentially the same; thus, the survival and proliferation of species of the small-bodied, barrel-chested, burrowing dicynodont *Lystrosaurus*—approximately 90 percent of all land vertebrates alive in the Early Triassic were lystrosaurs (Sahney and Benton 2008)—could just as well be due to selection for low-oxygen tolerance as for high-carbon-dioxide tolerance, and is probably due to both selective effects acting in concert (fig. 6.9). Land plants, however, present a more complex problem. The dominant and more recently evolved pinophyte spermatophytes of the Permian were replaced by more ancient, spore-reproducing isoetalean lycophytes in the terrestrial flora of the earliest Triassic (see table 3.1 for the phylogeny of Paleozoic land plants). Only in the late Olenekian Age of the Early Triassic did the pinophytes regain their dominance in the terrestrial flora, and then only in the higher-latitude regions of the Earth, whereas the lycophytes remained dominant in the equatorial regions until the beginning of the Middle Triassic (Sun et al. 2012). This latitudinal pattern of retreat from the tropics by the pinophytes, then gradual return to lower and lower latitudes until the tropics were once again reached by the Middle Triassic, is exactly the pattern one would expect to be produced by the onset of hot temperatures at the Permian/Triassic boundary, persistent high temperatures in the Early Triassic, and a final cooling down of the Earth by the Middle Triassic. This latitudinal-shift pattern in land plants is not a predicted result of the onset of low-oxygen, high-carbon-dioxide conditions in the Earth's atmosphere.

In fact, one might predict that the land plants would flourish in such an atmosphere—if temperatures were within livable ranges. To land plants, oxygen is a waste product. Unlike terrestrial arthropods or vertebrates, which need oxygen to survive, plants need carbon dioxide to synthesize their food. In the fossil record we repeatedly see that land plants flourish during times of low oxygen content, and high carbon-dioxide content, in the atmosphere. But did land plants flourish in the Early Triassic? No, they did the exact opposite. Rather than flourishing, we see a six-million-year global "coal gap" in which there was insufficient plant growth in peat swamps to form any coal deposits (Retallack et al. 1996). In the hills and highlands, there was insufficient plant growth to stabilize and bind the upland soils in the Early Triassic, and downstream, low-energy meandering rivers were replaced by high-energy braided rivers on a worldwide basis (Ward et al. 2000; Sahney and Benton 2008). Still, Gregory Retallack argues that the coal gap could also have been produced—at least in part—by "soil hypoxia," because low oxygen levels within soils will interfere with plant-root respiration and nutrient uptake (Retallack and Krull 2006).

Kill Mechanism 4: Radiation Poisoning

Land plants do show evidence of an additional kill mechanism not demonstrated for the Emeishan LIP eruption: radiation poisoning (table 6.2). The predicted consequences of the injection into the atmosphere of trillions of tonnes of methyl chloride and billions of tonnes of methyl bromide were new to the Siberian LIP eruption (table 6.1), as the Emeishan LIP magmas did not encounter buried evaporite strata. Chlorine and bromine catalytically destroy ozone,[23] as we know from modern times with our efforts to stop the destruction of the Earth's ozone layer at the poles of the planet caused by pollutant chlorofluorocarbons from our older coolant systems. The Siberian LIP eruption produced trillions of tonnes of hydrochloric acid, but atmospheric-photochemical computer simulations by the University of Sheffield biologist David Beerling and his colleagues suggest that that amount of hydrochloric acid, although huge, would only destroy 33–55 percent of the stratospheric ozone shield in the high-latitude Northern Hemisphere, where the Siberian LIP is located. However, when additional trillions of tonnes of methyl chloride are added to the mix, the simulations predict that 70–85 percent of the Northern Hemisphere ozone shield would be destroyed, and that 55–80 percent of the Southern Hemisphere ozone shield would be destroyed as well. And, to make matters worse, the addition of methyl bromide to the mix speeds up the catalytic process of ozone destruction (Beerling et al. 2007; Svensen et al. 2009).

With the destruction or serious thinning of the ozone layer, the radiation flux of high-energy ultraviolet light in the shortwave 290 to 315 nanometer spectrum[24]—deadly UV-B radiation—increases at the Earth's surface. The computer simulations of Beerling and colleagues predict that the UV-B radiation flux could increase by as much as 50 to 100 kilojoules (12 to 24 kilocalories) per square meter (square yard) per day (Beerling et al. 2007). That increased radiation flux would have mutagenic effects on land plants, as plants cannot flee the Earth's surface and hide underground as burrowing animals can. Thus, to the selective effects of hypercapnia and hypoxia, we can add the selective effect of radiation poisoning in favoring the survival of the burrowing lystrosaurs (table 6.2).

The Utrecht University paleobotanist Henk Visscher and his colleagues argue that that bleak radiation-poisoning scenario did indeed happen in the end-Permian mass extinction. They report finding mutation-induced changes in land-plant reproductive structures in Late Permian strata—anomalous variation in shape, size, wall thickness, and arrangement of pollen and spores. Numerous spores are found still bound together in groups of four, indicating that the normal process of spore development in reproduction was never completed. Lycophytes in particular are able to continue to reproduce asexually if their sexual reproductive structures—spores—are damaged or rendered sterile.

It has been proposed that radiation-induced damage to their pollen- and seed-producing structures contributed to the decline of the large woody conifers and the vanishing of closed-canopied forests in the Late Permian. The conifers were replaced by isoetalean lycophytes, the quillworts (table 3.1), and the great majority of the lycophyte species belonged to a single genus of isoetaleans, *Pleuromeia*. Thus the diverse Late Permian forests of large, woody conifers was replaced by a monotonous terrestrial plant cover of *Pleuromeia*, polelike lycophytes covered by a sleeve of tiny leaves and topped by sporangia, standing only 1.5 to two meters (five to 6.6 feet) high (Pfefferkorn 2004). The latitudinal pattern of global deforestation in the Late Permian fits well with a kill mechanism of radiation poisoning and hypothermia, perhaps assisted by acidosis, as discussed in the previous sections of the chapter. The subsequent invasion of herbaceous lycophytes fits well with an ecological scenario of forest destruction and replacement by fast-growing, opportunistic weedy plants that reproduced asexually. This scenario also explains why many of these lycophytes are found to have colonized and grown in drought-prone habitats that are usually off limits to the sexually reproducing lycophytes that require water in the reproductive process (Visscher et al. 2004).

The question then becomes: what sterilized the Late Permian land plants and caused their reproductive structures to mutate? Visscher and colleagues argue that anomalous changes in land-plant reproductive structures were produced by exposure to high-energy UV-B radiation, and that that exposure could only have been produced by the vanishing or serious thinning of the Earth's ozone shield in the Late Permian. And it did not happen just once: fossil layers containing mutant spores and pollen are also found in the Early Triassic, indicating a repeated cycle of collapse and recovery of the ozone layer. Such a cyclic pattern is consistent with the continued Siberian LIP eruptions and lethal gas emissions during the Early Triassic (Visscher et al. 2004; Sun et al. 2012; Black et al. 2014).

In summary, the worst mass extinction in Earth history appears to have been the result of a series of very bad events happening concurrently: first, the eruption in Siberia of the largest continental LIP in the Phanerozoic Eon; second, the penetration of the rising Siberian super-plume through the vast deposits of buried hydrocarbons (petroleum, coal) and evaporites containing chlorine, bromine, and sulfur (halite, anhydrite) in the Tunguska Basin (table 6.1); and third, the explosive nature of the catastrophic phase of the eruptions, in which volcanic gases were vented into the Earth's atmosphere through the formation of over 6,400 explosive pipes in the Tunguska Basin (figs. 6.7–6.8). As summarized by the Russian Academy of Sciences geologist Alexander Polozov and his colleagues, the "direct relationship between the Siberian Trap LIP and the abundance and style of pipes is unlike any other LIP encountered through the Earth's history, and points to the thick sequences of salt and carbonate rich sediments beneath the province playing a vital role in making the end-Permian event so extreme compared to many of the other LIP/extinction relationships" (Polozov et al. 2016).

The catastrophic phase of the Siberian LIP eruption took place in only 200,000 years, from 252.3 million years ago (the very latest Permian[25]) to 252.1 million years ago (the earliest Triassic) (Shen et al. 2011; Payne and Clapham 2012; Wang et al. 2014). On a geologic timescale, a mere 200,000 years is a blink of the eye, yet decades of research have progressively confirmed the extremely short time span during which the largest loss of animal biodiversity in Earth history occurred—and future research may narrow that time interval even further.[26] The mass extinction itself is now estimated to have occurred in only 60,000 ± 48,000 years, based on the radiometric dating of volcanic ash beds in

South China that lie above and below the extinction strata (Erwin 2015, xi, xiv). That is, the greatest die-off of animal life in Earth history may have taken place in as little as 12,000 years, and no more than 108,000 years, during the 200,000-year span of the catastrophic phase of the Siberian LIP eruption.

The Siberian LIP continued to erupt episodically all the way through the Early Triassic (Bryan et al. 2010), and the southern part of the LIP in Chelyabinsk experienced eruptive activity up to 243 million years ago, in the Anisian Age of the Middle Triassic (Reichow et al. 2009). Thus there was no respite from the end-Permian catastrophe—hot global climates and low levels of oxygen in the atmosphere persisted throughout the five million years of the Early Triassic (Retallack 1999; Sun et al. 2012). The end-Permian ecological catastrophe and the long duration of the Hot Earth environmental conditions of the Early Triassic proved to be too much for many of the typical life-forms of the Paleozoic to survive, and by the dawn of the Middle Triassic the 200-million-year old ecological structure of the Paleozoic world was gone forever.

THE EARTH POISONS ITS CHILDREN

We now know that in the end-Permian mass extinction the ultimate horror in geologic history occurred: the Earth poisoned its own children. Back in the 1970s, the once-popular "Gaia hypothesis" maintained that the Earth was similar to a living organism in that its geochemical and biochemical cycles seemed to act together to buffer and nurture the presence of life on the planet (Ruse 2013). For example, for the past 4,560 million years, major fluctuations have occurred in the temperature of the Earth, in the carbon dioxide and oxygen content of its atmosphere, in the salinity and other chemical components of its seas—yet all these fluctuations have occurred within upper and lower limits that have never exceeded the ability of life to continue to survive on the planet. This planetary physical-biological-homeostasis hypothesis was given the name Gaia by its formulator, the British scientist James Lovelock, after the ancient Greek goddess of the Earth.[27] In the decades that followed the Gaia proposal, some argued that the hypothesis was redundant in that everyone already knew that the evolution of life has had major effects on the physical characteristics of the Earth and that these changes in the physical characteristics of the Earth have had major effects on the subsequent evolution of life. Others have branded the hypothesis as pseudoscience, particularly the more extreme proposal by some that the planet Earth itself was a living organism (Ruse 2013).

Others have pointed out that life has had some extremely close calls to annihilation in the past 4,560 million years—and the end-Permian mass extinction was one of them. Rather than Gaia, the nurturing Greek goddess, the planet Earth in the Late Permian acted more like the Hindu god Shiva—a god that both creates and destroys entire worlds. Shiva is usually depicted as a male humanoid figure with two legs but with six arms, dancing barefoot on the back of a prostrate and squalling human baby. In the case of the Late Permian, the world that was destroyed was the Paleozoic world, the very world that had been created earlier in the Cambrian Explosion and the Great Ordovician Biodiversification Event. In the next two sections of the chapter, we will examine in detail—an obituary?—the destruction of the Paleozoic world: first the end of the world in the oceans, and then the end of the world on land.

The End of the Paleozoic Marine World

What was the Paleozoic world like in the oceans? In a series of classic papers in the 1980s and early 1990s, the University of Chicago[28] paleontologist J. John (Jack) Sepkoski Jr. argued that the Phanerozoic marine biosphere can be divided into three evolutionary faunas: the Cambrian evolutionary fauna, which dominated the Cambrian Period; the Paleozoic evolutionary fauna, which was dominant from the Ordovician through the Permian; and the modern evolutionary fauna, which was dominant in the Mesozoic and Cenozoic (fig. 6.10) (Sepkoski 1981, 1984, 1990; Sepkoski and Miller 1985). His research demonstrated that each evolutionary fauna possessed distinctive evolutionary traits; for example, each successive evolutionary fauna had a slower rate of diversification and species turnover but a higher level of maximum diversity. Jack Sepkoski and his student Arnie Miller demonstrated that these same three evolutionary faunas can be reconstructed from ecological community data (rather than just the taxonomic diversity data that Sepkoski had used in his initial works), and also that the three evolutionary faunas have characteristic ecological distributions along an onshore-offshore environmental gradient—thus the evolutionary faunas are

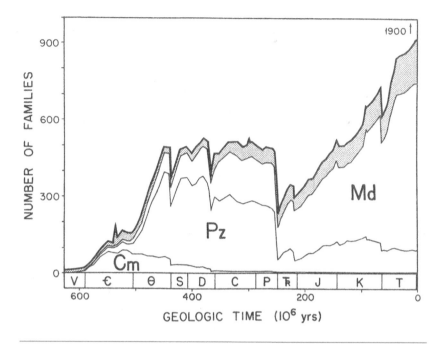

FIGURE 6.10 Sepkoski's classic plot of the familial diversity of marine animals in geologic time, showing the temporal diversities of the Cambrian Evolutionary Fauna (Cm), the Paleozoic Evolutionary Fauna (Pz), and the Modern Evolutionary Fauna (Md); see text for discussion. Geologic timescale abbreviations: V, Ediacaran; barred-C, Cambrian; Θ, Ordovician; S, Silurian; D, Devonian; C, Carboniferous; P, Permian; T_R, Triassic; J, Jurassic; K, Cretaceous; T, Tertiary (Paleogene and Neogene).

Source: From *Paleobiology*, volume 10, pp. 246–267, by J. J. Sepkoski Jr., "A Kinematic Model of Phanerozoic Taxonomic Diversity: III. Post-Paleozoic Families and Mass Extinctions," copyright © 1984 The Paleontological Society. Reprinted with permission of Cambridge University Press.

recurrent associations in space as well as in time (Sepkoski and Miller 1985). Subsequent research by other workers has demonstrated that each of the three evolutionary faunas has a unique and characteristic guild structure, with each successive evolutionary fauna occupying more "ecospace" and having more guilds in total (Bambach 1983, 1985); that each evolutionary fauna has its own tiering structure, both above and below the sediment-water interface (Bottjer and Ausich 1986); and that each of the evolutionary faunas is characterized by a unique series of periods of ecological structural stability (Sheehan 1996).

Sepkoski's three evolutionary faunas raised a few eyebrows at the time they were published because their time designations were not standard. First, the Cambrian Period is a part of the Paleozoic Era. Sepkoski's mathematical analyses showed that the Cambrian organisms are evolutionarily and ecologically demonstrably different from the Ordovician-Permian organisms, and subsequent research has shown that the Ordovician-Permian evolutionary fauna originated in the Great Ordovician Biodiversification Event (GOBE) (Webby et al. 2004; Servais et al. 2010), hence postdating the Cambrian fauna, but how can one speak of a "Paleozoic" evolutionary fauna without including the Cambrian organisms (see the top and middle graphs in figure 6.11)? Second, he lumped the Mesozoic marine fauna and the Cenozoic marine fauna together as one "modern" evolutionary fauna, and most people do not think of the extinct marine reptile *Ichthyosaurus* as a modern animal (see animal-illustration number 23 in the bottom graph in figure 6.11). Yet the point that Sepkoski argued was that the ecological structures of the Mesozoic and Cenozoic organisms were the same; for example, a Mesozoic ichthyosaur (a reptile) is ecologically equivalent to the convergently[29] evolved Cenozoic porpoise (a mammal; see animal-illustration number 24 in the bottom graph in figure 6.11).

In this book I consider the Paleozoic marine world to be just that—the ecological world of the marine organisms characteristic of the Paleozoic Era—and that that world consisted of two evolutionary phases: the initial Cambrian Explosion diversification that produced the Cambrian evolutionary fauna and the subsequent Great Ordovician Biodiversification Event that produced the Ordovician-Permian evolutionary fauna (top and middle graphs in figure 6.11). Thus, in table 6.4, I have combined Sepkoski's two evolutionary faunas of Paleozoic age into a single list of marine organisms characteristic of the Paleozoic marine world, and contrast those with the marine organisms characteristic of Sepkoski's modern evolutionary fauna. Note that many of the characteristic organisms of the Paleozoic world are extinct—but that even the modern fauna have extinct components, such as the Mesozoic marine reptiles like *Ichthyosaurus*.

At first glance (table 6.4) it might seem that the end-Permian mass extinction triggered the extinction of older, less well adapted, clades of organisms in favor in newly evolved, better adapted, clades. The actual pattern of survival is more subtle than a simplistic older-clade-versus-younger-clade scenario. If we take the list of characteristic organisms of the Paleozoic world and of the modern

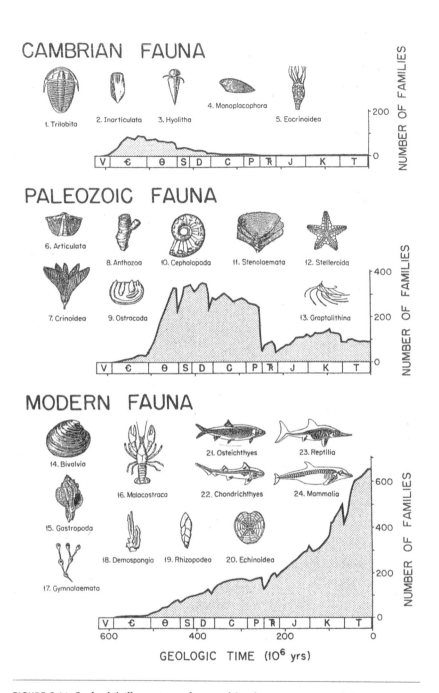

FIGURE 6.11 Sepkoski's illustrations of some of the characteristic animals belonging to the Cambrian, Paleozoic, and Modern Evolutionary Faunas. For geologic timescale abbreviations, see figure 6.10.

Source: From *Paleobiology*, volume 10, pp. 246–267, by J. J. Sepkoski Jr., "A Kinematic Model of Phanerozoic Taxonomic Diversity: III. Post-Paleozoic Families and Mass Extinctions," copyright © 1984 The Paleontological Society. Reprinted with permission of Cambridge University Press.

TABLE 6.4 Characteristic marine organisms of the Paleozoic world and of the modern evolutionary fauna.

Paleozoic World Fauna	Modern Evolutionary Fauna
1. Rugose† corals	1. Rhizarian unicells
2. Tabulate† corals	2. Demosponges
3. Stenolaemate bryozoans	3. Scleractinian corals
4. Brachiopods	4. Gymnolaemate bryozoans
5. Monoplacophoran molluscs	5. Gastropod molluscs
6. Hyolithan† lophophorates	6. Bivalve molluscs
7. Ammonoid† cephalopods	7. Malacostracan crustaceans
8. Trilobite† arthropods	8. Echinoid echinoderms
9. Ostracod crustaceans	9. Chondrichthyan fishes
10. Eocrinoid† echinoderms	10. Osteichthyan fishes
11. Crinoid echinoderms	11. Marine reptiles† (Mesozoic)
12. Asteroid echinoderms	12. Marine mammals (Cenozoic)
13. Graptolite† hemichordates	

Source: Modified from the classic evolutionary fauna analysis of Sepkoski (1984, 1990).

Note: Extinct taxa are marked with a dagger (†).

evolutionary fauna (table 6.4) and arrange those organisms in terms of their phylogenetic relationships (table 6.5), then we quickly see that in many cases the Paleozoic world and modern fauna organisms *belong to the same clades*. That is, the global ecological restructuring that the end-Permian mass extinction triggered was more of a within-clade than a between-clade phenomenon.

Let us examine the pattern of the collapse of the Paleozoic world in its evolutionary context (table 6.5). To begin, the clade of rhizarian unicellular organisms contains organisms that perished in the end-Permian mass extinction—namely, the calcareous bottom-dwelling foraminifera that died in the acidification of the late Permian oceans. However, this clade also contains elements of the modern evolutionary fauna—the highly successful siliceous radiolarians, which float up in the water column, and the bottom-dwelling agglutinated foraminifera, both of which groups were largely immune to the effects of oceanic acidification because neither has shells composed of calcium carbonate—as well as the calcareous floating foraminifera that evolved after the end-Permian mass extinction was over. Skipping down the list, the clade of the zoantharian corals contains

TABLE 6.5 Phylogenetic relationships of the marine organisms of the Paleozoic world (italicized) and of the modern evolutionary fauna (underlined).

Eukarya (eukaryote cells)

– Bikonta

– – Rhizaria

– – – Actinopoda

– – – – *Radiolaria*

– – – Foraminifera

– – – – *calcareous benthic foraminifera*

– – – – <u>agglutinated benthic foraminifera</u>

– – – – <u>planktic calcareous foraminifera</u>

– Unikonta

– – Opisthokonta

– – – Choanozoa

– – – – Metazoa (animals)

– – – – – Demospongiae

– – – – – – *Stromatoporoidea†*

– – – – – – <u>Sclerospongea</u>

– – – – – <u>Hexactinellida</u> (glass sponges)

– – – – – Eumetazoa

– – – – – – Cnidaria (jellyfish, corals, and kin)

– – – – – – – Zoantharia (corals)

– – – – – – – – *Tabulata†* (tabulate corals)

– – – – – – – – *Rugosa†* (horn corals)

– – – – – – – – *Heterocorallia†*

– – – – – – – – <u>Scleractinia</u>

– – – – – – Bilateria

– – – – – – – Protostomia (animals with protostome development)

– – – – – – – – Lophotrochozoa

– – – – – – – – – Lophophorata

– – – – – – – – – – Bryozoa (moss animals)

– – – – – – – – – – – *Stenolaemata*

– – – – – – – – – – – Gymnolaemata

– – – – – – – – – – – – <u>Ctenostomata</u>

– – – – – – – – – Phoronozoa

– – – – – – – – – – Phoronida

– – – – – – – – – – *Hyolitha†*

– – – – – – – – – – Brachiopoda (lampshells)

– – – – – – – – – – – *Linguliformea*

– – – – – – – – – – – Rhynchonelliformea

– – – – – – – – – – – – *Strophomenata†*

(continued)

```
– – – – – – – – – – – – Rhynchonellata
– – – – – – – – – – – – – Orthida†
– – – – – – – – – – – – – Pentamerida†
– – – – – – – – – – – – – Atrypida†
– – – – – – – – – – – – – Athyridida†
– – – – – – – – – – – – – Spiriferida†
– – – – – – – – – – – – – Spiriferinida†
– – – – – – – – – – – – – Rhynchonellida
– – – – – – – – – – – – – Terebratulida
– – – – – – – Eutrochozoa
– – – – – – – – Spiralia
– – – – – – – – – Annelida (segmented worms)
– – – – – – – – – – Polychaeta
– – – – – – – – – Mollusca
– – – – – – – – – – Eumollusca
– – – – – – – – – – – Polyplacophora (chitons)
– – – – – – – – – – – Conchifera
– – – – – – – – – – – – Monoplacophora
– – – – – – – – – – – – Ganglioneura
– – – – – – – – – – – – – Visceroconcha
– – – – – – – – – – – – – ?Tentaculitoidea†
– – – – – – – – – – – – – Cephalopoda (octopus, squid, and kin)
– – – – – – – – – – – – – – Ammonoidea† (shelled cephalopods)
– – – – – – – – – – – – – – Nautiloidea (shelled cephalopods)
– – – – – – – – – – – – – – Coleoidea (octopus, squid, and kin)
– – – – – – – – – – – – – – Gastropoda (snails)
– – – – – – – – – – – – – Diasoma
– – – – – – – – – – – – – – Rostroconcha†
– – – – – – – – – – – – – – Bivalvia (clams, mussels, and kin)
– – – – – – – Cuticulata
– – – – – – – – Ecdysozoa
– – – – – – – – – Panarthropoda
– – – – – – – – – – Arthropoda
– – – – – – – – – – – Cheliceriformes
– – – – – – – – – – – – Eurypterida† (sea scorpions)
– – – – – – – – – – – – Trilobita†
– – – – – – – – – – – – Merostomata (horseshoe crabs)
– – – – – – – – – – – – Pycnogonida (sea spiders)
```

```
– – – – – – – – – – – Mandibulata
– – – – – – – – – – – – Myriapoda
– – – – – – – – – – – – Pancrustacea (crustaceans)
– – – – – – – – – – – – – Maxillopoda
– – – – – – – – – – – – – – Ostracoda
– – – – – – – – – – – – – – Copepoda
– – – – – – – – – – – – – – Cirripedia (barnacles)
– – – – – – – – – – – – – – Malacostraca (crabs, lobsters, krill, and kin)
– – – – – – Deuterostomia (animals with deuterostome development)
– – – – – – – Echinodermata
– – – – – – – – Eocrinoidea†
– – – – – – – – Crinoidea (sea lilies)
– – – – – – – – Blastoidea†
– – – – – – – – Asteroidea (starfish)
– – – – – – – – Echinoidea (sea urchins)
– – – – – – – – Holothuroidea (sea cucumbers)
– – – – – – – – Pharyngotremata
– – – – – – – – Hemichordata
– – – – – – – – – Graptolithina†
– – – – – – – – Chordata
– – – – – – – – – Myomerozoa
– – – – – – – – – – Craniata
– – – – – – – – – – – Conodonta†
– – – – – – – – – – – Vertebrata
– – – – – – – – – – – – Gnathostomata (jawed vertebrates)
– – – – – – – – – – – – – Placodermi† (armored fishes)
– – – – – – – – – – – – – Chondrichthyes (sharks, rays, and kin)
– – – – – – – – – – – – – Acanthodii† (spine-fin fishes)
– – – – – – – – – – – – – Osteichthyes (bony fishes)
– – – – – – – – – – – – – – Actinopterygii (ray-fin fishes)
– – – – – – – – – – – – – – Sarcopterygii (lobe-fin fishes)
– – – – – – – – – – – – – – – Actinistia (coelacanths)
– – – – – – – – – – – – – – – Dipnoi (lung fishes)
– – – – – – – – – – – – – – – Tetrapodomorpha
– – – – – – – – – – – – – – – – Tetrapoda (limbed vertebrates; see table 6.8)
```

Source: Phylogenetic classification modified from Lecointre and Le Guyader (2006) and Benton (2015).

Note: Extinct taxa are marked with a dagger (†).

both the Paleozoic world heterocorals, tabulates, and rugosans that did not survive into the Mesozoic and the scleractinian corals that evolved in the Triassic, which are the main components of coral reefs today. The clade of the bryozoans, the tiny colonial moss animals, contains both the highly calcified stenolaemates that perished in the acidic late Permian seas and the uncalcified ctenostomate gymnolaemates that survived, along with the calcareous cheilostome gymnolaemates that would evolve later in the Jurassic. Skipping further down the list, the clade of the cephalopod molluscs contains both the numerous ammonoid shelled cephalopods that died from hypercapnia in the carbon-dioxide-poisoned terminal Permian oceans and the nautiloid shelled cephalopods that, with their better-buffered respiratory physiologies, survived (Knoll et al. 2007). Likewise, in the clade of the deuterostome echinoderms, the burrowing echinoids and holothurians, with their high metabolic activity levels and their tolerance of high carbon-dioxide levels within the sediment, survived differentially over the immobile or slow-moving eocrinoids, blastoids, and numerous crinoids with poorly-buffered respiratory physiologies.

Thus, in many cases, the effects of the end-Permian environmental catastrophe triggered within-clade ecological replacements, not wholesale replacement of one evolutionary clade by another. However, some clades, such as the brachiopod lampshells, were almost universal losers in the collapse of the Paleozoic world fauna (table 6.5). Each biodiversity crisis in the Paleozoic triggered diversity losses in the brachiopods, and even some of the rhynchonelliform groups that managed to survive the end-Permian mass extinction—the athyridids and spiriferinids—later succumbed to extinction in the Mesozoic and are not alive today. Only the rhynchonellids and terebratulids survive today, a small remnant of the once highly diverse brachiopod shellfish fauna of the Paleozoic world.[30]

Another way of understanding the magnitude of the collapse of the Paleozoic world is to examine the pattern of ecological replacements triggered by the end-Permian mass extinction. Table 6.6 shows nine major ecological roles of life, or megaguilds, for marine life (Droser et al. 2000). Within each megaguild are listed the organisms of the Paleozoic world that lived those ecological roles and the organisms of the modern evolutionary fauna that replaced the Paleozoic world fauna in those ecological roles following the end-Permian mass extinction.

In the oceans of the Earth, the pelagic ecological structure of the Paleozoic world's oceanic waters was totally destroyed—essentially all of the characteristic

TABLE 6.6 Ecological replacement of the fauna of the Paleozoic world by the modern evolutionary fauna in the marine realm following the end-Permian mass extinction.

I. *Pelagic Mode of Life* (living within oceanic waters)

Detritivores	*Paleozoic World:* chitinozoans†; tentaculitoid† molluscs; trilobite† arthropods; conodont† craniates; graptolite† hemichordates
	Modern Fauna: radiolarian and foraminiferan rhizarians; pteropod gastropods; krill and copepod crustaceans
Carnivores	*Paleozoic World:* ammonoid† cephalopods; eurypterid† arthropods; armored†, spine-fin†, and lobe-fin fishes
	Modern Fauna: nautiloid and squid cephalopods; sharks and rays; ray-fin fishes; marine reptiles†

II. *Epibenthic Mode of Life* (living on the surface of the sea bottom)

Sessile Detritivores	*Paleozoic World:* stromatoporoid† demosponges; stenolaemate bryozoans; acrotretid†, strophomenate†, orthid†, pentamerid†, atrypid†, athyridid†, spiriferid†, and spiriferinid† brachiopods; hyolithan† lophophorates; bivalve molluscs (low diversity); eocrinoid†, blastoid† and crinoid echinoderms
	Modern Fauna: sclerosponge demosponges; glass sponges; ctenostomate bryozoans; bivalve molluscs (high diversity); barnacle crustaceans
Sessile Carnivores	*Paleozoic World:* heterocoral†, rugose†, and tabulate† corals
	Modern Fauna: scleractinian corals
Mobile Detritivores	*Paleozoic World:* monoplacophoran and gastropod molluscs, trilobite† arthropods; ostracod crustaceans; armored† fishes
	Modern Fauna: gastropod and bivalve molluscs; malacostracan crustaceans; echinoid echinoderms; ray-fin fishes
Mobile Herbivores	*Paleozoic World:* monoplacophoran and gastropod molluscs; merostomate arthropods; ostracod and malacostracan crustaceans
	Modern Fauna: polychaete annelids; polyplacophoran and gastropod molluscs; malacostracan crustaceans; echinoid echinoderms; ray-fin fishes
Mobile Carnivores	*Paleozoic World:* ammonoid† cephalopods; malacostracan crustaceans; asteroid echinoderms
	Modern Fauna: polychaete annelids; gastropod molluscs; octopus cephalopods; pycnogonid arthropods; malacostracan crustaceans; asteroid echinoderms

---

III. *Endobenthic Mode of Life* (living within the sediment on the sea bottom)

Mobile Detritivores	*Paleozoic World:* linguliformean brachiopods; rostroconch† and bivalve molluscs (low diversity); trilobite† arthropods; conodont† craniates
	Modern Fauna: polychaete annelids; bivalve molluscs (high diversity); echinoid and holothurian echinoderms
Mobile Carnivores	*Paleozoic World:* polychaete annelids; merostomate arthropods
	Modern Fauna: polychaete annelids; gastropod molluscs; malacostracan crustaceans

---

Source: Data from Levinton (1982), Bambach (1983), and McGhee (2011).

Note: Extinct taxa are marked with a dagger (†).

organisms of that world are extinct (table 6.6). The collapse of the Paleozoic world ecological trophic pyramid in the oceans was total: both the zooplankton—the tiny detritivores—and the larger swimming predatory fauna were essentially eliminated. Only one extremely rare remnant of the lobe-fin fishes survives in our oceans today—the species *Latimeria chalumnae* (the famous "living fossil" coelacanth, which is an actinistian sarcopterygian; see table 6.5). In contrast to the Paleozoic, our modern oceans are filled with a zooplankton of rhizarians, gastropods, and crustaceans and a predatory fauna of squids, sharks, and ray-fin fishes. Only the marine reptiles of the Mesozoic are today extinct—but their ecological niche has been convergently filled by the evolution of modern-day marine mammals like killer whales and porpoises.[31]

On the sea bottom, the ecological structure of immobile, sessile benthic organisms of the Paleozoic world was totally destroyed—essentially all of the characteristic organisms of that world are also extinct (table 6.6). Again, both the detritivore and carnivore levels of the trophic structure of the Paleozoic ecosystem were destroyed. Instead of a bottom-dwelling fauna almost totally comprised of vast expanses of sessile brachiopod lampshells (some of them giants; see figure 4.9) and stalked echinoderms, and huge reefs comprised of massive calcareous stromatoporoid demosponges, tabulate corals, and horn corals, our modern ocean bottoms are populated by survivors of the latest Permian acid seawaters—sclerosponges, glass sponges, and ctenostome bryozoans—and active organisms with well-buffered respiratory physiologies—bivalve molluscs and crustaceans. Our modern reefs are dominantly comprised of scleractinian corals, corals that evolved from soft-bodied, skeletonless survivors of the Permian acid seas.

The ecological structure of the mobile bottom-dwelling organisms of the Paleozoic world was not as severely affected as that of the immobile organisms, probably because these organisms, being mobile, could at least try to flee late Permian toxic water conditions as they developed—an option the sessile organisms did not have (table 6.6). Thus, more of the ancient Paleozoic mobile bottom-dwelling fauna have survived to the present day, including the monoplacophoran molluscs, horseshoe crabs, ostracod and Paleozoic malacostracan crustaceans, and the starfish. Still, these surviving elements of the Paleozoic world are today vastly outnumbered by modern mollusc faunas of chitons, snails, bivalves, and octopuses, sea spider and modern crustacean arthropods, sea urchins, and ray-fin fishes (compare tables 6.5 and 6.6).

Living within the sediment on the sea bottom is a more restrictive mode of life, so fewer organisms have evolved adaptations to actively burrow in bottom muds and silts. Yet even here the ecological structure of burrowing organisms of the Paleozoic world was almost totally destroyed (table 6.6). The burrowing rostroconch molluscs, trilobites, and conodonts of the Paleozoic are all extinct, and the living linguliform lampshells are rare. Instead, our modern seas are dominated by burrowing polychaete worms, numerous bivalve molluscs and snails, and echinoderm sea urchins and sea cucumbers. All of these animals are descendants of active ancestors with a high tolerance for high carbon-dioxide levels, and low oxygen levels, within the marine muds of the latest Permian seas.

Although the end-Permian eruption of the Siberian LIP and the environmental catastrophe that followed sealed the fate of the Paleozoic marine world, the ecological structure of the Paleozoic world had already suffered several major shocks prior to its final collapse. The first biodiversity crisis to strike the Paleozoic world was the end-Ordovician (Hirnantian Age, table 1.3) mass extinction, which was triggered by the Late Ordovician glaciation (table 1.2). Sepkoski's data revealed that the Paleozoic world lost 22 percent—almost a quarter—of the animal families alive in the world's oceans in the Late Ordovician, a loss that is clearly visible in the sharp drop in the standing diversity curves shown in figures 6.10 and 6.11 at the end of the Ordovician. However, the Ordovician-Permian evolutionary fauna not only survived the extinction, it also rebounded in diversity in the Silurian back to a level similar to that present before the end-Ordovician mass extinction. This pattern of survival and recovery was not seen in the Cambrian evolutionary fauna, where this older Paleozoic fauna never recovered its previous diversity level (top graph in figure 6.11); thus the Paleozoic world permanently lost a small amount of its marine diversity at the end of the Ordovician.

The next biodiversity crisis in the Paleozoic world occurred in the Late Silurian (Ludfordian Age, table 1.3) and triggered a much smaller drop in marine diversity (figs. 6.10, 6.11). As in the end-Ordovician crisis, the Ordovician-Permian evolutionary fauna survived and recovered its diversity levels postcrisis in the Early Devonian. Although both the end-Ordovician and Late Silurian crises triggered extinctions and diversity losses, both had minimal ecological impact in the Paleozoic marine world (table 1.3) (McGhee et al. 2013).

The same is not true of the Late Devonian (Frasnian Age, table 1.3) biodiversity crisis. This next crisis struck the Paleozoic world hard, and although Sepkoski's data show that it triggered a loss of marine animal families very similar to the end-Ordovician crisis (figs. 6.10, 6.11), the ecological impact of the crisis was devastating (McGhee 1996, 2013; McGhee et al. 2004, 2013). The Paleozoic marine world never fully recovered from the marine diversity losses it suffered at the end of the Frasnian Age. A rediversification phase began in the following Famennian Age of the Late Devonian, but it was terminated by the end-Devonian crisis before previous diversity levels could be attained (middle graph in figure 6.11). All in all, the effects of the twin Late Devonian crises eliminated the chitinozoans[32] and tentaculitoids in the zooplankton, the atrypid brachiopods on the seafloor, the stromatoporoid sponges in the reef ecosystems, and the armored fishes. A smaller and slower rediversification followed the end-Devonian crisis, but it also was terminated by the Early Carboniferous biodiversity crisis (Serpukhovian Age, table 1.3). It is known that both the Famennian and Serpukhovian crises were triggered by the stepwise onset of the Late Paleozoic Ice Age and, as I argued in chapter 1, it is likely that the Frasnian biodiversity crisis was also.

In summary, the Paleozoic marine world suffered a series of ecological shocks largely triggered by the various glaciation phases of the Late Paleozoic Ice Age. Still, Sepkoski's mathematical analyses showed that the Ordovician-Permian evolutionary fauna of the Paleozoic world constituted a stable ecological structure that had persisted in the oceanic realm for over 200 million years—until the end-Permian mass extinction (middle graph in figure 6.11). The ancient tabulate and rugose reef coral faunas had suffered major setbacks in the Frasnian and Serpukhovian crises, but they survived. These hardy corals now finally encountered an environmental crisis too severe even for them to survive, and they perished. The trilobites are found in marine rocks from the Cambrian through the Permian, and are virtually synonymous with the word "Paleozoic." They, and their close cousins the sea scorpions, were driven to extinction at the close of the Permian and would evolve no further. The diverse strophomenate, orthid, and spiriferid brachiopod lampshell groups were wiped out, along with their conical-shell lophophorate cousins, the hyoliths (Moysiuk et al. 2017), eliminating most of the typical shellfish of the Paleozoic seas. Our modern seas are dominated by molluscan shellfish, but even the molluscs suffered the extinction of the long-lived rostroconchs, and the highly active ammonoid predators just barely survived. Among the deuterostome animals, the benthic gardens of blastoid echinoderms perished while their close cousins the crinoids, the sea lilies, came very close to extinction. In the chordates, the pelagic graptolites and spine-fin fishes died, but the tiny conodonts managed to survive the Changhsing, only to succumb to extinction in the Triassic.

The magnitude of the end-Permian crisis was so great that it reset the global pattern of both extinction rates (Van Valen 1984) and speciation rates (Sepkoski 1998) in the marine ecosystems of the Earth. The hot, acid, and poisonous seawaters of the latest Permian oceans were too much for the Paleozoic fauna, and the 200-million-year-old ecological structure of the Ordovician-Permian evolutionary fauna of the Paleozoic marine world collapsed—and this time it never recovered (middle graph in figure 6.11; table 6.6).

The End of the Paleozoic Terrestrial World

At present, there exists no mathematical division of the terrestrial animal biosphere into evolutionary faunas equivalent to those produced for the marine animal biosphere by Sepkoski (Sepkoski 1984, 1990). However, Jack Sepkoski did write in 1990[33] that he considered the University of Bristol vertebrate paleontologist Mike Benton's 1985 "terrestrial tetrapod assemblages" (Benton 1985) to be similar in ecological nature to his marine evolutionary faunas. Benton argued that three distinct ecological-evolutionary assemblages of terrestrial tetrapod vertebrates could be recognized in geologic time: first, a Paleozoic labyrinthodont-synapsid-parareptile assemblage that was terminated in the end-Permian extinction; second, a Mesozoic diapsid-pterosaur-dinosaur assemblage that was terminated in the end-Cretaceous extinction; and third, the Cenozoic assemblage of modern lissamphibians, turtles, lizards, crocodiles, birds, and mammals.

Who were these animals? Similar to tables 6.4 and 6.5 for marine animals, in tables 6.7 and 6.8 I have listed the characteristic land-dwelling vertebrates of the Paleozoic and Mesozoic worlds. The labyrinthodont component of Benton's 1985 Paleozoic tetrapod assemblage referred to the ancient amphibian-like animals that existed in the Carboniferous and Permian, which consisted of older paraphyletic or "partial" evolutionary groupings that we now recognize as basal batrachomorphs (then called temnospondyls, and ancestral to our modern amphibians), basal lepospondyls (microsaurs), and basal reptiliomorphs (anthracosaurs; compare tables 6.7 and 6.8). These ancient animals were joined by the early amniote animals—basal synapsids (pelycosaurs), advanced therapsid synapsids, and the diapsid reptiles—to constitute the typical fauna that one would have encountered on the continents of the Paleozoic world.

In contrast to the Paleozoic, the Mesozoic world land-dwelling vertebrate fauna is comprised almost entirely of reptiles (tables 6.7 and 6.8). Only the cynodonts and their descendants, the mammals, remained from the once ecologically diverse and numerous land-dwelling synapsid faunas of the Paleozoic world (Abdala and Ribeiro 2010). What happened? The synapsid vertebrates (a group that includes us, the modern humans) were the dominant and most advanced land animals of the world in the Paleozoic. In the Mesozoic world, the synapsids had been ecologically replaced by reptiles—and very advanced reptiles indeed. Not only did the reptiles take over the continents of the Earth, they also convergently re-evolved fins and other aquatic adaptations[34] and invaded the marine realm (the ichthyosaurs, plesiosaurs, and kin; table 6.8) and even modified their forelimbs into wings and invaded the aerial realm as well (the pterosaurs; table 6.8). The first flying vertebrate animals in Earth history were reptiles, not synapsids (which only managed to evolve wings much later in the Cretaceous[35]).

TABLE 6.7 Characteristic land-dwelling vertebrate faunas of the Paleozoic world and of the Mesozoic world.

Paleozoic World Fauna	Mesozoic World Fauna
1. "temnospondyls" (batrachomorphs)	1. cynodont/mammal synapsids
2. "microsaurs" (lepospondyls)	2. marine reptiles
3. "anthracosaurs" (reptiliomorphs)	3. archosaurian reptiles
4. "pelycosaurs" (synapsids)	4. pterosaurian reptiles
5. therapsids (synapsids)	5. dinosaurian reptiles
6. parareptiles (reptiles)	

Note: Older paraphyletic tetrapod group names are in quotation marks; see table 6.8 for a phylogenetic classification.

TABLE 6.8 Phylogenetic relationships of the land-dwelling vertebrates of the Paleozoic world (italicized) and of the Mesozoic world (underlined).

TETRAPODA (limbed vertebrates)

– *basal tetrapods*

– Neotetrapoda

– – BATRACHOMORPHA (ancestors of amphibians)

– – – *basal batrachomorphs ("temnospondyls")*

– – – *Capitosauria†*

– – – *Trematosauria†*

– – – *Eryopidae†* (giants)

– – – <u>Lissamphibia</u> (living amphibians)

– – Lepospondyli†

– – – *basal lepospondyls ("microsaurs")*

– – – *Nectridea†*

– – – *Aistopoda†*

– – REPTILIOMORPHA (ancestors of amniote tetrapods)

– – – *basal reptiliomorphs ("anthracosaurs")*

– – – *Seymouriamorpha†*

– – – *Diadectomorpha†* (giants)

– – – AMNIOTA (amniote tetrapods, conquerors of the land)

– – – – SYNAPSIDA (ancient ancestors of mammals)

– – – – – *basal synapsids ("pelycosaurs")*

– – – – – *Eothyrididae†*

– – – – – *Caseidae†* (giants)

– – – – – *Varanopidae†*

– – – – – *Ophiacodontidae†* (giants)

– – – – – *Edaphosauridae†* (giants)

– – – – – *Sphenacodontidae†* (giants)

– – – – – THERAPSIDA (closer ancestors of mammals)

– – – – – – *Biarmosuchia† (basal therapsids)*

– – – – – – *Dinocephalia†* (giants)

– – – – – – *Dicynodontia†*

– – – – – – *Gorgonopsia†*

– – – – – – *Therocephalia*

– – – – – – – Cynodontia

– – – – – – – – <u>MAMMALIA</u> (mammals)

– – – – REPTILIA (ancestors of reptiles)

– – – – – *basal reptiles*

– – – – – PARAREPTILIA

– – – – – – *Mesosauridae†*

– – – – – – *Procolophonidae†*

(*continued*)

TABLE 6.8 Phylogenetic relationships of the land-dwelling vertebrates of the Paleozoic world (italicized) and of the Mesozoic world (underlined). (*continued*)

```
– – – – – – – Pareiasauridae† (giants)
– – – – – – EUREPTILIA
– – – – – – – Captorhinidae† (giants)
– – – – – – – DIAPSIDA
– – – – – – – – Weigeltisauridae†
– – – – – – – – Testudinata (turtles)
– – – – – – – – LEPIDOSAUROMORPHA
– – – – – – – – – Ichthyosauria† (marine reptiles; ichthyosaurs)
– – – – – – – – – Sauropterygia† (marine reptiles; plesiosaurs and placodonts)
– – – – – – – – – Lepidosauria (living lizards and snakes)
– – – – – – – – ARCHOSAUROMORPHA
– – – – – – – – – Rhynchosauria†
– – – – – – – – – Archosauria
– – – – – – – – – – Proterosuchidae†
– – – – – – – – – – Crurotarsi (ancestors of living crocodiles)
– – – – – – – – – – Phytosauridae†
– – – – – – – – – – Stagonolepididae† (aetosaurs)
– – – – – – – – – – Rauisuchia†
– – – – – – – – – – Crocodylomorpha (crocodiles)
– – – – – – – – – – Ornithodira
– – – – – – – – – – Pterosauria† (flying reptiles; pterodactyls and rhamphorhynchids)
– – – – – – – – – – DINOSAURIA (dinosaurs, including living birds)
– – – – – – – – – – – Saurischia
– – – – – – – – – – – – Sauropodomorpha†
– – – – – – – – – – – – Theropoda
– – – – – – – – – – – Ornithischia†
```

Source: Phylogenetic classification modified from Benton (2015).

Note: Extinct taxa are marked with a dagger (†). Older paraphyletic tetrapod group names are in quotation marks; major clades are in capitals.

The evolutionary and ecological innovativeness of the clade of the reptiles (table 6.8) in the Mesozoic terrestrial world is better revealed when we examine the ecological structure of that world in table 6.9. We have previously considered the ecological replacements that took place in the pelagic realm when the Paleozoic marine world ended (table 6.6). Note at the top of table 6.9 that the "marine reptiles" listed in table 6.6 are now identified as two groups of reptiles: the diapsid ichthyoperytians and the lepidosaurian sauroptergyians (compare tables 6.8 and 6.9). At the bottom of table 6.9, note that the reptiles had already produced some evolutionary experiments in flight even within the Paleozoic terrestrial world—namely, the gliding weigeltisaurid diapsids. These interesting animals had bodies that were dorsoventrally flattened and laterally widened. Within their bodies their ribs were flattened and greatly elongated laterally away from the anteroposterior axis of the vertebral column. In midair these animals looked like a circular discus or frisbee in flight—but with a pair of forelimbs and a head attached on one side of the discus and a

TABLE 6.9 Ecological replacement of the vertebrate fauna of the Paleozoic world by the Mesozoic world fauna in the terrestrial realm in the Late Triassic.

I. Vertebrates in the Oceans

Carnivores	*Paleozoic World:* armored†, spine-fin†, and lobe-fin fishes
	Mesozoic World: sharks and rays; ray-fin fishes; ichthyosaurian† and sauropterygian† lepidosauromorphs

II. Vertebrates in Freshwater Rivers and Lakes

Detritivores	*Paleozoic World:* armored† fishes
	Mesozoic World: ray-fin fishes
Carnivores	*Paleozoic World:* armored† and lobe-fin fishes, capitosaurian† and trematosaurian† batrachomorphs; nectridean lepospondyls†; seymouriamorph† reptiliomorphs; mesosaurian† parareptiles
	Mesozoic World: ray-fin fishes; phytosaurid† archosaurs; spinosaurid† theropod dinosaurs

III. Vertebrates on Land

Herbivores	*Paleozoic World:* diadectomorph† reptiliomorphs; caseid†, edaphosaur†, dinocephalian†, dicynodont† and cynodont synapsids; procolophonid† and pareiasaurid† parareptiles; captorhinid† reptiles
	Mesozoic World: rhynchosaur† archosauromorphs; aetosaur† archosaurs; sauropodomorph† and ornithischian† dinosaurs
Carnivores	*Paleozoic World:* eryopid† batrachomorphs; varanopid†, ophiacodont†, sphenacodont†, and gorgonopsid† synapsids
	Mesozoic World: cynodont and mammalian synapsids (low diversity); proterosuchid† and rauisuchian† archosaurs; theropod dinosaurs (high diversity)

IV. Vertebrates in the Air

	Paleozoic World: gliding weigeltisaurid† diapsids
	Mesozoic World: flying pterosaurs†; avian dinosaurs (birds)

Note: Extinct taxa are marked with a dagger (†).

pair of hindlimbs and a tail on the other (fig. 6.12). This identical morphology has been repeatedly, convergently, evolved within the diapsid reptiles and can be seen today in the modern gliding lizard *Draco melanopogon*.[36]Although innovative, the weigeltisaurid gliders apparently were not innovative enough because they still perished at the end of the Changhsingian in the end-Permian mass extinction.

The great flying reptiles of the Mesozoic world—the pterosaurs (table 6.9)—were not descendants of the early weigeltisaurid gliders of the Permian. These animals were ornithodirans, highly derived archosaurian reptiles closely related to the famous dinosaurs (table 6.8). Unlike their distant cousins the weigeltisaurid gliders, the pterosaurs had true wings and could fly. Their wings were modified forelimbs, originally used solely for walking but now used also as

FIGURE 6.12 A Paleozoic world gliding reptile, the weigeltisaurid *Coelurosauravus*.

Source: Illustration by Mary Persis Williams. Modified and redrawn from Steyer (2012).

FIGURE 6.13 A Mesozoic world flying reptile, the pterosaur *Eudimorphodon*.
Source: Illustration by Mary Persis Williams. Modified and redrawn from Palmer et al. (2012).

lift-generating structures, that consisted of a single finger—the fourth, or ring, finger—that was vastly elongated and to which a membrane of skin was attached and stretched all the way to the side of the animal's body (fig. 6.13). Their membranous wings were thus somewhat similar to those of modern bats, although a bat's wing is made by elongating all five fingers, not just the fourth. Later in the Mesozoic world, the theropod dinosaur predators would produce a flying competitor to their pterosaur cousins—the animals we know today as birds (table 6.8). The wing of a bird, an avian dinosaur, differs even more from the membranous wings of pterosaurs and bats. It consists of numerous elongated flight feathers that are attached to its arm, but still the basic structure is a modification of the forelimb of the animal that was originally used for walking.

In freshwater ecosystems on land, the rivers and lakes of the Paleozoic world contained a diverse fauna of ancient fishes, ancient ancestors of the amphibians (batrachomorphs), amphibian-like lepospondyls, and advanced seymouriamorph reptiliomorphs that were very closely related to the amniotes (table 6.8). The only major group of reptiles in freshwater ecosystems in the Paleozoic world were the mesosaur parareptiles (tables 6.8 and 6.9). In contrast, the rivers and lakes of the Mesozoic world were filled with ray-fin fishes, which exploded in diversity in the Triassic, the fish-eating phytosaurs, a group of advanced archosaurs related to modern-day crocodiles, and the fish-eating theropod dinosaurs—the spinosaurs (tables 6.8 and 6.9).

The megaguild of herbivores in the Paleozoic terrestrial world comprised a diverse assemblage of diadectomorph reptiliomorphs (the giant *Diadectes maximus* we considered in chapter 4; see fig. 4.5) and, in the synapsid clade, the caseids, edaphosaurs, dinocephalians (the giant *Moschops* dinocephalians were the largest land vertebrates in the Paleozoic; fig. 4.7), and the advanced dicynodonts and cynodont herbivores. The reptile clade was also diverse in the herbivore megaguild of the Paleozoic terrestrial world, with procolophonid and pareiasaur parareptiles (relatives of the mesosaurs in freshwater ecosystems) and captorhinids (table 6.9). In contrast, the herbivore megaguild of the early Mesozoic terrestrial world was much less diverse, dominated by rhynchosaur archosauromorphs, more advanced aetosaur archosaurs, and the early dinosaurs—both sauropodomorphs (later to become the largest land animals in Earth history) and ornithischians (table 6.9). The diverse synapsid herbivores of the Paleozoic world were gone.

The Paleozoic carnivore megaguild was composed of eryopid early amphibians (including the giant *Eryops megacephalus*; fig. 4.4), pelycosaurian synapsids (varanopids, ophiacondonts, and sphenacodonts, some of which were giants; see fig. 4.6), and advanced gorgonopsid synapsids (fig. 6.14). The gorgonopsid

FIGURE 6.14 A Paleozoic world synapsid carnivore, the gorgonopsid *Lycaenops*.

Source: Illustration by Mary Persis Williams. Modified and redrawn from Steyer (2012).

predators in particular were lithe, catlike animals with elongated fang teeth very similar to the saber teeth to be evolved much later by the true cats (table 6.9). In the Mesozoic world, the dominant land-dwelling carnivores were proterosuchid and rauisuchian archosaurs (the latter a relative of the phytosaurian carnivores in freshwater ecosystems; tables 6.8 and 6.9) and the theropod dinosaurs (later to produce the famous *Tyrannosaurus rex* and other very large carnivores). The once-dominant synapsid carnivores were rare—only the cynodonts and, near the end of the Triassic, the first of the true mammals.

Just as with the end of the Paleozoic marine world, the end of the Paleozoic terrestrial world was not a single instantaneous event confined to the end of the Changhsingian Age in the Permian. The Paleozoic marine world experienced a series of biodiversity crises before the coup de grâce of the end-Permian mass extinction. The Paleozoic marine ecological structure was older than the Late Paleozoic Ice Age, and the various glacial phases of the ice age usually triggered extinctions and biodiversity losses in the marine fauna. In essence, the Late Paleozoic Ice Age presented a series of environmental and ecological challenges that the Paleozoic world marine fauna encountered and survived, even while losing overall diversity with time (middle graph in figure 6.11). Thus we can hypothesize that the Paleozoic marine world might have continued to survive past the end of the Late Paleozoic Ice Age if it had not been for the global environmental catastrophe produced by the Emeishan and Siberian LIP eruptions.

In contrast, the Paleozoic terrestrial world fauna was largely the product of climatic conditions produced by the Late Paleozoic Ice Age, and land vertebrates flourished in the equatorial rainforests of the Earth during the Carboniferous and Permian. However, the Late Paleozoic Ice Age began to wane in the Early Permian, as discussed in the last chapter, and the P2-to-P3 interpulse greenhouse period triggered a major contraction of the ever-warm, ever-wet, equatorial zone during the late Artinskian–early Roadian interval of the Permian. By the beginning of the Middle Permian Epoch, some elements of the typical Paleozoic terrestrial world fauna had died out, including the eryopid basal amphibians, the nectridean and aistopod groups of the amphibian-like lepospondyls, the seymouriamorph and diadectomorph basal reptiliomorphs, the eothyridid, ophiacodont, edaphosaur, and sphenacodont basal synapsids, and the mesosaurs in the reptile clade (tables 6.8 and 6.9)—all in "Olson's extinction" that we considered in chapter 5. Then came the

Emeishan LIP eruption, and the surviving non-therapsid synapsids, the caseids and varanopids, were eliminated, along with the biarmosuchian basal therapsids (table 6.8) in the Capitanian. The catastrophe of the Siberian LIP eruption killed the great dinocephalian herbivores and lithe gorgonopsid predators in the synapsid clade, and in the reptile clade the pareiasaurs, captorhinids, and gliding weigeltisaurids also perished (tables 6.8 and 6.9) in the Changhsingian. At the dawn of the Early Triassic, in the synapsid clade, only the dicynodonts, therocephalians, and cynodonts had survived (Sahney and Benton 2008). In the reptile clade, only the procolophonid parareptiles, marine lepidosauromorphs (the ichthyosaurs), and basal archosauromorphs lived on (Sahney and Benton 2008).

The fate of the Paleozoic terrestrial world flora presents a somewhat more complex picture than that of the fauna. As mentioned at the beginning of this section, the terrestrial paleozoologists have not produced a formal division of the terrestrial animal biosphere into evolutionary faunas using the analytical techniques that Sepkoski used for the marine biosphere. However, the terrestrial paleobotanists have: Christopher Cleal and Borja Cascales-Miñana explicitly "adopted exactly the same mathematical approach as used by Sepkoski (1981) and applied it to the most recent compilation of families and classes of tracheophytes (vascular plants) as revealed in the fossil record" (Cleal and Cascales-Miñana 2014, 469). Their analyses reveal the existence of five evolutionary floras in the terrestrial realm during the Phanerozoic, rather than the three evolutionary faunas that existed in the oceans.

Of particular interest here is that Cleal and Cascales-Miñana recognize a Paleophytic flora in the interval of time from the Devonian through the Permian, a flora dominated by lycophytes, filicophytes, and equisetophytes (lycopsids, pteropsids, and equisetopsids in their terminology; see Cleal and Cascales-Miñana 2014, table 2). This floral ecological assemblage collapsed following the end-Permian catastrophe in a pattern remarkably similar to that shown by the Paleozoic marine fauna (middle graph in figure 6.11) (Cleal and Cascales-Miñana 2014, fig. 4, factor-loading graph for factor 3). The Paleophytic flora was later replaced in the Triassic by the Mesophytic flora, a flora dominated by pinophytes and ginkgophytes (pinopsids and ginkgoopsids in their terminology; see Cleal and Cascales-Miñana 2014, table 2), and Cascales-Miñana and colleagues conclude that the end-Permian climatic catastrophe had a "profound effect on plant life; vegetation changed fundamentally, in effect resetting plant evolutionary history and marking the appearance of what has become today's vegetation" (Cascales-Miñana et al. 2015, 1072)—a fate similar to that of the terrestrial faunas and the marine faunas of the Earth in this interval of time. Thus the end-Permian mass extinction was a completely global event, of maximum ecological severity in both the oceans and on the land, and for both animal and plant life of the Earth (McGhee et al. 2004; Cascales-Miñana et al. 2015).

We now enter the "post-apocalyptic greenhouse" (Retallack 1999) world of the Early Triassic—and that world was very weird indeed. As discussed previously, Yadong Sun and colleagues have presented evidence that the equatorial tropical zones of the Earth were lethally hot for the entire five-million-year duration of the Early Triassic Epoch. The surviving land vertebrates were confined to latitudes higher than 30° in the Northern Hemisphere and 40° in the Southern Hemisphere—the tropics were barren. The same was true of the hot oceans, with marine fishes and marine reptiles, the ichthyosaurs, occurring only in the cooler waters of the high latitudes of the Earth (Sun et al. 2012).

On the land, the tetrapod faunas of the entire Earth were dominated by species of a single genus: the barrel-chested, burrowing dicynodont therapsid *Lystrosaurus*. All of the continents of the Earth were joined together in the single supercontinent of Pangaea, and in the Northern Hemisphere you could have traveled from western North America to Europe to eastern Asia without ever leaving dry land—and just about all of the land animals you would have seen would have been lystrosaurs. Flying over the barren lands of the superhot tropics to the cooler high latitudes of the Southern Hemisphere, you would have encountered scenes eerily similar to those of the Northern Hemisphere—lumbering groups of lystrosaurs wandering across the landscapes, from South America in the west to Africa to India to Australia in the east.

Lystrosaurs were smaller herbivores (fig. 6.15) that also were burrowers, and this behavioral trait probably contributed to their surviving the end-Permian mass extinction, as discussed above (Retallack et al. 2006; Retallack and Krull 2006). They had large, blocky heads with massive jaw muscles for chewing (Benton 2015) the tough vegetation present on land in the hothouse Early Triassic world—for land plants had also suffered major extinction in the end-Permian catastrophe. The Late Permian expanses of conifer trees were gone, replaced by low-growing lycophyte club mosses, equisetophyte horsetail rushes, and ferns (Payne and Clapham 2012; Sun et al. 2012; Irmis and Whiteside 2012).

FIGURE 6.15 A synapsid disaster species that briefly flourished in the hothouse Early Triassic world, the dicynodont herbivore *Lystrosaurus*.

Source: Illustration by Mary Persis Williams.

All of these plants are of low nutrient content and are tough to chew—the equisetophytes in particular, as these plants concentrate silica in their tissues in order to discourage herbivores from eating them—so many of the lystrosaurs, as burrowing animals, may have concentrated on digging up the roots of the sparse Early Permian land-plant vegetation and eating this more palatable plant material instead.

Other tetrapod groups were present on land in the Early Triassic, but in very low diversity, as 90 percent of the land vertebrates alive in the Early Triassic were lystrosaurs (Sahney and Benton 2008). The synapsid lithe, catlike gorgonopsid predators of the Permian were gone, replaced by the smaller, short-legged, and more crocodile-like archosaur *Proterosuchus*, a reptile. The synapsids (our ancestors) had lost their ecological position as the top carnivores to the archosaurs in the Early Triassic hothouse Earth, and their fate would become even bleaker in the Late Triassic. The diversifying archosaurs produced the freshwater phytosaur piscivores and the land-dwelling rauisuchian carnivores (table 6.9), which further competed with the few predatory cynodont synapsids, such as *Galesaurus* and *Thrinaxodon*, that lived in the Early Triassic. But even these competing archosaurian groups would become extinct by the end of the Triassic following the evolution and spread of an entirely new type of archosaurian predator—the theropod dinosaurs (tables 6.8 and 6.9). The oldest known dinosaurs are Carnian in age—predators like *Herrarasaurus* and *Coelophysis*—and they rapidly diversified in the Norian Age of the Late Triassic (Benton 2015) until theropods became the top carnivores in terrestrial ecosystems around the planet and the Mesozoic world ecological structure was firmly established. For the next 150 million years, the theropod dinosaurs would remain the lords of the predator megaguild, and the cynodonts and their new descendants—the mammals—would remain small and cryptic, hiding underground and foraging for food in the dark of night.

At first glance, it might seem that the synapsids had retained their dominance as the major herbivores in terrestrial ecosystems, given the ubiquitous planetwide distribution of the lystrosaurs in the Early Triassic. However, the lystrosaurs were opportunistic "disaster species"—lucky survivor species of an ecological catastrophe that rapidly spread into the ecological space vacated by the unlucky victims of the end-Permian mass extinction (Sun et al. 2012; Irmis and Whiteside 2012). The proliferation of a few disaster species, both animal and plant, following a major environmental catastrophe results in post-catastrophe ecosystems that are very low in species diversity but, of the species that are present, high population abundances and widespread geographic distribution. Such was the Early Triassic world, in which 90 percent of the individuals in the tetrapod populations of the world were lystrosaurs (fig. 6.15) (Sahney and Benton 2008).

The synapsid disaster species of the Early Triassic world were rapidly replaced by more diverse tetrapod faunas as the global terrestrial ecosystem began to recover in the Middle Triassic (Sahney and Benton 2008; Irmis and Whiteside 2012). In particular, the reptiles progressively moved into the terrestrial herbivore niche; by the Late Triassic, they would displace the last of the herbivorous synapsids. First it was the surviving procolonophorid parareptiles, themselves a remnant of the Paleozoic world fauna (tables 6.8 and 6.9), that managed to achieve large population densities in South Africa in the hothouse Early Triassic. Then, in the Middle and Late Triassic, the more advanced rhynchosaur archosauromorphs began to ecologically displace the remaining dicynodont and cynodont synapsid herbivores. From small beginnings, by the Carnian Age of the Late Triassic, the rhynchosaurs had evolved body sizes up to two meters (6.6 feet) long and were the dominant herbivores in terrestrial ecosystems, constituting up to 60 percent of the herbivore populations (Benton 2015). These animals had massive heads—the back of the skull was wider than the skull itself was long, and

FIGURE 6.16 A Mesozoic world reptile that flourished and displaced the synapsid disaster species in the Late Triassic, the rhynchosaur herbivore *Hyperodapedon*.

Source: Illustration by Mary Persis Williams.

the front part of the skull was laterally compressed and had a parrot-like beak (fig. 6.16). They possessed massive muscles for chewing in the back part of the skull, and their jaws moved from side to side with no sliding backwards or forwards (Benton 2015). Even though they were well adapted to eating tough vegetation, they still were driven to extinction—along with the last of the herbivorous synapsids—at the end of the Carnian Age, and the herbivore megaguild in terrestrial ecosystems was firmly taken over by the lords of the Mesozoic world—the plant-eating dinosaurs (tables 6.8 and 6.9).

The Paleozoic terrestrial world came to a final end at the end of the Carnian Age of the Late Triassic. It was replaced by the ecological structure of the Mesozoic terrestrial world, with the land floras dominated by the Mesophytic flora and the land faunas by the ruling dinosaurs. The dinosaurian ecosystem would persist for some 150 million years—until it too came to an end, when the Chixulub asteroid impacted the Earth and the end-Cretaceous mass extinction began. Very late in the Carnian Age[37] of the Late Triassic, a small, furry, shrew-like animal rustled in the underbrush, hunting for insects to eat. This tiny animal was a new descendant of the cynodont synapsid carnivores that had been the dominant predators on Earth in the Permian, during the time of the Paleozoic terrestrial world. It was *Adelobasileus cromptoni*, the first true mammal (Benton 2005, 2015) and a basal member of the synapsid lineage that would eventually include the evolution of *Homo sapiens*, modern humans. Mammals would eventually come to rule the terrestrial realm—but not for 150 million years. Back in the Late Triassic, the long night for the mammals—the Age of Dinosaurs—had just begun.

7 | The Legacy of the Late Paleozoic Ice Age

Palaeozoic tropical rainforests and their effect on global climates: is the past the key to the present? . . . The Palaeozoic evidence clearly confirms that there is a correlation between levels of atmospheric CO_2 and global climates. However, care must be taken in extrapolating this evidence to the present-day tropical forests, which do not act as a comparable unsaturated carbon sink.

—Cleal and Thomas (2005, 13)

THE EFFECT OF THE LATE PALEOZOIC ICE AGE IN EARTH HISTORY

What were the effects of the Late Paleozoic Ice Age in Earth history? First, the climatic conditions created by the ice age—a planet with cold poles, arid temperate zones, and an ever-wet, ever-warm tropical zone—were perfect for forming what Stephen Greb and his colleagues, as noted in chapter 3, called the "largest tropical peat mires in Earth history" (Greb et al. 2003, 127), perfect habitats in which the peculiar lycophyte trees flourished. Those peat mires and lycophyte tropical rainforests later fossilized to produce the world-wide distribution of Carboniferous coal deposits that would be used by humans over 250 million years later as the principal source of energy in the process of industrialization, first in Europe and then around the world—continuing to the present day in developing countries like China and India. In this chapter, we will examine how the climatic and biological processes that created these massive coal deposits during the Late Paleozoic Ice Age have affected our modern world.

Legacy 1: The Industrial Revolution?

Humans learned how to create fire and burn wood in prehistory. In fact, the use of fire in the human lineage is much older than the 300,000-year-old species *Homo sapiens*—we know from the fossil record that the older hominin species *Homo erectus* also learned how to used fire some 1.5 million years ago. We also know from historical records that our ancestors knew about coal, the strange black rocks that they found lying on the ground in some regions of the Earth—black rocks that could ignite and burn. However, for the great majority of our history, our ancestors burned wood for energy. Later humans learned how to make charcoal from wood by heating wood in kilns, cooking out volatiles like water and partially oxidizing the remaining carbon. Charcoal is a more concentrated form of burnable carbon; that is, charcoal has a higher energy content per unit mass than ordinary wood. Everyone who has ever used charcoal to grill hamburgers or other food on a summer day is familiar with this fact, recognizing that it would take a much larger mass of wood to accomplish the same amount of cooking (the amount of wood required would not fit in your average backyard grill!).

In Europe, people first began to burn coal to create hotter fires in order to melt metal ores from their rock matrix and to soften metal for working it into tools and ornaments. People rapidly discovered that burning less concentrated-carbon forms of coal, such as bituminous coal, created irritating and unhealthy air pollution and thus began to use more concentrated-carbon forms of coal, such as anthracite, for domestic purposes like heating their homes.

Then the first primitive steam engines were invented. It was discovered that the pressure produced by boiling water into steam could be controlled and used to power machines that could, for example, turn the massive heavy millstones that were used to grind grain into flour. These steam engines were much more powerful than traditional sources of energy for grinding grain, such as using the energy of flowing water over waterwheels or connecting the millstones to a team of horses that walked around and around in circles. More experimentation produced more efficient steam engines, until the Scot James Watt patented the first efficient, modern-type steam engine in 1769. Producing the steam power in these engines required a very hot fire—and coal was the perfect fuel for that fire. Even today, it is amazing to ride on a train powered by a now-antique steam engine. The steam engine produces so much power that it not only propels itself, but it also pulls along its source of energy—a second car loaded with coal—as well as a whole series of cars connected to the train behind the coal car. Those cars are heavy, and they can be filled with even heavier loads of coal ore from coal mines in freight cars (fig. 7.1) or loads of people in passenger cars. Yet the steam engine at the front of the train pulls them all, powered by the energy released by burning the coal in the coal car.

Next, just as people discovered that charcoal could be made by cooking off the volatiles in wood, people discovered that coke could be made by cooking coal. Burning coke could then produce extremely hot fires in very large blast furnaces, and those huge blast furnaces could be used to produce much less expensive, high-quality iron in massive amounts—cheap iron that could be used to make more machines, particularly steam-driven machines.

Thus was born the technology needed to drive the European Industrial Revolution, beginning in Britain in the mid-1700s. The Industrial Revolution probably began first in Britain because, by geologic coincidence, large amounts of

FIGURE 7.1 Coal-powered, steam-engine train hauling freight cars loaded with coal ore in Australia.

Source: Photograph courtesy of Wikipedia/Christchurch. Reprinted under permission of the GNU Free Documentation License.

coal-containing strata and metallic-ore-containing rocks were both present. And those coal strata in England, Wales, and Scotland are almost all Carboniferous in age. Then the Industrial Revolution spread to continental Europe—particularly to southern Belgium and the middle region of Germany, where coal-bearing strata were also present. And those coal strata in the Ardennes of Belgium and the Ruhrgebiet of Germany are almost all Carboniferous in age. The Industrial Revolution spread around the world in the 1700s and 1800s, following the geologic outcrops of Carboniferous coal. As Nick Lane pointed out (see chapter 2), 90 percent of all the coal strata on Earth were deposited during the height of the Late Paleozoic Ice Age—in the time interval from the Serpukhovian Age in the Early Carboniferous to the Wuchiapingian Age in the Late Permian (Lane 2002, 84). Coal was originally present on the surface of the Earth, where early humans discovered that these peculiar black rocks could burn. With the need for coal to fuel the Industrial Revolution, intensive mining of coal spread around the world. Coal strata are relatively shallow and could be mined by open-pit excavations or shallow mine tunnels underground. As the shallow deposits of coal were mined out, deeper underground mines were dug, and entire mountain tops were removed in some regions of the Earth to reach the coal buried at depth. The mining and burning of coal for energy continues to the present day, but most of the coal burned today is used to produce electricity, not to power steam engines.

The powerful steam engines of the 1700s, 1800s, and early 1900s have been replaced today by an even more powerful, more efficient type of engine—the internal combustion engine. The invention and near-universal use of internal combustion engines today was made possible by a new type of energy source—petroleum—and its refined products such as gasoline and jet fuel. Unlike coal, large quantities of petroleum are not present on the surface of the Earth or at shallow depths underground. It was only after people perfected deep-drilling machines—the early ones powered by steam engines—that abundant petroleum became available as an energy source.

Legacy 2: Modern Global Climate Change?

It is no accident that climate scientists divide the recent history of carbon-dioxide levels in the Earth's atmosphere into "preindustrial" and "industrial" phases, in recognition of the atmospheric effect of the tons of coal that has been burned since the beginning of the Industrial Revolution. The preindustrial amount of carbon dioxide in the atmosphere was 0.03 percent—that is, in the entire recorded human history of burning wood and charcoal, over 6,000 years, the carbon dioxide content of the atmosphere did not exceed 0.03 percent. At the beginning of the Industrial Revolution, humans began to burn coal in vast quantities, and the present-day level of carbon dioxide in the atmosphere has risen to 0.04 percent in less than 200 years.

It is an empirical observation, a fact, that the Earth is heating up. Combining that fact with the fact that the amount of the greenhouse gas carbon dioxide in the atmosphere has vastly increased in a tiny amount of time on geologic time-scales has led climate scientists to conclude cause and effect—that is, that the industrially driven[1] increase in the carbon-dioxide content of the atmosphere has caused the Earth to retain more heat in the atmosphere instead of losing it to space and, as a result, the planet has become hotter. As 90 percent of the coal in the Earth's strata that has been burned since the beginning of the Industrial Revolution was deposited during the Late Paleozoic Ice Age, by extension one could blame the Late Paleozoic Ice Age for modern global climate change!

However, as one might expect, the situation is more complicated than that, and one cannot blame the Late Paleozoic peat bogs and tropical mires for everything. The industrial burning of coal is clearly a major culprit in the increase of carbon dioxide in our atmosphere, but since about the mid-1900s, vast quantities of petroleum have been burned, and the burning of petroleum has also added tonnes of carbon dioxide to the atmosphere. It is usually impossible to determine how old that petroleum was—was it formed from organics deposited during the Late Paleozoic Ice Age or from some other geologic time period? Petroleum is a liquid, and a slippery liquid at that, and it easily migrates within rocks under the influence of gravity and differential pressure within the rock. Petroleum can accumulate in vast pools deep underground that are very distant from the original strata in which the petroleum formed.

Another complicating factor is that not only has the amount of carbon dioxide in the atmosphere increased rapidly in less than 200 years, but the number of human beings on the Earth has also increased rapidly in the same period of time. In the early 1800s, early in the Industrial Revolution, there were about one billion human beings present on the Earth. Two hundred years later, at the beginning of the third millennium, the number of human beings on the Earth has increased to 7.6 billion. All of those people have depended primarily on energy from burning hydrocarbons—first coal, then petroleum—and that burning of hydrocarbons has added carbon dioxide to the Earth's atmosphere. If the Earth's human population had remained stable at around one billion individuals throughout those 200 years, clearly only a small fraction of the coal and petroleum that has been burned during that time interval would have been burned—and the Earth would be a much cooler (and less crowded) place.

Still, the fact remains that all of that Carboniferous coal (and petroleum, whatever age it happens to be) was burned for energy—and continues to be burned today. Where is the Earth headed in the future? Are we headed for a hot-house world like the Earth that existed in the Early Triassic—a world, described in detail in chapter 6, so hot that the tropical zones were essentially lethal, a world in which complex plant and animal life was confined to latitudes higher than 30° in the Northern Hemisphere and 40° in the Southern Hemisphere?

Legacy 3: The Successful Invasion of Land by Animals?

A twin result of the huge size and the peculiar growth pattern of the trees in the ancient lycophyte rainforests was the massive removal of carbon dioxide from the Earth's atmosphere, the fixation of the carbon from carbon-dioxide molecules into plant tissues, and the burial of those carbon-rich plant tissues in bogs where they would later be fossilized into coal. Simultaneously, the oxygen released in fixing carbon during plant photosynthesis resulted in a massive injection of free oxygen into the Earth's atmosphere during the Late Paleozoic Ice Age. That abundance of atmospheric oxygen led to the evolution of gigantism in both marine and land animals and, on land, in both arthropod and vertebrate clades of animals, as we saw in chapter 4.

In the vertebrate clade of animals, the initial boost of atmospheric oxygen, it is argued, was a major assist in both the final invasion of land by the tetrapods in the Visean Age of the Carboniferous and in the evolution of the first amniotes (Graham et al. 1995, 1997; McGhee 2013). As outlined in chapter 4, Jeffrey Graham and his colleagues have pointed out that the hyperoxic Carboniferous atmosphere enabled the previously aquatic tetrapods of the Devonian to obtain more oxygen with their primitive lungs, to decrease the amount dehydration they experienced in air breathing, and to boost their metabolic rates so they could move more energetically and endure the constant pull of gravity experienced in moving on dry land.

The key innovation in the successful invasion of land by vertebrates was the evolution of the amniote egg and the anatomical antidehydration adaptations that characterize the amniote animals. The hyperoxic Carboniferous atmosphere would have allowed tetrapods to develop larger eggs and, at the same time, to minimize water loss from those eggs (Graham et al. 1997). With an atmosphere rich in oxygen, the egg could absorb more oxygen per unit surface area of the egg. A larger egg has a larger internal volume (hence more space for extra tissue layers and fluid in addition to the embryo) with a relatively smaller external surface area (thus less water loss across that surface area to the outside world) than a small egg. The amniote egg is an innovative, two-layer adaptation to protect the developing embryo from dehydration in the harsh, dry-air environments of the terrestrial realm. First, the embryo is enclosed in a water-filled region contained by the surrounding amniotic membrane. In essence, rather than floating in an actual pond of water as amphibian embryos do, the amniote embryo floats within its own private amniotic pond. Second, the outside layer of the egg is a tough shell rather than the soft, gelatinous outer covering of the amphibian egg. Both of these layers help protect the embryo from serious water loss and death by dehydration. Thus the amniotes can lay their eggs on dry land and have them survive. This is not true of most amphibians. Their eggs will rapidly dehydrate if exposed to dry air, shriveling up and shrinking as moisture is lost to the atmosphere, and the embryo will die.

Finally, a large container of food for the embryo is enclosed within the amniote egg—the yolk sac. Thus the embryo can develop to a considerable degree within the egg itself, feeding on nutrients from the yolk sac, rather than hatching out of the egg at an early growth stage and foraging for food in a free-swimming larval stage, as in many amphibians. In consequence, the amniotes have direct development—there is no larval stage. What emerges from an amniote egg looks like a small, scaled-down version of the adult animal. In contrast, what emerges from a frog egg—a tadpole—looks like a fish, not a frog. Fish, of course, need water to swim in; a hatchling tetrapod amniote does not.

The amniotes were the victorious conquerors of land in the clade of the vertebrate animals.[2] That victory was achieved by the evolution of the amniote egg, which freed the vertebrates from the constraint of having to reproduce in or near a body of water such as a river or lake. Prior to this innovation, tetrapods still laid their eggs in water, much like fish—and many of the non-amniote tetrapods, the amphibians, still do so today. Free from this constraint, the amniotes invaded the highlands and dry areas of the Earth far from standing bodies of water.

Just as the amniotes were the final conquerors of land in the vertebrate clade, the winged insects were the victorious conquerors of land in the clade of the arthropods (McGhee 2013). The hyperoxic atmosphere of the Carboniferous, it is argued, was also a major assist in the evolution of the winged insects, as we considered in chapter 4. Flying is a highly energetic activity that requires a lot of oxygen. The arthropods invaded land back in the Silurian, long before the vertebrates emerged from the water and the start of the Late Paleozoic Ice Age (McGhee 2013). Why, then, did it take them so long to take the next evolutionary step, to develop flight? The development of a hyperoxic atmosphere in the Carboniferous and the evolutionary diversification and spread of the flying insects during the Late Carboniferous may not be a coincidence (Graham et al. 1995, 119, fig. 2). With the evolution of flight, the insects could disperse over huge areas of land in a very short period of time—the world was theirs for the taking. Only much later, in the Late Triassic, would the vertebrates evolve powered flight in the clade of the reptiles with the first pterosaurs.

Legacy 4: The Evolution of Mammals and Dinosaurs?

The decline and depletion of free oxygen in the atmosphere following the end of the Late Paleozoic Ice Age and the catastrophic Emeishan and Siberian LIP eruptions, it is argued, was a major impetus for the evolution of yet more efficient respiratory metabolisms—and eventually the evolution of endothermic metabolisms—in both the synapsid and reptilian clades of vertebrates, and the eventual evolution of both mammals and dinosaurs (Graham et al. 1995, 1997). Geologic history—the very existence of the Mesozoic and Cenozoic

Eras of geologic time—would not have developed the way it did if these two major groups of land animals had not evolved. Every schoolchild knows that the Mesozoic was the Age of Dinosaurs and the Cenozoic is the Age of Mammals. They may not be aware, however, that both the dinosaurs and the mammals evolved in the Late Triassic, and that the ecological conditions on land that resulted from the end of the Late Paleozoic Ice Age may have contributed to the evolution of both groups.

Some may argue that the evolution of dinosaurs and mammals is not really a legacy of the Late Paleozoic Ice Age but rather a legacy of the Emeishan and Siberian LIP eruptions, which produced the conditions that resulted in oxygen depletion in the Earth's atmosphere at the end of the Permian. Still, the hyperoxic atmosphere created by the Late Paleozoic Ice Age set the stage for the environmental selective pressures that would result from the collapse of that atmosphere, and for the onset of the hypoxic and poisonous atmospheric conditions in the latest Permian and early Triassic. As we considered in detail in chapter 4, Graham and colleagues have argued that the onset of hypoxia in the latest Permian and earliest Triassic triggered the evolution of the four-chambered heart, and perhaps full endothermy, in the therapsids within the synapsid clade (Graham et al. 1997). In the reptilian clade, we know that four-chambered hearts evolved in the Early Triassic, as the crocodilian archosaurs appear in the fossil record at this time and they possess four-chambered hearts. Still, as discussed in chapter 4, the oldest known (as yet) true mammals and dinosaurs appear in the fossil record in the Late Triassic—at least 15 million years after the Late Permian–Early Triassic hypoxic interval of time.

Legacy 5: The Destruction of the Paleozoic World?

The ecological shocks of the successive phases of glaciation during the Late Paleozoic Ice Age triggered a series of extinctions that progressively winnowed and depleted Paleozoic marine world animals in the Earth's oceans. The Frasnian, Famennian, and Serpukhovian extinctions all preferentially eliminated marine animals that belonged to the Paleozoic world fauna, and preferentially spared marine animals that belong to the modern world fauna (Sepkoski 1984, 1990, 1996; Stanley 2007). We know that the Famennian and Serpukhovian extinctions were triggered by Late Paleozoic Ice Age glaciations, and, as discussed in chapter 1, evidence exists that the Frasnian extinction was as well. In essence, the winnowing effect of the Late Paleozoic Ice Age glacially induced extinctions resulted in a Paleozoic marine world fauna that was progressively weakened and eventually could not withstand the catastrophic environmental changes triggered by the Emeishan and Siberian LIP eruptions. As discussed in chapter 6, the hot, acid, poisonous seawaters of the latest Permian oceans were too much for the Paleozoic fauna, and the 200-million-year old ecological structure of the Paleozoic marine world collapsed, never to recover (figs. 6.10, 6.11; table 6.6).

In contrast, the Paleozoic terrestrial world was largely the product of the terrestrial climatic conditions created by Late Paleozoic Ice Age, as discussed in chapter 6; how, then, can that world's demise be attributed to its creator? It is the end of the Late Paleozoic Ice Age that was the harbinger of the end of the Paleozoic terrestrial world, and thus one may argue that both the existence of and the end of the Paleozoic terrestrial world were a legacy of the Late Paleozoic Ice Age. This causal relationship can be seen most clearly in Olson's extinction, which we considered in detail in chapter 5. The waning of the Late Paleozoic Ice Age in the Early Permian triggered major contractions in the ever-warm, ever-wet equatorial zone of the Earth, and major elements of the terrestrial vertebrate fauna of the Paleozoic terrestrial world died out as a result. One could argue that the waning of the Late Paleozoic Ice Age produced its own winnowing effect in preferentially eliminating species of the Paleozoic terrestrial world fauna, resulting in a weakened fauna that could not withstand the catastrophic environmental changes triggered by the Emeishan and Siberian LIP eruptions in the Late Permian.

On the other hand, it can be counterargued that the destruction of the Paleozoic World is not really a legacy of the Late Paleozoic Ice Age but rather a legacy of the Emeishan and Siberian LIP eruptions; those eruptions produced the catastrophic environmental conditions that lethally heated and poisoned vast areas of both the seas and the land of the Earth, and the very atmosphere of the entire planet. The counterargument would be that had it not been for the Emeishan and Siberian LIP eruptions, the characteristic faunas of the Paleozoic world might have survived. This hypothetical possibility is but one of many that we will explore in the next sections of the chapter.

What If the Late Paleozoic Ice Age Had Never Happened?

From the perspective of human history, this question has major implications. If there had been no Late Paleozoic Ice Age, then there would have been no Earth with a tropical zone with huge expanses of lycophyte rainforests from 326 to 254 million years ago (tables 3.2 and 5.1). If there had been no lycophyte rainforests, then there would have been no massive coal deposits in Carboniferous strata—indeed, there would have been no "Carboniferous" at all in the geologic timescale. If the Carboniferous coal strata had never existed, would there ever have been an industrial revolution?

From prehistoric times we have burned wood for heat—heat to keep our habitats warm and heat to cook our food. Wood is readily available at the Earth's surface, and it is also a renewable resource, as new trees can be grown to replace old ones that have been cut and used for fuel. However, neither wood nor charcoal made from wood contains enough energy per unit mass to efficiently fuel a steam engine—for that we need coal. In human history, coal also was initially readily available at the surface of the Earth, or only shallowly buried beneath that surface. Yet 90 percent of the coal that was present in the Earth's strata was formed during the Late Paleozoic Ice Age, and if the Late Paleozoic Ice Age had never occurred, then that coal would not have existed. How long would a nascent industrial revolution have lasted that was fueled by only 10 percent of the coal that was used to fuel the historical Industrial Revolution? In an alternative world in which the Late Paleozoic Ice Age had never occurred, could humankind have quickly made the jump—while those meager 10 percent of coal supplies still lasted—from steam engines to the use of petroleum and the invention of the internal combustion engine?

The problem with petroleum is that it is usually not readily available at the Earth's surface—or even shallowly below that surface. It takes powerful drilling machines to reach the petroleum pools located deep beneath the surface of the Earth. In an alternative world in which the Late Paleozoic Ice Age had never occurred, would humankind have used its coal-fired steam engines to drill for oil before the coal ran out? Or would that alternative humankind have remained in the preindustrial stage of our own world, forever existing in small cities and predominantly agrarian communities that used wood for fuel, unaware of the existence of the vast pools of oil buried deep beneath the surface of the Earth?

Clearly our modern dilemma of carbon-dioxide-induced global warming would not have occurred if the Late Paleozoic Ice Age had not occurred, as the vast deposits of coal that formed in that ice age would never have existed to be burned by humans. If the Industrial Revolution had never occurred, the human population itself might not have exploded as it did; in an alternative world in which the Late Paleozoic Ice Age had never occurred the human population might have grown at a much slower rate, fueled solely by our intellectual advances in more efficient agricultural methods to produce more food and our advances in more efficient medical methods to preserve and prolong life.

The question "What if the Late Paleozoic Ice Age had never happened?" also has major implications for Earth history and evolutionary ecology. If the great lycophyte tropical rainforests had never existed, the hyperoxic atmosphere of the Earth during much of the Carboniferous–Permian interval of geologic time also would never have existed, as those peculiar rainforests not only acted as massive carbon sinks but also released tons of free molecular oxygen into the atmosphere. In the absence of a hyperoxic atmosphere, the convergent evolution of the numerous giant animals that we considered in detail in chapter 4 would never have happened. Giant griffenflies would never have soared through the skies of the Earth, alligator-size millipedes would never have crawled through the peat bogs and rainforests below, and so on.

However, there is a more serious side to this question than the potential absence of animal gigantism in the Carboniferous and Permian. Without the energetic boost of that oxygen-enriched atmosphere, would the tetrapods have finally managed to emerge from the rivers and lakes and become fully terrestrial in the Early Carboniferous? Without that hyperoxic atmosphere, would the amniotes have evolved? Without extra oxygen, would the tracheal-breathing insects have finally managed to perfect the highly energetic activity of flight? Or, in an alternative world in which the Late Paleozoic Ice Age had never occurred, would the tetrapods have remained in the aquatic evolutionary stage, and if the amniotes had never evolved, would the land areas of the Earth have been populated by amphibian vertebrates only? Within the terrestrial arthropods, would the insects have remained grounded in the mode of life in which arthropods had previously existed for some 100 million years since their initial invasion of land back in the Silurian?

As discussed in the first section of the chapter, we also know that the Famennian and Serpukhovian extinctions were triggered by the Late Paleozoic Ice Age glaciations, and evidence exists that the Frasnian extinction was as well. If the Late Paleozoic Ice Age had never happened, then the glacially induced Famennian and Serpukhovian—and possibly the Frasnian—extinctions would never have happened. These events were the seventh, sixth, and fourth most ecologically severe biodiversity crises in Earth history (table 1.3). How would our Earth have been different if these crises had never happened?

In the oceans, the Frasnian crisis destroyed the largest reefs in Earth history—the massive Paleozoic-style skeletal reefs of stromatoporoid sponges, tabuate corals, and rugose corals (fig. 7.2). The stromatoporoids just barely survived the Frasnian extinction, only to be exterminated by the Famennian crisis. The tabulate corals never recovered their previous diversity, but the rugosans did begin to recover and diversify in the Early Carboniferous, only to be decimated by the Serpukhovian crisis (McGhee et al. 2012). One hundred thirty million years were to pass before the corals once again became the major skeletal-building element in reefs—in the Middle Triassic, with the evolution of the scleractinian corals (Flügel and Stanley 1984; Fois and Gaetani 1984). If the Late Paleozoic Ice Age had never happened, would Paleozoic-style massive skeletal reefs have survived to the present? Would the scleractinian corals and our modern-style reefs have never evolved?

The Devonian is also known as the "Age of Armored Fishes," the great placoderms (table 7.1) (Young 2010). As discussed in chapter 2, some of the great armored fishes were as big as modern-day killer whales, but unlike killer whales, the armored fishes had no teeth. Instead, they had sharp bone blades in their mouths that functioned in a way similar to the sharp bone beaks of modern-day snapping turtles. And even though they possessed massive armor plates of bone around their heads, internally these peculiar fishes had skeletons made of cartilage, rather than bone, like modern-day sharks. Imagine a huge shark with a head like a snapping turtle and you have a fish similar to the great placoderm predators (see fig. 2.1).

The Frasnian extinctions eliminated half of the world's diversity of armored fishes (table 7.1); the Famennian extinctions eliminated them all (table 7.1). Our modern ecologically and numerically dominant fishes, the sharks (the chondrichthyans, table 7.1) and ray-fin fishes (the actinopterygians, table 7.1),

TABLE 7.1 Detailed phylogeny of placoderm and sarcopterygian vertebrates showing the effect of the twin Late Devonian extinctions (modified from tables 6.5 and 6.8).

VERTEBRATA (animals with vertebrae)
– Gnathostomata (jawed vertebrates)
– – PLACODERMI (armored fishes) †**Famennian**
– – – Acanthothoraci †**Frasnian**
– – – unnamed clade †**Famennian**
– – – – unnamed clade †**Famennian**
– – – – – Rhenanida †**Frasnian**
– – – – – Antiarchi †**Famennian**
– – – – unnamed clade †**Famennian**
– – – – – Arthrodira †**Famennian**
– – – – – unnamed clade †**Famennian**
– – – – – – Petalichthyida †**Frasnian**
– – – – – – Ptychtodontida †**Famennian**
– – CHONDRICHTHYES (sharks, rays, and kin)
– – Acanthodii (spine fin fishes)
– – Osteichthyes (bony fishes)
– – – ACTINOPTERYGII (ray fin fishes)
– – – SARCOPTERYGII (lobe fin fishes + descendants)
– – – – CROSSOPTERYGII
– – – – – Porolepiformes †**Famennian**
– – – – – unnamed clade
– – – – – – Onychodontida †**Famennian**
– – – – – – Actinista (living coelacanths)
– – – – DIPNOI (lung fishes)
– – – – TETRAPODOMORPHA (tetrapod-like fishes + descendants)
– – – – – Rhizodontia
– – – – – Osteolepidiformes
– – – – – – Osteolepididae †**Famennian**
– – – – – – Megalichthyidae
– – – – – – Eotetrapodiformes
– – – – – – – Tristichopteridae †**Famennian**
– – – – – – – unnamed clade
– – – – – – – – Elpistostegalia †**Frasnian**
– – – – – – – – TETRAPODA (limbed vertebrates)
– – – – – – – – – Family Elginerpetontidae †**Frasnian**
– – – – – – – – – – *Elginerpeton pancheni* †**Frasnian**
– – – – – – – – – – *Obruchevichthys gracilis* †**Frasnian**
– – – – – – – – – Family incertae sedis †**Frasnian**
– – – – – – – – – – *Sinostega pani* †**Frasnian**
– – – – – – – – – unnamed clade
– – – – – – – – – – Family incertae sedis †**Famennian**
– – – – – – – – – – – *Densignathus rowei* †**Famennian**

```
— — — — — — — — — unnamed clade
— — — — — — — — — — Family incertae sedis †Famennian
— — — — — — — — — — Ventastega curonica †Famennian
— — — — — — — — — — unnamed clade †Frasnian
— — — — — — — — — — — Family incertae sedis †Frasnian
— — — — — — — — — — — Metaxygnathus denticulus †Frasnian
— — — — — — — — — — Family incertae sedis †Famennian
— — — — — — — — — — Jakubsonia livnensis †Famennian
— — — — — — — — — — unnamed clade
— — — — — — — — — — — Family Acanthostegidae †Famennian
— — — — — — — — — — — Acanthostega gunnari †Famennian
— — — — — — — — — — — unnamed clade
— — — — — — — — — — — Family incertae sedis †Famennian
— — — — — — — — — — — Ymeria denticulata †Famennian
— — — — — — — — — — — unnamed clade
— — — — — — — — — — — Family Ichthyostegidae †Famennian
— — — — — — — — — — — — Ichthyostega stensioei †Famennian
— — — — — — — — — — — — Ichthyostega watsoni †Famennian
— — — — — — — — — — — — Ichthyostega eigili †Famennian
— — — — — — — — — — — unnamed clade
— — — — — — — — — — — Family incertae sedis †Famennian
— — — — — — — — — — — — Hynerpeton bassetti †Famennian
— — — — — — — — — — — — unnamed clade
— — — — — — — — — — — — Family Tulerpetontidae †Famennian
— — — — — — — — — — — — — Tulerpeton curtum †Famennian
— — — — — — — — — — — — unnamed clade
— — — — — — — — — — — — — Family Colosteidae
— — — — — — — — — — — — — unnamed clade
— — — — — — — — — — — — — — Family Crassigyrinidae
— — — — — — — — — — — — — — unnamed clade
— — — — — — — — — — — — — — Family Whatcheeriidae
— — — — — — — — — — — — — — unnamed clade
— — — — — — — — — — — — — — — Family Baphetidae
— — — — — — — — — — — — — — — unnamed clade
— — — — — — — — — — — — — — — Neotetrapoda
— — — — — — — — — — — — — — — BATRACHOMORPHA (ancestors of amphibians)
— — — — — — — — — — — — — — — Lepospondyli
— — — — — — — — — — — — — — — REPTILIOMORPHA (ancestors of amniote tetrapods)
```

Source: Phylogenetic data modified from Lecointre and Le Guyader (2006), Ahlberg et al. (2008), Clack et al. (2012), and Benton (2015).

Note: Lineages that went extinct in either the Frasnian or Famennian are marked with a dagger (†); the age of their extinction is in bold; major clades are in capitals.

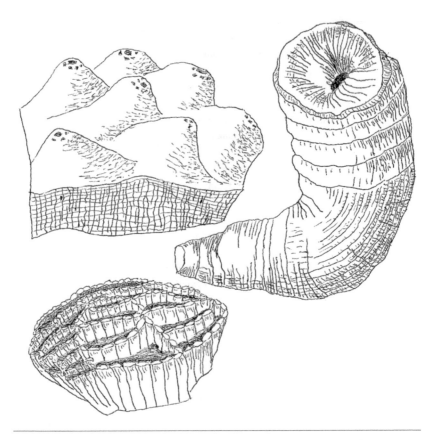

FIGURE 7.2 The skeletal-building elements of the massive Paleozoic-style reefs were chiefly stromatoporoid sponges like *Stromatopora polyostiolata* (top left), tabulate corals like *Halysites catenularia* (bottom left), and rugosan horn-corals like *Caninia torquia* (right).

Source: *Stromatopora*, *Halysites*, and *Caninia* modified and redrawn from Tasch (1973); Hoskins, Inners, and Harper (1983); and Moore, Lalicker, and Fischer (1952), respectively.

diversified only in the Early Carboniferous, filling the ecological void left by the vanished armored fishes (Sallan and Coates 2010; McGhee 2013). If the Late Paleozoic Ice Age had never happened, would our rivers, lakes, and seas still be populated by giant dinichthyids and titanichthyids—the "terrible fishes" and "titanic fishes"—that were seven meters (23 feet) long, with massive bony head armor and jaws like a snapping turtle? Would sharks and ray-fin fishes never have diversified, but only continued to exist in the shadow of the great armored fishes if the Late Paleozoic Ice Age had never happened?

The Late Devonian extinctions were particularly severe in the clade of the sarcopterygian vertebrates (table 7.1), where they acted as evolutionary bottlenecks. As discussed in chapter 2, an evolutionary or genetic bottleneck is produced when the population sizes of a given species shrink almost to the critical minimum level from which a species cannot recover. One of the consequences of a species' surviving an evolutionary bottleneck is the sharp reduction of morphological or genetic diversity in the population of survivors. Another is a reduction in the geographic range of a species: species that may once have had large populations spread across a continent may be found in only a few regional patches with small numbers of individuals after an evolutionary bottleneck. Finally, as implied in the descriptive name "bottleneck," only a few lineages manage to survive past that evolutionary constriction. All of these classic bottleneck phenomena are seen in the lineage of the sarcopterygians in the twin Late Devonian extinctions; hence I call these two evolutionary events the End-Frasnian Bottleneck and the End-Famennian Bottleneck (McGhee 2013).

The sarcopterygian lobe-fin fishes are not familiar to us today, but they greatly outnumbered our modern, familiar ray-fin fishes (the actinopterygians, table 7.1) back in the Devonian. The sarcopterygian clade was hit hard by the Late Devonian extinctions, and today they are represented by only one genus of coelacanth fish, three groups of lung fish, and, of course, ourselves—the modern living tetrapods.

The End-Famennian Bottleneck eliminated two-thirds of the lineages of the crossopterygian lobe-fin fishes (table 7.1). For many years, paleontologists thought that the crossopterygians were extinct—that is, that they had encountered an evolutionary dead end rather than a bottleneck. Then, in 1938, off the coast of South Africa, fishermen unexpectedly captured a very odd-looking fish. Unlike a ray-fin fish, whose fins appears to attach directly to its body, this fish had peculiar stumpy lobes—lobes that contained bones—protruding from its body, and its fins were attached to these lobes. To their astonishment, paleontologists recognized the fish as a living coelacanth crossopterygian and gave it the name *Latimeria chalumnae*. Since that time, other individuals of *Latimeria chalumnae* have been sighted and photographed in the waters off South Africa and Madagascar, have appeared in *National Geographic* magazine and television programs, and so on.

The crossopterygians had survived the End-Famennian Bottleneck after all, and are still alive today.

Only three families of lungfishes, the dipnoian sarcopterygians (table 7.1), survive today—the Neoceratodontidae in Australia, the Lepidosirenidae in South America, and the Protopteridae in Africa. These widely scattered lungfish groups are rare, the only survivors of a group of fishes that were very numerous in the Devonian. What if lobe-fin crossopterygian and dipnoian fishes were common today, as they were in the Devonian, and ray-fin fishes were rare? The idea that we evolved from fish might not seem so strange to people if every time they bought a fish at the supermarket, it had four stumpy limblike fins, complete with internal bones, as in a crossopterygian fish. What if it were not at all unusual to go down to a river and see several different types of fishes crawling about on the banks, breathing air? If, like common pet turtles in aquaria, modern-day children kept air-breathing fish as pets at home, would their parents still be skeptical than land vertebrates are the descendants of fish?

Next we encounter the tetrapodomorph fishes—the ancestors of the living tetrapods (table 7.1). No fewer than three groups of lobe-fin, tetrapodomorph fishes were independently evolving tetrapod-like morphologies in the Devonian— the tristichopterids like *Eusthenopteron* (fig. 7.3), the elpistostegalians like *Tiktaalik* (fig. 7.3), and, of the course, the tetrapods themselves (table 7.1) (Ahlberg and Johanson 1998; Coates et al. 2008; McGhee 2013). The End-Frasnian Bottleneck eliminated the elpistostegalians, and the End-Famennian Bottleneck eliminated the tristichopterids (table 7.1). The extinction of the elpistostegalian, tristichopterid, and osteolepidid tetrapodomorphs created a large phylogenetic gap between the advanced tetrapods and the basal rhizodont and megalichthyid tetrapodomorphs (table 7.1). The rhizodonts would not survive the Carboniferous, and the megalichthyids would not survive the Permian. By the end of the Paleozoic, only the tetrapod clade remained of the once diverse tetrapodomoph lineages. As a result of the Late Devonian extinctions and bottlenecks, a large phylogenetic gap exists today between living limbed vertebrates (the tetrapods; table 7.1) and living finned vertebrates (the ray-fin fishes, the actinopterygians; table 7.1). All of our intermediate cousins are gone, which is why a modern-day lizard seems to us to be such a radically different animal from a trout.

FIGURE 7.3 From lobed fins to limbs (from top to bottom): comparison of the bones in the fin of the tristichopterid fish *Eustenopteron foordi*, in the fin of the elpistostegalian tetrapodomorph *Tiktaalik roseae*, in the forelimb of the aquatic tetrapod *Acanthostega gunnari*, in the forelimb of the cynodont synapsid *Thrinaxodon liorhinus*, in the forelimb of the mammalian synapsid mountain lion *Felis concolor*, and in the arm of the human *Homo sapiens*.

Source: Illustration by Kalliopi Monoyios © 2013. Reprinted with permission.

If these two bottlenecks in vertebrate evolution had not occurred, might further parallel evolution in both the elpistostegalian and tristichopterid lineages have produced animals with four limbs? That is, could the condition of possessing four limbs have evolved independently in three separate vertebrate lineages, instead of just one? This is not an outlandish idea: it is a fact that three other groups of vertebrates have convergently, independently, evolved wings from their forelimbs—the pterosaurs, the birds, and the bats. The pterosaurs are ornithodiran reptiles, and they evolved wings from their forelimbs in the Triassic. The birds are also reptiles, but they belong to a much more derived lineage of reptiles—the dinosaurs. They independently evolved wings from their forelimbs in the Jurassic. The bats are very different from the pterosaurs and birds as they are synapsids, not reptiles. They are also are very highly derived synapsids—they are mammals—and yet they also independently evolved wings from their fore-limbs in the Paleocene, at the beginning of the Cenozoic (Benton 2005). If three separate groups of vertebrates could independently, convergently, accomplish the seemingly highly unlikely feat of evolving wings from modified forelimbs, it is equally likely that three separate groups of advanced lobe-fin fishes could have independently, convergently, evolved four limbs from their four ventral lobe fins. Today the skies of the Earth are the territory of birds and bats—two very different kinds of winged animals that have no common winged ancestor. What if the land areas of the Earth were also home to two, or even three, different kinds of limbed vertebrates that also had no common limbed ancestor but had independently evolved their four limbs? We will never know, as the Late Devonian bottlenecks eliminated that evolutionary possibility forever.[3]

The first of the tetrapods—the true limbed vertebrates (table 7.1)—were aquatic animals, not land dwellers. Their limbs were short, and the manus and pes (hand and foot) morphologies on the limb, used as swimming and crawling paddles, were very different from those of modern tetrapods (fig. 7.3). Only with subsequent evolution would the paddlelike limbs of the early aquatic tetrapods be modified to limbs that would bear the weight of the animal as it walked on dry land. The Devonian aquatic tetrapods like *Acanthostega* (fig. 7.3) were the classic "fish with feet" that are featured today on cars with Darwin bumper stickers.

The End-Frasnian Bottleneck hit the tetrapods hard (table 7.1). Before the bottleneck, these early aquatic tetrapods had a worldwide distribution—from Scotland (*Elginerpeton pancheni*) and Latvia (*Obruchevichthys gracilis*) in Europe in the west to China (*Sinostega pani*) and Australia (*Metaxygnathus denticulus*) in the east. After the bottleneck, surviving early Famennian tetrapods are found only in Europe. Before the bottleneck, tetrapods had a diversity of body sizes, from large individuals up to 2.5 meters (8.2 feet) long to small individuals about a 0.5 meters (1.6 feet) long. After the bottleneck, surviving Famennian tetrapods are about all the same size—about a meter (3.3 feet) long. The End-Frasnian Bottleneck triggered a sharp reduction in morphological variance and geographic range in surviving tetrapod faunas, reductions that are classic evolutionary bottleneck phenomena. What diverse Famennian morphological novelties and body sizes might our cousins in the other Frasnian tetrapod lineages have produced if they had not been eliminated by the End-Frasnian Bottleneck? How would the future course of tetrapod evolution have changed if these lineages had survived?[4]

Following the Famennian Gap cold period (table 1.7), the tetrapods began to recover their species diversity and began to lose many of their fishlike traits as they became more adapted to life on dry land. Then they encountered the End-Famennian Bottleneck, and it triggered an even more severe diversity contraction than had the End-Frasnian Bottleneck (table 1.7). Not only were almost all of the new Famennian clades eliminated, two lineages of Frasnian survivors (the lineages of *Densignathus rowei* and *Ventastega curonica*) were also driven to extinction. Only three groups are thought to have survived the bottleneck—the colosteids, the crassigyrinids, and the whatcheeriids (table 1.7). Only the whatcheeriids are known from fossils in the Famennian (Daeschler et al. 2009; McGhee 2013), and the presumed survival of the colosteids and crassigyrinids is based solely on phylogenetic analyses that indicate that they are more primitive than the whatcheeriids and thus evolved before the whatcheeriids.[5]

As discussed in chapter 2, following the Famennian glaciations and the Tournaisian Gap cold period (table 1.7), the tetrapods began once again to recover. The whatcheeriids (see fig. 2.2) in particular blossomed in diversity and achieved a worldwide distribution, with new species being found in North America, Europe, and Australia. By the Visean Age, the tetrapod invasion of land was in full progress, and the two major clades of the batrachomorphs (ancestors of our modern amphibians) and the reptiliomorphs (ancestors of the amniotes—the living reptiles, dinosaurs, and mammals) had evolved (tables 2.4 and 7.1).

At the very least, the End-Frasnian Bottleneck set back and delayed the successful invasion of land by tetrapods by some six million years (the duration of the Famennian Gap, table 1.7), and the End-Famennian Bottleneck likewise generated another delay of some ten million years (the duration of the Tournaisian Gap). At the worst, the bottlenecks changed the direction of tetrapod evolution forever—we will never know how the exterminated lineages would have evolved or what ecological diversifications and evolutionary innovations they would have produced. In our impoverished world, only one lineage of four-limbed vertebrates exists—a sad legacy of the Late Paleozoic Ice Age.

On the other hand, as discussed at the beginning of this section, the hyperoxic atmosphere produced by the Late Paleozoic Ice Age was a major boost to that surviving tetrapod lineage in its final and successful invasion of land. The other major clade of terrestrial invaders, the arthropods, likewise received that energetic boost—a boost that sent them into the skies. In a world where the hyperoxic atmosphere of the Late Paleozoic Ice Age had never existed, perhaps both of those evolutionary events might still have happened—but they might also have taken much longer than the rapid burst of evolutionary events that occurred in the Serpukhovian–Bashkirian interval of Carboniferous time.

WHAT IF THE SIBERIAN MANTLE PLUME HAD MISSED THE TUNGUSKA BASIN?

The continental glaciations of the Late Paleozoic Ice Age eventually would have ended even if the global warming produced by the Emeishan LIP eruption and then the extreme global heating produced by the Siberian LIP eruption had never happened. After its long 115-million-year trek, the supercontinental mass of Gondwana was finally moving off the South Pole in the latest Permian, as discussed in chapter 1 (fig. 1.3). By the Middle Triassic, about 240 million years ago, the South Pole was open ocean.[6] The giant landmass of Gondwana began to break up in the Middle Jurassic: the plate fragment holding South America and Africa began to move north, and the plate fragment holding India, Antarctica, and Australia began to move south, back toward the South Pole. By the Early Cretaceous, the continental region of Antarctica was once again located over the South Pole, and then first India and later Australia began to split away from it. By the Eocene Epoch of the Cenozoic Era, Antarctica was a totally separate continent—a small fraction of the original giant continent Gondwana—sitting directly on the South Pole, and the initial phases of the continental glaciation that would produce the Cenozoic Ice Age had begun.[7]

In itself, eruption of a mantle plume the size of the Siberian one would have affected the climate of the entire planet, but not to the extent that it did when combined with the vaporization and explosive venting of gases from the huge volume of Tunguska Basin organic-rich and evaporite strata that the Siberian mantle plume burned through on its way to the surface of the Earth. The Siberian mantle plume alone erupted enough lava to cover five million square kilometers (1.93 million square miles) of land; as discussed in chapter 6, that is enough lava to cover 62 percent of the area of the United States between the Atlantic and Pacific coasts. The Siberian flood-basalt eruption is estimated to have been 8.3 to 20 times larger than the Emeishan LIP, producing some 45 to 108 trillion tonnes (50 to 119 trillion tons) of sulfur-dioxide pollutant and its byproduct, sulfuric acid. The acidification of the world's oceans and acid rain poisoning of the world's land areas would have proceed just as they did at the end of the Permian in the absence of any further gaseous input from the Tunguska Basin strata—but it can be argued that the environmental impacts would have been measured on a smaller scale. This is because the massive additional injection of sulfur dioxide into the atmosphere from the Tunguska Basin anhydrite deposits—strata rich in calcium sulfate—would not have occurred if the Siberian mantle plume had missed the Tunguska Basin.

Beyond the effects of sulfur dioxide, the environmental consequences of the Siberian LIP eruption would have been radically different if the hot plume had missed the Tunguska Basin. At the end of the Permian, between 100 and 160 trillion tonnes (110 to 176 trillion tons) of carbon dioxide and between 14 and 42 trillion tonnes (16 to 46 trillion tons) of methane were injected into the Earth's atmosphere from the organic-rich strata in the Tunguska Basin (Svensen et al. 2009; Payne et al. 2010), as discussed in chapter 6. What if this had never happened? If the Earth's atmosphere had never become vastly enriched and polluted with those two greenhouse gases, the horrific Hot Earth of the latest Permian and Early Triassic would never have occurred. The global depletion of oxygen from the Earth's atmosphere that occurred at the end of the Permian—caused by the oxidation of those trillions of tonnes of methane—would never have happened. The additional acidification of the world's oceans by carbonic acid would never have occurred; ocean acidification would have been the product of the mantle plume's own sulfur dioxide without any extra boost from the Tunguska Basin carbon deposits. Coal fly ash from the burning Tunguska Basin coal strata would never have been explosively injected into the atmosphere, oceanic euxinia would not have occurred, and unknown trillions of marine organisms would never have suffocated from the oxygen-depleting effects of eutrophication. Thus three of the end-Permian kill mechanisms that we considered in chapter 6—heat death, carbon-dioxide poisoning, and suffocation—would either not have occurred at all or would have been vastly reduced in their lethality.

Finally, we have the evaporite strata of the Tunguska Basin. At the end of the Permian, between five and 15 trillion tonnes (six to 17 trillion tons) of methyl chloride and between 87 and 255 billion tonnes (96 to 281 billion tons) of methyl bromide were injected into the Earth's atmosphere from the burning of the Tunguska Basin evaporite deposits (Svensen et al. 2009). If the Siberian mantle plume had missed the Tunguska Basin, those gases never would have been injected into the atmosphere, the ozone layer of the atmosphere never would have been destroyed, and the vast increase in the radiation flux of ultraviolet at the Earth's surface never would have occurred. Yet another of the end-Permian kill mechanisms—radiation poisoning—would either not have occurred at all or would have been vastly reduced in its lethality.

In chapter 6 we also considered the alternative mantle-plume-chemistry model of Sobolev and colleagues (Sobolev et al. 2011), who argued that the Siberian LIP itself could have produced as much as 175 trillion tonnes (193 trillion tons) of carbon dioxide and as much as 18 trillion tonnes (20 trillion tons) of hydrochloric acid if the melting and recycling of oceanic crustal rock located beneath Siberia had produced a magma that was more gaseous and carbon rich than typical basaltic-type lavas of other LIP eruptions like those of Laki, Iceland. Thus, in the case of each kill mechanism discussed above, I have included the possibility that even if the Siberian mantle plume had missed the Tunguska Basin, the kill mechanisms of heat death, carbon-dioxide poisoning, suffocation, and radiation poisoning still might have occurred *but would have been vastly reduced in their lethality.* Even if the magma-chemistry model of Sobolov and colleagues is eventually shown to be correct, it is a known fact that the Siberian mantle plume did indeed burn through the organic-rich and evaporitic strata of the Tunguska Basin. Thus, whatever the lethality of the kill mechanisms produced by the Sobolev and colleagues model for the Siberian mantle plume alone, all of the intensification of those kill mechanisms produced by the vast tonnages of carbon dioxide, methane, methyl chloride, and methyl bromide that were injected into the Earth's atmosphere at the end of the Permian would have to be subtracted if the mantle plume had missed the Tunguska Basin.

Would the end-Permian mass extinction have happened at all if the kill mechanisms of heat death, carbon-dioxide poisoning, suffocation, and radiation poisoning had never occurred, or occurred with much reduced intensities? If the Siberian mantle plume had missed the Tunguska Basin, would the only major kill mechanism triggered have been acidification resulting from the plume's venting of sulfur dioxide? The Siberian volcanic eruptions were the largest continental

LIP in Earth history (Reichow et al. 2009), but would sulfur-dioxide-induced acidification of the world's oceans and land areas have been lethal enough to cause the largest loss of biodiversity and the severest ecological catastrophe in Earth history, the end-Permian mass extinction (table 1.3)?

If an extinction event of much reduced severity had occurred at the end of the Permian—in essence, if there had been no end-Permian mass extinction—would the Paleozoic world have never come to an end (table 6.6)? Jack Sepkoski would have argued that the Paleozoic marine world would still have ended, even in the absence of the end-Permian mass extinction—it just would have taken longer and not been as abrupt in an alternative world in which the Siberian mantle plume missed the Tunguska Basin. Sepkoski's analyses of the interactive evolutionary dynamics (Sepkoski 1981, 1984, 1990; Sepkoski and Miller 1985) of his Paleozoic evolutionary fauna and modern evolutionary fauna supported the conclusion that the modern evolutionary fauna would have actively, competitively, ecologically replaced the Paleozoic evolutionary fauna with time, just as his Paleozoic evolutionary fauna competitively replaced the Cambrian evolutionary fauna (figs. 6.10, 6.11). Sepkoski's analyses showed that every major crisis in the Paleozoic had differentially eliminated Paleozoic marine world fauna and favored the modern evolutionary fauna—in essence, that the Paleozoic marine world was doomed even in the absence of the catastrophic effects of the end-Permian mass extinction. The winnowing effect of the Frasnian, Famennian, Serpukhovian, and Captanian extinctions on the Paleozoic marine world fauna would still have occurred in a world where there had been no end-Permian mass extinction, and Sepkoski would have argued that each future biodiversity crisis in the marine realm would have continued to eliminate Paleozoic marine world fauna in favor of the modern evolutionary fauna in an inexorable process of natural selection.

However, the story might have been very different for the Paleozoic terrestrial world fauna in a world in which the end-Permian mass extinction had never happened. In such an alternative world, in the absence of the catastrophic effects of the Siberian LIP, would the diverse land-dwelling vertebrate faunas of the Paleozoic world have been replaced almost entirely by reptiles (tables 6.7 and 6.8)? And if the synapsids had not been weakened by the end-Permian mass extinction, might they not have been replaced by dinosaurs in the Late Triassic (table 6.9)—might the synapsids have continued to be the numerically and ecologically dominant land vertebrates? Might the synapsids have evolved endothermic mammals before the dinosaurs, and might the dominant mammals have kept the dinosaurs in check for the next 150 million years—rather than the dinosaurs keeping the mammals in check, as happened in our world in the Mesozoic?

The dinosaurs evolved a robust ecosystem structure that persisted for 150 million years before being destroyed in the environmental catastrophe triggered by the Chicxulub asteroid strike at the end of the Cretaceous. After the collapse of the dinosaur ecosystem, the same ecosystem structure was convergently evolved by the mammals in the Cenozoic (table 7.2). It is important to note that the structure of the 19 ecological roles, or niches, listed in table 7.2

TABLE 7.2 Convergent evolution of ecological-analog compositions of Mesozoic dinosaurian-dominated ecosystems and Cenozoic mammalian-dominated ecosystems.

Convergent Ecological Analog	Mesozoic Ecosystem	Cenozoic Ecosystem
1. Large ambush predator	Allosaur (Dinosauria: Allosauridae; *Allosaurus fragilis* †Jurassic)	Lion (Mammalia: Felidae; *Panthera leo*)
2. Small ambush predator	Coelophysis (Dinosauria: Coelophysidae; *Coelophysis rhodesiensis* †Triassic)	Wild cat (Mammalia: Felidae; *Felis sylvestris*)
3. Pursuit pack predator	Velociraptor (Dinosauria: Dromaeosauridae; *Velociraptor mongoliensis* †Cretaceous)	Wolf (Mammalia: Canidae; *Canis lupus*)
4. Nocturnal foraging predator	Troodont (Dinosauria: Troodontidae; *Saurornithoides mongoliensis* †Cretaceous)	Wolverine (Mammalia: Mustelidae; *Gulo gulo*)
5. Ant eater	"Desert bird" alvarezsaurid (Dinosauria: Alvarezsauridae; *Shuvuuia deserti* †Cretaceous)	Anteater (Mammalia: Myrmecophagidae; *Myrmecophaga tridactyla*)
6. Large, herding, browsing herbivore	Sauropod (Dinosauria: Titanosauridae; *Titanosaurus madagascariensis* †Cretaceous)	Elephant (Mammalia: Elephantidae; *Loxodonta africana*)
7. Midsize, fast-running, browsing herbivore	Hypsilophodont (Dinosauria: Hypsilophodontidae; *Hypsilophodon foxii* †Cretaceous)	Deer (Mammalia: Cervidae; *Cervus elaphus*)

8. Large, herding, mixed-feeding herbivore	Hadrosaur (Dinosauria: Hadrosauridae; *Parasaurolophus walkeri* +Cretaceous)	Bison (Eutheria: Bovidae; *Bison bison*)
9. Large, horned, grazing herbivore	Triceratops (Dinosauria: Ceratopsidae; *Triceratops albertensis* +Cretaceous)	Rhinoceros (Mammalia: Rhinocerotidae; *Rhinoceros unicornus*)
10. Large, armored, grazing herbivore	Ankylosaur (Dinosauria: Ankylosauridae; *Ankylosaurus magniventris* +Cretaceous)	Glyptodont (Mammalia: Glyptodontidae; *Doedicurus clavicaudatus* +Pleistocene)
11. Midsize ruminant grazer	Pachycephalosaur (Dinosauria: Pachycephalosauridae; *Pachycephalosaurus wyomingensis* +Cretaceous)	Goat (Mammalia: Bovidae; *Capra hircus*)
12. Giant frugivore (fruit eater)	"Sloth-claw" therizinosaur (Dinosauria: Therizinosauridae; *Nothronychus mckinleyi* +Cretaceous)	Giant ground sloth (Mammalia: Megatheriidae; *Nothrotheriops shastensis* +Pleistocene)
13. Fluvial piscivore (fish eater)	Spinosaur (Dinosauria: Spinosauridae; *Spinosaurus maroccanus* +Cretaceous)	Gavialid crocodile (Archosauria: Crocodilidae; *Gavialis gangeticus*)
14. Shallow-marine piscivore	Plesiosaur (Diapsida: Sauropterygia: Cryptocleididae; *Cryptocleidus oxoniensis* +Jurassic)	Sea lion (Mammalia: Otariidae; *Zalophus californianus*)
15. Small open-ocean carnivore	Ichthyosaur (Diapsida: Ichthyosauria: Ichthyosauridae; *Ichthyosaurus platyodon* +Jurassic)	Porpoise (Mammalia: Phocaenidae; *Phocaena phocaena*)
16. Large open-ocean carnivore	Pliosaur (Diapsida: Sauropterygia: Rhomaleosauridae; *Rhomaleosaurus megacephalus* +Jurassic)	Killer whale (Mammalia: Delphinidae; *Orcinus orca*)
17. Flying insectivore (insect eater)	"Frog-jaw" pterosaur (Ornithodira: Pterosauria: Aneurognathidae; *Batrachognathus volans* +Jurassic)	Bat (Mammalia: Vespertilionidae; *Myotis myotis*)
18. Flying frugivore (fruit eater)	"Bakony dragon" pterosaur (Ornithodira: Pterosauria: Azhdarchidae; *Bakonydraco galaczi* +Cretaceous)	Fruit bat (Mammalia: Pteropodidae; *Rousettus aegyptiacus*)
19. Flying piscivore	Pteranodon (Ornithodira: Pterosauria: Pteranodontidae; *Pteranodon longiceps* +Cretaceous)	Pelican (Aves: Pelecanidae; *Pelicanus occidentalis*)

Source: Modified from McGhee (2011).

Note: The age of extinct species is marked with a dagger and a geologic date (e.g., +Jurassic).

first appeared in the evolution of the dinosaurian-dominated ecosystems of the Mesozoic world, not in the mammalian-dominated ecosystem structure of the Cenozoic world. Mesozoic dinosaurs did not independently evolve the ecological roles of lions, wild cats, wolves, wolverines, anteaters, elephants, deer, bison, rhinoceroses, glyptodonts, goats, and ground sloths—it is the Cenozoic mammal fauna that has convergently refilled the ecological niches of allosaurs, coelophysises, velociraptors, troodonts, alvarezsaurids, sauropods, hypsilophodonts, hadrosaurs, triceratopses, ankylosaurs, pachycephalosaurs, and therizinosaurs! Some of these two groups of animals resemble each other morphologically (triceratopses and rhinoceroses, ankylosaurs and glyptodonts, therizinosaurs and ground sloths), whereas the others appear radically different—the mammalian predators are all quadrupeds, whereas the dinosaurian predators walked on their hind legs only. Yet they all were ecological analogs of each other, making their living in the roughly same way. The goats, for example, have a ruminant stomach system to process plant material, whereas the pachycephalosaurs had a gizzard-like gastric mill to accomplish the same purpose. The similarity in the herbivorous ecological roles of many of the ornithopod dinosaurs, such as the hypsilophodonts and hadrosaurs, to ungulate mammals prompted paleobiologists Matthew Carrano, Christine Janis, and Jack Sepkoski to conclude: "Although late Mesozoic and late Cenozoic terrestrial ecosystems were profoundly different in terms of both animal and plant taxa, there may be universal constraints on the ecological roles played by large herbivores, resulting in convergence in morphology (and, by implication, behavioral ecology) between groups as taxonomically distinct as dinosaurs and mammals" (Carrano et al. 1999, 256). Thus Carrano and colleagues suggest that Mesozoic hadrosaurs and Cenozoic ungulates may provide an example of both ecological and behavioral convergent evolution.[8]

Not all dinosaurian ecological roles are filled by Cenozoic mammals, and not all mammalian roles were filled by dinosaurs. Although the otter hunts fish in rivers, the gavialid crocodile is a much closer modern ecological analog than the otter to the toothy spinosaur (table 7.2). The dinosaurs were exclusively terrestrial animals, yet if we examine marine and oceanic habitats, we find that other Mesozoic animals independently evolved ecological roles that are now played by Cenozoic mammals: the plesiosaur, ichthyosaur, and pliosaur ecological niches have been convergently refilled by Cenozoic sea lions, porpoises, and killer whales. In the air, the insect-eating pterosaur and fruit-eating pterosaur niches have been independently re-evolved by the insect-eating bats and the fruit-eating bats. And the modern pelican, itself an avian dinosaur, convergently refilled the niche of the Mesozoic pteranodon and even resembles a small pteranodon in appearance. Mammals also have convergently evolved flying fish-eaters, such as the greater bulldog bat, *Noctilio leporinus*, although it looks nothing like a pteranodon (excepting the wings).

In these 19 ecological roles (table 7.2), the Cenozoic ecosystem play is scripted in the same way as the Mesozoic ecosystem play. The roles of the ecological play have not changed, although the cast of actors changed dramatically following the end-Cretaceous mass extinction. Would this same ecosystem structure have evolved if the end-Permian mass extinction had never happened, if the synapsids have evolved endothermic mammals before the dinosaurs, and if the dinosaur-dominated ecosystem structure of our Mesozoic world had never evolved? That is, in an alternative world in which the end-Permian mass extinction had never happened, would the mammals have evolved the same ecosystem structure (table 7.2) back in the Late Triassic? Would the dinosaurs have existed only as minor elements in that ecosystem structure, coexisting with the mammals but being held in check by the mammalian ecological dominants—the reverse of what happened in our world in the Mesozoic?

If the mammals had "got there first," so to speak, and established a mammalian-dominated ecosystem structure that persisted throughout the Mesozoic, what would have happened when the Chicxulub asteroid impacted the Earth at the end of the Cretaceous? That cosmic event would still have happened, whether there had ever been an end-Permian mass extinction or not, as that event had an extraterrestrial cause. In our world, that impact-generated global environmental catastrophe triggered the collapse of the dinosaurian ecosystem and the extinction of all of the nonavian dinosaurs. In an alternative world in which the end-Permian mass extinction had never happened, would the Chicxulub asteroid impact have destroyed the mammalian ecosystem instead, and would the dinosaurs have then convergently refilled the ecological niches vacated by the extinct mammals? Would that alternative world have had a Mesozoic fauna dominated by mammals, and a Cenozoic fauna dominated by dinosaurs?

The answer to that hypothetical question is . . . probably not. One of the traits that enabled the mammals to differentially survive the Chicxulub asteroid strike was their ability to burrow underground and hibernate. That trait is an ancient synapsid trait, one that predates the evolution of the mammals—and one that also enabled the synapsids to differentially survive the end-Permian mass extinction, as discussed in chapter 6. It was no accident that the small-bodied, barrel-chested, burrowing dicynodont synapsid *Lystrosaurus* survived the catastrophic environmental conditions triggered by the Siberian LIP, and that approximately 90 percent of all land vertebrates alive in the Early Triassic were lystrosaur synapsids (Sahney and Benton 2008). (However, remember from chapter 6 that that synapsid population resurgence was short and not sweet, as the reptilian rhynchosaur and rauisuchian competitors were soon to evolve in the Middle Triassic, and the deadly dinosaurs proliferated in the Late Triassic; table 6.9). It is a curious quirk of evolution that this trait is found in both highly derived, high-metabolic-rate endothermic mammals and ancient, primitive, ectothermic animals like amphibians—animals that are not even amniotes. The phylogenetic distribution of this trait explains some of the great mystery of why almost all of the highly derived, ecologically advanced, high-metabolic-rate dinosaurs were exterminated by the end-Cretaceous mass extinction, whereas very ancient ectothermic animals like frogs and snakes survived: the dinosaurs did not burrow underground and they did not hibernate. The dinosaurs, to our knowledge, never evolved the ecological equivalents of mammalian moles, woodchucks, rabbits, and the like.

Large body size is always a survival liability in times of ecological crisis—large-bodied animals need too much food—and in a hypothetical world in which the mammals were ecologically dominant when the Chicxulub asteroid struck the Earth, all of the large-bodied mammals would no doubt have perished, just as all of the large-bodied dinosaurs did in our world. But—just as actually happened in our world—diverse lineages of mammals that could burrow and hibernate would have survived to rediversify in the postextinction hypothetical world. The small-bodied dinosaurs did not have this capability, and they also perished, along with their large-bodied kin, in the end-Cretaceous mass extinction. Only the avian dinosaurs survived, probably as a result of both small body size and their ability to fly: they could cover large areas in search of food when food was in very short supply in the post-asteroid-impact world. No doubt vast, uncountable numbers of birds starved to death at the end of the

Mesozoic terrestrial world, but enough passed through the evolutionary bottle-neck to continue the survival of the dinosaurian lineage to this day.

In summary, rather than the end-Permian mass extinction, it was the end-Cretaceous mass extinction that created our modern Cenozoic terrestrial world with its mammal-dominated ecosystems (McGhee et al. 2004). Even in a hypothetical world in which the end-Permian mass extinction never happened, in which the dinosaurian ecosystem never evolved, and in which the mammals became dominant in the Late Triassic, the selective effect of the end-Cretaceous mass extinction would still have resulted in a postapocalyptic terrestrial world that continued to be dominated by mammals, not by dinosaurs.

However, if the end-Permian mass extinction had never happened and the dinosaurian ecosystem had never evolved, then the mammals would have escaped being ecologically dominated and preyed upon by the dinosaurs. Mammals never would have been forced to remain small, mouse- or chipmunk-size creatures during the Jurassic and Cretaceous, digging underground to hide from the terrible small coelurosaurian dinosaur predators, forced to remain noctur-nal creatures scurrying furtively for food in the dark of the night in hopes of not being noticed by nocturnally hunting troodont dinosaurs. The 150-million-year-long nightmare of our mammalian ancestors might never have happened.

Notes

1. HARBINGERS OF THE LATE PALEOZOIC ICE AGE

1. The percentage of carbon dioxide in the atmosphere is now 0.04 and steadily rising because of the burning of fossil hydrocarbons—many of Carboniferous age.
2. Anaerobic photosynthesizing bacteria: purple sulfur, purple nonsulfur, green sulfur, green nonsulfur, and heliobacteria. Aerobic photosynthesizing bacteria: cyanobacteria.
3. $CO_2 + H_2O \rightarrow CH_2O + O_2\uparrow$.
4. $CH_4 + 2O_2 \rightarrow CO_2 + 2H_2O$.
5. Five to 18 percent of the atmosphere's present 21 percent level of oxygen (Lane 2002).
6. Κρυος + γενής.
7. Given its short duration, some question whether the Gaskiers was a snowball-Earth glaciation at all—it may instead have been the first of the Phanerozoic-style glaciations (Pu et al. 2016).
8. Φανερός + ζωον.
9. The discovery of soft-tissue fossils of the enigmatic hyoliths has now demonstrated that the hyoliths were lophophorates (Moysiuk et al. 2017).
10. In the Oligocene Epoch.
11. Aerobic: $CO_2 + H_2O \rightarrow CH_2O + O_2$; anaerobic: $CO_2 + 2H_2S \rightarrow CH_2O + H_2O + 2S$.
12. $CO_2 + CaSiO_3 \rightarrow CaCO_3 + SiO_2$.
13. $CH_4 + 2O_2 \rightarrow CO_2 + 2H_2O$.
14. A mineral formed at very high pressures and temperatures deep in the Earth.
15. Measuring the total duration of the Late Paleozoic Ice Age from an onset in the late Frasnian Age of the Late Devonian, 375 million years ago, to an ending in the Wuchiapingian Age of the Late Permian, about 260 million years ago.
16. Another significant accumulation of coal-rich strata also occurred in the Paleogene (Nelsen et al. 2016).

17. Zachos et al. (2001) estimate the size of the Oi-1 ice sheet to have been 50 percent of the present-day ice sheet, or 7.0×10^6 km², which is here taken as a minimum estimate. Pusz et al. (2011) give an average estimate of the Oi-1 ice sheet that is 85 percent of the present-day ice sheet, or 11.9×10^6 km², which is here taken as a maximum estimate.

18. By 12 million years ago in the Miocene, sea level had fallen to the level present in our modern-day state of glaciation (Westerhold et al. 2005); thus the initial estimate of the size of the Miocene ice sheet in Antarctica is 14×10^6 km², the same as today, and this value is here taken as a minimum estimate. However, Wilson and Luyendyk (2009) have argued that the Transantarctic Mountains in Antarctica are a small remnant of a previous highland and that West Antarctica once was exposed above sea level, increasing the total land area of Antarctica by 10 to 20 percent in the earlier Cenozoic. The Miocene ice sheet may thus have been as large as 15.4 to 16.8×10^6 km², and the upper value is here taken as a maximum estimate.

19. 250 μm and larger.

20. See note 18.

21. The Famennian Gap spanned ten conodont zones and was estimated to have been seven million years in duration based on the Gradstein et al. (2004) geologic timescale. However, the duration of the Famennian Age has been shortened from 15.3 to 13.3 million years in the new Gradstein et al. (2012) timescale, leading to a new estimate of 0.6 million years' duration for each of the 22 Famennian conodont zones. As the Famennian Gap spanned ten of those zones, ten zones times 0.6 million-years-per-zone yields the new duration estimate of six million years for the Famennian Gap.

22. The duration of the Tournaisian Gap was once thought to have been some 14 million years—the duration of the entire Tournaisian Age in the Gradstein et al. (2004) timescale—and twice the duration of the original seven-million-year estimate of the duration of the Famennian Gap (see tables 6.2 and 6.7 in McGhee 2013). However, while McGhee (2013) was in press, Smithson et al. (2012) argued that the Tournaisian Gap (which they called Romer's Gap) lasted for ten million years, not 14 million, before land vertebrate faunas began to recover from the Lilliput-effect small-body-size constraint imposed by "adverse conditions, such as aridity of other climatic conditions" (Smithson et al. 2012, 4535). It should be noted that this new ten-million-year estimate of the duration of the Tournaisian Gap is based solely on the land vertebrate fossil record and has yet to be corroborated by data from the marine fossil record and the land-plant fossil record, both of which were also affected by the climatic conditions of the Tournaisian Gap (see McGhee 2013, 184–188).

23. Additional geochemical similarities between the onset of the Cenozoic and Late Paleozoic Ice Ages also exist; for example, both the Miocene and Famennian glaciations were characterized by positive carbon-isotope anomalies of +0.8‰ δ^{13}C and +1.2‰ δ^{13}C, respectively (Zachos et al. 2001; Kaiser et al. 2006, 2016); oxygen-isotope increases of 0.3–1.0‰ δ^{18}O and 0.8–1.2‰ δ^{18}O, respectively (Flower and Kennett 1995; Kaiser et al. 2006); and estimated drops in sea-surface temperature of 6–7°C in the Miocene and at least 2–4°C in the Famennian, based on partial data (Shevenell et al. 2004; Kaiser et al. 2006). Likewise, both the Oligocene and proposed Frasnian glaciations were characterized by positive carbon-isotope anomalies of +0.8‰ δ^{13}C and +3.0‰ δ^{13}C, respectively (Zachos et al. 2001; Joachimski and Buggisch 2002); oxygen-isotope increases of 0.5–1.0‰ δ^{18}O and 1.0–1.5‰ δ^{18}O, respectively (Pusz et al. 2011; Joachimski and Buggisch 2002); and estimated drops in sea-surface temperature of 3–4°C in the Oligocene, based on partial data, and 5–7°C in the Frasnian (Wade et al. 2012; Joachimski and Buggisch 2002).

2. THE BIG CHILL

1. For a detailed examination of the aftermath of the Devonian extinctions, see McGhee (2013).
2. A lyginopterid spermatophyte, an extinct group of seed plants.
3. Lophophorate lophotrochozoans. Brachiopods still exist today, but in very low diversity in modern oceans.
4. Eutrochozoan lophotrochozoans.
5. It was originally argued that the Tournaisian Gap lasted 14 million years (McGhee 2013, 184–188), but a more recent estimate is ten million years (Smithson et al. 2012, 4535). For a more detailed discussion of the data, see note 22 in chapter 1.
6. For a detailed examination of the effect of the Devonian extinctions on the tetrapod faunas, see McGhee (2013).
7. For a detailed examination of the post-Tournaisian-Gap fauna, see McGhee (2013).
8. Tentative evidence from bone fragments (Daeschler et al. 2009) suggest that a whatcheeriid-like tetrapod may have been present in Pennsylvania, USA, in the latest Famennian. If true, this makes the enigma of the lack of further evolution in the family during the ten million years of the Tournaisian Gap even more perplexing.
9. Dated from the Lower *crenulata* through the *isosticha* conodont zones; see discussion in McGhee (2013, 196–199).
10. Possible further geochemical corroboration of this assessment is the documented presence of a large-magnitude positive anomaly in carbon-isotope ratios in strata dating from the middle Tournaisian in both North America and Europe. We saw in chapter 1 that the onset of glaciations in both the Late Paleozoic and Cenozoic ice ages was accompanied by positive carbon-isotope anomalies (see note 23 in chapter 1), and the mid-Tournaisian anomaly is "one of the largest known Phanerozoic $\delta^{13}C$ events" (Saltzman et al. 2000).
11. Dated from the Upper *commutata* through the Middle *bilineatus* conodont zones; see discussion in McGhee (2013, 196–199).
12. The midpoint of the Visean, which spans in geologic time from 330.9 million to 346.7 million years ago, is 338.8 million years.
13. $\delta^{18}O = 1.8‰$ (Mii et al. 2001).
14. See Epshteyn (1981a), who argues that Ustritsky's (1973) Early Permian (Sakmarian) glacial strata are Bashkirian-Muscovian in age.
15. However, the marine fauna of high-latitude regions did continue to experience fluctuations in extinction rates during the waxing and waning of glaciers during the Late Carboniferous; see the analysis of Balseiro (2016).
16. A United States quarter-dollar coin is 24 millimeters in diameter.
17. Benton (1989) documents the extinction of only four families of tetrapods in the Serpukhovian, and all four were of low species diversity.
18. See the extensive discussion of the evolution of amniotes, and of the evolution of flight in insects, in McGhee (2013).
19. Both are found in the famous Joggins tree-stump *Lagerstätte* in Nova Scotia, Canada; see discussion in McGhee (2013).
20. The question of the evolution of the first winged insects and their subsequent diversification becomes even more enigmatic if molecular analyses are included. Molecular phylogenies with

divergence-date estimates have predicted the evolution of wings, hence the first Pterygota, to have occurred 406 million years ago in the Emsian Age of the Early Devonian (Misof et al. 2014)! If true, this would mean that insects possessing the key adaptation of wings still did not diversify and did not achieve large population sizes until the Bashkirian Age of the Late Carboniferous, some 83 *million* years later. Ecologically, that scenario does not make sense.

3. THE LATE CARBONIFEROUS ICE WORLD

1. Based on molecular phylogenies; see discussion in Bell et al. (2010).
2. Not all calamitean trees grew from rhizomes; see Rößler et al. (2012).
3. However, some cordaitean trees grew as tall as 30 meters; see Stewart (1983).
4. $CO_2 + H_2O \rightarrow CH_2O + O_2\uparrow$.
5. More precisely, the strata are dated to the Middle *expansa* conodont zone, the VH spore zone, and the Fa2c chronozone of older timescales; see McGhee (2013).
6. The late Robert Berner (1935–2015) did modify the Geocarbsulf model in a brief, three-page paper in 2009, bringing its predicted oxygen values closer to the older Rock-Abundance model. Still, the revised Geocarbsulf model predicts that oxygen levels did not increase to 30 percent or higher until the latest Carboniferous and, more significantly, actually predicts a *lower* oxygen content in the atmosphere in the Early Carboniferous than the original Geocarbsulf model! The Early Carboniferous is the same time interval in which animal gigantism evolved in numerous independent species lineages, a biological phenomenon that has been used to argue for a hyperoxic atmosphere on the Earth, as will be discussed in chapter 4.
7. Some peat fires do occur today in the tropics, during the dry season.
8. To help the reader, here is a rough correlation of some of the major older Carboniferous geologic time divisions with the modern time scale: Namurian A (pars) = Serpukhovian; Namurian A (pars), B, C, and Westphalian A, B = Bashkirian; Westphalian C, D, and Stephanian (Cantabrian pars) = Moscovian; Stephanian (Cantabrian pars) A, B = Kasimovian; and Stephanian C, Autunian (pars) = Gzhelian.
9. Παν + γαια.
10. See also DiMichele et al. (2009, 210), in which differences in interpretation of the wettest part of cyclothemic rhythms are discussed—one interpretation holding that the wettest periods occurred during periods of sea-level lowstand, another holding that the wettest period occurred during sea-level highstand.

4. GIANTS IN THE EARTH . . .

1. Some have argued that *Megarachne servinei* was a giant land-dwelling eurypterid water scorpion, not a spider. If true, this would simply mean that this giant predatory arthropod was a merostome chelicerate rather than an arachnid chelicerate (table 4.1).
2. Giant arthropleuran fossils are reported from strata ranging in age from the Carboniferous Westphalian C [= Moscovian] to the Permian Lower Rotliegend [= Asselian].

3. Data from Clapham and Kerr (2012; supplemental tables S1, wing length, and S2, body width). Total wingspan is calculated by the formula: Wingspan = 2(wing length) + (body width).

4. The basal batrachomorphs are often called "temnospondyls" in the older literature, and the basal reptiliomorphs are called "anthracosaurs." Both of these older taxonomic groupings are paraphyletic.

5. A United States quarter-dollar coin is 24 millimeters in diameter.

6. Specifically, the Cryogenian Period of the Neoproterozoic Era; see Erwin et al. (2011).

7. The most common symbionts are species of the dinoflagellate genus *Symbiodinium*; see table 4.3 for dinoflagellate evolutionary relationships.

8. Specifically, species in the fusulinacean families Neoschwagerinidae and Verbeekinidae; see discussion in McGhee et al. (2013).

9. For a discussion of the convergent evolution of ecological niches in Cenozoic mammals and Mesozoic dinosaurs, see McGhee (2011).

10. Named for the vertebrate paleontologist Everett C. Olson, who spent his life studying these vertebrates; Sahney and Benton (2008).

11. Specifically, eruption of the Emeishan Large Igneous Province in China, which we will examine in detail in chapter 6; see also McGhee et al. (2013).

12. These animals were the first of many vertebrate lineages to convergently evolve the ability to glide; see McGhee (2011).

13. These animals were the first of many vertebrate lineages to convergently evolve bipedalism; see McGhee (2011).

5. THE END OF THE LATE PALEOZOIC ICE AGE

1. See the paleogeographic map given in figure 5 of the paper by Isbell et al. (2016). However, these same authors argue that the region was ice free in the Permian.

2. See the new dating of the P3 and P4 glacial pulses given in Metcalfe et al. (2015).

3. See note 1.

4. Παν + θάλασσα.

5. Παν + γαια; see discussion of the tectonic effects of the assembly of Pangaea in chapter 3.

6. The Ophiacodontidae, Edaphosauridae, Eothyrididae, and Sphenacodontidae; see Kemp (2006).

6. THE END OF THE PALEOZOIC WORLD

1. LIP is pronounced like the word "lip," and not spelled out L-I-P like an acronym.

2. The term "large igneous province" was introduced by Coffin and Eldholm (1991, 1994); see discussion in Saunders (2005). For comprehensive discussions of the LIPs that have occurred in geologic time, see Bryan and Ernst (2008) and Bryan et al. (2010).

3. A LIP geographic region sometimes also is called the "Traps," as in the "Emeishan Traps." The designation "trap" comes from "traprock," an informal name commonly used by miners for basaltic rocks.

4. For extensive eyewitness accounts of this period of crisis in Europe, see tables A1 and A2 in Thordarson and Self (2003).

5. Some researchers have questioned the causal link between the Laki eruption and the abnormally cold winter of 1783–1784; see, for example, D'Arrigo et al. (2011); Lanciki et al. (2012). For a counterargument to these studies, see Schmidt et al. (2012).

6. (500 eruptions/100,000 years) × (20 km³ lava/eruption) = 10,000 km³ lava/100,000 years; 1,500,000 years/100,000 years = 15; thus 15 × (10,000 km³ lava)/ 15 × (100,000 years) = 150,000 km³ lava/1,500,000 years.

7. The degree of doming produced by the Emeishan super plume continues to be debated; Ukstins-Peate and Bryan (2008) argue that no doming at all occurred. For a detailed discussion of the debate, see Shellnutt (2013).

8. $CH_4 + 2O_2 \rightarrow CO_2 + 2H_2O$.

9. For every molecule of CH_4 oxidized, two molecules of O_2 are removed from the atmosphere and one molecule of CO_2 is released into the atmosphere; see note 8.

10. Families Neoschsagerinidae and Verbeedinidae; see Vachard et al. (2010).

11. Staffelids and schubertellids; see Vachard et al. (2010).

12. The Alatoconchidae; see Aljinović et al. (2008); Bond, Wignall, et al. (2010).

13. The model of Ganino and Arndt (2009) estimates a total emission of 78.4 to 162.4 trillion tonnes of CO_2 into the atmosphere, based on the assumption of a 1×10^6 km³ LIP volume. Our current maximum estimates of the original Emeishan LIP volume are about a third smaller, at 0.6×10^6 km³. Thus I have reduced the original Ganino and Arndt (2009) CO_2 injection-mass estimates by one-third.

14. The model of Self et al. (2005) estimates a total emission of 11.7 billion tonnes of SO_2 into the atmosphere from a small LIP magma volume of 1.3×10^3 km³. Scaling this estimate up to an Emeishan LIP magma volume of 0.6×10^6 km³ results in a SO_2 injection mass of 5.4 trillion tonnes.

15. $^{12}CO_2 + H_2O \rightarrow {}^{12}CH_2O + O_2$.

16. $Ca^{12}CO_3$.

17. The index is calculated as follows: $\delta^{13}C = [({}^{13}C/{}^{12}C)_{sample} / ({}^{13}C/{}^{12}C)_{standard} - 1] \times 10^3$, where values are reported as per mille (‰) relative to a standard. For many biostratigraphic studies, the standard taken is the isotopic ratio of a fossil belemnite from the Cretaceous Pee Dee Formation of the Carolinas, the "Pee Dee Belemnite" standard or simply $\delta^{13}C_{(pdb)}$.

18. Methane clathrates.

19. Dolerite is a coarse-crystalline form of basalt formed at depth within the Earth.

20. The Olenekian Age is subdivided, from oldest to youngest, into the Dienerian, Smithian, and Spathian sub-ages; the highest temperatures occur in the Smithian (Sun et al. 2012).

21. Just as the photosynthetic process produces oxygen in the production of hydrocarbons, $CO_2 + H_2O \rightarrow CH_2O + O_2$, the burning of those hydrocarbons consumes oxygen, $CH_2O + O_2 \rightarrow CO_2 + H_2O$.

22. $CH_4 + 2O_2 \rightarrow CO_2 + 2H_2O$.

23. The most general depletion cycle is $Cl + O_3 \rightarrow ClO + O_2$, which destroys one ozone molecule (O_3), and then $ClO + O \rightarrow Cl + O_2$, which simultaneously prevents atomic oxygen (O) from producing a new ozone molecule $(O + O_2 \rightarrow O_3)$ and produces a free chlorine atom to start the cycle all over again; see Beerling et al. (2007).

24. For contrast, humans see light in the 400 (violet) to 700 (red) nanometer spectrum.

25. In the timescale of Gradstein et al. (2012).

26. For example, in 1998 Bowring et al. narrowed the time interval down to 900,000 years, from 252.3 to 251.4 Ma; in 2004 Mundil et al. narrowed the time interval down to 400,000 years, from 252.8 to 252.4 Ma; and in 2011 Shen et al. narrowed the time interval down to 200,000 years, from 252.3 to 252.1 Ma.

27. Γαια.

28. However, it was at the University of Rochester in upstate New York that Jack Sepkoski collected most of the massive amount of data that he used in his analyses. For a brief glimpse into graduate student life and paleontological research conducted at the University of Rochester in those years, see McGhee (1996, 168–171).

29. For a detailed discussion of convergent ecological evolution, see McGhee (2011).

30. For an analysis of the differential fates of the rhynchonelliform brachiopods in the acidic oceans of the end-Permian, see Garbelli et al. (2017).

31. For an extensive discussion of the phenomenon of ecological convergence, see McGhee (2011).

32. The chitinozoans were the egglike reproductive phase of as yet unknown small invertebrate animals; see Grahn and Paris (2011).

33. Jack Sepkoski tragically died at age 50 in the year 1999.

34. For a discussion of the convergent evolution of swimming morphologies, see McGhee (2011).

35. The volaticotherians. For a discussion of the convergent evolution of flight morphologies, see McGhee (2011).

36. For a discussion of the convergent evolution of gliding morphologies in animals, see McGhee (2011).

37. Unfortunately, too late, as the predatory dinosaurs had already evolved at the beginning of the Carnian and diversified rapidly.

7. THE LEGACY OF THE LATE PALEOZOIC ICE AGE

1. The excess carbon dioxide is overwhelmingly the lighter isotope of carbon, $^{12}CO_2$, which would be released by the burning of coal and other hydrocarbons. As we are not living at the end of the Permian, when coal strata and petroleum-rich strata were burned by mantle-plume magma (chapter 6), the only other source of large excesses of light-isotope carbon dioxide in the atmosphere is anthropogenic.

2. For a detailed examination of the process of the invasion of land by vertebrates, see McGhee (2013).

3. See the more extensive discussion of this evolutionary possibility in McGhee (2013).

4. For a more extensive discussion of the effects of the End-Frasnian Bottleneck, see McGhee (2013).

5. For a more extensive discussion of the effects of the End-Famennian Bottleneck, see McGhee (2013).

6. See the detailed paleogeographic map sequences in Blakey (2008).

7. See the detailed paleogeographic map sequences in Blakey (2008).

8. For an extensive discussion of the phenomenon of behavioral convergent evolution, see McGhee (2011).

References

Abdala, F., and A. M. Ribeiro. 2010. Distribution and diversity patterns of Triassic cynodonts (Therapsida, Cynodontia) in Gondwana. *Palaeogeography, Palaeoclimatology, Palaeoecology* 286:202–217.

Ahlberg, P. E., J. A. Clack, E. Lukševičs, H. Blom, and I. Zupiņš. 2008. *Ventastega curonica* and the origin of tetrapod morphology. *Nature* 453:1199–1204.

Ahlbert, P. E. and Z. Johanson. 1998. Osteolepiforms and the ancestry of tetrapods. *Nature* 395:792–794.

Algeo, T. J., R. A. Berner, J. B. Maynard, and S. E. Scheckler. 1995. Late Devonian oceanic anoxic events and biotic crises: "Rooted" in the evolution of vascular land plants? *GSA Today* 5:63–66.

Algeo, T. J., S. E. Scheckler, and J. B. Maynard. 2001. Effects of the Middle to Late Devonian spread of vascular land plants on weathering regimes, marine biotas, and global climate. In *Plants invade the land*, ed. P. G. Gensel and D. Edwards, 213–236. New York: Columbia University Press.

Ali, J. R., G. M. Thompson, M.-F. Zhou, and X. Song. 2005. Emeishan large igneous province, SW China. *Lithos* 79:475–489.

Aljinovic, D., Y. Isozaki, and J. Sremac. 2008. The occurrence of giant bivalve Alatoconchidae from the *Yabeina* zone (Upper Guadalupian, Permian) in European Tethys. *Gondwana Research* 13:275–287.

Archer, D., H. Kheshgi, and E. Maier-Reimer. 1997. Multiple timescales for neutralization of fossil fuel CO_2. *Geophysical Research Letters* 24:405–408.

Averbuch, O., N. Tribovillard, X. Devleeschouwer, L. Riquier, B. Mistiaen, and B. van Vliet-Lanoe. 2005. Mountain building-enhanced continental weathering and organic carbon burial as major causes for climatic cooling at the Frasnian-Famennian boundary (c. 376 Ma)? *Terra Nova* 17:25–34.

Babarro, J. M. F., and A. De Zwaan. 2008. Anaerobic survival potential of four bivalves from different habitats: A comparative survey. *Comparative Biochemistry and Physiology, Part A* 151:108–113.

Baker, R. A., and W. A. DiMichele. 1997. Biomass allocation in Late Pennsylvanian coal-swamp plants. *Palaios* 12:127–132.

Balseiro, D. 2016. Compositional turnover and ecological changes related to the waxing and waning of glaciers during the late Paleozoic ice age in ice-proximal regions (Pennsylvanian, western Argentina). *Paleobiology* 42:335–357.

Bambach, R. K. 1983. Ecospace utilization and guilds in marine communities through the Phanerozoic. In *Biotic interactions in recent and fossil benthic communities*, ed. M. J. S. Tevesz and P. L. McCall, 719–746. New York: Plenum.

Bambach, R. K. 1985. Classes and adaptive variety: The ecology of diversification in marine faunas through the Phanerozoic. In *Phanerozoic diversity patterns*, ed. J. W. Valentine, 191–253. Princeton: Princeton University Press.

Bambach, R. K., A. H. Knoll, and S. C. Wang. 2004. Origination, extinction, and mass depletions of marine diversity. *Paleobiology* 30:522–542.

Barham, M., J. Murray, M. M. Joachimski, and D. M. Williams. 2012. The onset of the Permo-Carboniferous glaciation: Reconciling global stratigraphic evidence with biogenic apatite $\delta^{18}O$ records in the late Visean. *Journal of the Geological Society, London* 169:119–122.

Becker, R. T., M. Piecha, M. Gereke, and K. Spellbrink. 2016. The Frasnian/Famennian boundary in shelf basin facies north of Diemelsee-Adorf. *Münster Forschungen zur Geologie und Paläontologie* 108:220–231.

Beerling, D. J., and R. A. Berner. 2000. Impact of a Permo-Carboniferous high O_2 event on the terrestrial carbon cycle. *Proceedings of the National Academy of Sciences USA* 97:12428–12432.

Beerling, D. J., M. Harfoot, B. Lomax, and J. A. Pyle. 2007. The stability of the stratospheric ozone layer during the end-Permian eruption of the Siberian Traps. *Philosophical Transactions of the Royal Society (London)* A365:1843–1866.

Beerling, D. J., F. I. Woodward, M. R. Lomas, M. A. Wills, W. P. Quick, and P. J. Valdes. 1998. The influence of Carboniferous palaeo-atmospheres on plant function: An experimental and modelling assessment. *Philosophical Transactions of the Royal Society (London)* B353:131–140.

Bell, C. D., D. E. Soltis, and P. S. Soltis. 2010. The age and diversification of the angiosperms re-revisited. *American Journal of Botany* 97:1296–1303.

Benton, M. J. 1985. Patterns in the diversification of Mesozoic non-marine tetrapods and problems in historical diversity analysis. *Special Papers in Palaeontology* 33:185–202.

Benton, M. J. 1989. Mass extinctions among tetrapods and the quality of the fossil record. *Philosophical Transactions of the Royal Society (London)* B325:369–386.

Benton, M. J. 2003. *When life nearly died: The greatest mass extinction of all time.* London: Thames and Hudson.

Benton, M. J. 2005. *Vertebrate palaeontology.* 3d ed. Oxford: Blackwell.

Benton, M. J. 2015. *Vertebrate palaeontology.* 4th ed. Chichester: Wiley Blackwell.

Berner, R. A. 2002. Examination of hypotheses for the Permo-Triassic boundary extinction by carbon cycle modeling. *Proceedings of the National Academy of Sciences USA* 99:4172–4177.

Berner, R. A. 2006. GEOCARBSULF: A combined model for Phanerozoic atmosphere O_2 and CO_2. *Geochimica et Cosmochimica Acta* 70:5653–5664.

Berner, R. A. 2009. Phanerozoic atmospheric oxygen: New results using the GEOCARBSULF model. *American Journal of Science* 309:603–606.

Berner, R. A., D. J. Beerling, R. Dudley, J. M. Robinson, and R. A. Wildman. 2003. Phanerozoic atmospheric oxygen. *Annual Review of Earth and Planetary Sciences* 31:105–134.

Bishop, J. W., I. P. Montañez, E. L. Gulbranson, and P. L. Brenckle. 2009. The onset of mid-Carboniferous glacio-eustacy: Sedimentologic and diagenetic constraints, Arrow Canyon, Nevada. *Palaeogeography, Palaeoclimatology, Palaeoecology* 276:217–243.

Black, B. A., J.-F. Lamarque, C. A. Shields, L. T. Elkins-Tanton, and J. T. Kiehl. 2014. Acid rain and ozone depletion from pulsed Siberian Traps magmatism. 2014. *Geology* 42:67–70.

Blackburn, T. M., and K. J. Gaston. 1994. Animal body size distributions: Patterns, mechanisms and implications. *Trends in Ecology and Evolution* 9:471–474.

Blakey, R. C. 2008. Gondwana paleogeography from assembly to breakup—A 500 m.y. odyssey. In *Resolving the Late Paleozoic Ice Age in time and space* (Special Paper 441), ed. C. R. Fielding, T. D. Frank, and J. L. Isbell, 1–28. Boulder, Colo.: Geological Society of America.

Blieck, A. 2011. From adaptive radiations to biotic crises in Palaeozoic vertebrates: A geobiological approach. *Geologica Belgica* 14:203–227.

Bond, D. P. G., J. Hilton, P. B. Wignall, J. R. Ali, L. G. Stevens, Y. Sun, and X. Lai. 2010. The Middle Permian (Capitanian) mass extinction on land and in the oceans. *Earth-Science Reviews* 102:100–116.

Bond, D. P. G., and P. B. Wignall. 2014. Large igneous provinces and mass extinctions: An update. In *Volcanism, impacts, and mass extinctions: Causes and effects*, ed. G. Keller and A. C. Kerr, 29–55. Boulder, Colo.: Geological Society of America Special Paper 505.

Bond, D. P. G., P. B. Wignall, W. Wang, G. Izon, H.-S. Jiang, X.-L. Lai, Y.-D. Sun, R. J. Newton, L.-Y. Shao, S. Védrine, and H. Cope. 2010. The mid-Capitanian (Middle Permian) mass extinction and carbon isotope record of South China. *Palaeogeography, Palaeoclimatology, Palaeoecology* 292:282–294.

Bonelli, J. R., Jr., and M. E. Patzkowsky. 2011. Taxonomic and ecologic persistence across the onset of the late Paleozoic ice age: Evidence from the upper Mississippian (Chesterian Series), Illinois Basin, United States. *Palaios* 26:5–17.

Bottjer, D. J., and W. I. Ausich 1986. Phanerozoic development of tiering in soft substrata suspension-feeding communities. *Paleobiology* 12:400–420.

Bowring, S. A., D. H. Erwin, Y. G. Jin, M. W. Martin, D. Davidek, and W. Wang. 1998. U/Pb zircon geochronology and tempo of the end-Permian mass extinction. *Science* 280:1039–1045.

Boyce, C. K., and W. A. DiMichele. 2016. Arborescent lycopod productivity and lifespan: Constraining the possibilities. *Review of Palaeobotany and Palynology* 227:97–110.

Braddy, S. J., M. Poschmann, and O. E. Tetlie. 2007. Giant claw reveals the largest ever arthropod. *Biology Letters* 4:106–109.

Brezinski, D. K., C. B. Cecil, and V. W. Skema. 2010. Late Devonian glacigenic and associated facies from the central Appalachian Basin, eastern United States. *Geological Society of America Bulletin* 122:265–281.

Briggs, D. E. G. 1985. Gigantism in Palaeozoic arthropods. *Special Papers in Palaeontology* 33:1571.

Brunton, C. H. C., S. S. Lazarev, and R. E. Grant. 2000. Productida. In *Treatise on invertebrate paleontology, part H*, ed. R. L Kaesler, 350–643. Boulder, Colo: Geological Society of America and Lawrence, Kan.: University of Kansas.

Bryan, S. E., and R. E. Ernst. 2008. Revised definition of Large Igneous Provinces (LIPs). *Earth-Science Reviews* 86:175–202.

Bryan, S. E., I. Ukstins-Peate, D. W. Peate, S. Self, D. A. Jerram, M. R. Mawby, J. S. Marsh, and J. A. Miller. 2010. The largest volcanic eruptions on Earth. *Earth-Science Reviews* 102:207–229.

Caputo, M. V., J. H. G. Melo, M. Streel, and J. L. Isbell. 2008. Late Devonian and Early Carboniferous glacial records of South America. In *Resolving the Late Paleozoic Ice Age in time and space* (Special Paper 441), ed. C. R. Fielding, T. D. Frank, and J. L. Isbell, 161–173. Boulder, Colo.: Geological Society of America.

Carmichael, S. K., J. A. Waters, C. J. Batchelor, D. M. Coleman, T. J. Suttner, E. Kido, L. M. Moore, and L. Chadimová. 2016. Climate instability and tipping points in the Late Devonian: Detection of the Hangenberg Event in an open oceanic island arc in the Central Asian Orogenic Belt. *Gondwana Research* 32:213–231.

Carrano, M. T., C. M. Janis, and J. J. Sepkoski. 1999. Hadrosaurs as ungulate parallels: Lost lifestyles and deficient data. *Acta Palaeontological Polonica* 44:237–261.

Carroll, R. 2009. *The rise of amphibians: 365 million years of evolution.* Baltimore: Johns Hopkins University Press.

Cascales-Miñana, B., J. B. Diez, P. Gerrienne, and C. J. Cleal. 2015. A palaeobotanical perspective on the great end-Permian biotic crisis. *Historical Biology* 28:1066–1074.

Chumakov, N. M. 1994. Evidence of Late Permian glaciation in the Kolyma River Basin: A repercussion of the Gondwana glaciations in northeast Asia? *Stratigraphy and Geological Correlation* 2:426–444.

Clack, J. A. 2002. *Gaining ground: The origin and evolution of tetrapods.* Bloomington: Indiana University Press.

Clack, J. A. 2012. *Gaining ground: The origin and evolution of tetrapods.* 2d ed. Bloomington: Indiana University Press.

Clack, J. A., P. E. Ahlberg, H. Blom, and S. M. Finney. 2012. A new genus of Devonian tetrapod from north-east Greenland, with new information on the lower jaw of *Ichthyostega. Palaeontology* 55:73–86.

Clapham, M. E., and J. A. Kerr. 2012. Environmental and biotic controls on the evolutionary history of insect body size. *Proceedings of the National Academy of Sciences USA* 109:10927–10930.

Clapham, M. E., and J. L. Payne. 2011. Acidification, anoxia, and extinction: A multiple logistic regression analysis of extinction selectivity during the Middle and Late Permian. *Geology* 39:1059–1062.

Clapham, M. E., S. Z. Shen, and D. J. Bottjer. 2009. The double mass extinction revisited: Reassessing the severity, selectivity, and causes of the end-Guadalupian biotic crisis (Late Permian). *Paleobiology* 35:32–50.

Clarkson, E. N. K. 1998. *Invertebrate palaeontology and evolution.* Oxford: Blackwell Science.

Clarkson, M. O., S. A. Kasemann, R. A. Wood, T. M. Lenton, S. J. Daines, S. Richoz, F. Ohnemueller, A. Meixner, S. W. Poulton, and E. T. Tipper. 2015. Ocean acidification and the Permo-Triassic mass extinction. *Science* 348:229–232.

Cleal, C. J. 2010. Tiering on land—trees and forests (Late Palaeozoic). In *Encyclopedia of life sciences* (ELS). Chichester: Wiley. doi: 10.1002/9780470015902.a0001642.pub2.

Cleal, C. J., and B. Cascales-Miñana. 2014. Composition and dynamics of the great Phanerozoic Evolutionary Floras. *Lethaia* 47:469–484.

Cleal, C. J., R. M. James, and E. L. Zodrow. 1999. Variation in stomatal density in the Late Carboniferous gymnosperm frond *Neuropteris ovata. Palaios* 14:180–185.

Cleal, C. J., and B. A. Thomas. 2005. Palaeozoic tropical rainforests and their effect on global climates: Is the past the key to the present? *Geobiology* 3:13–31.

Coates, M. I., M. Ruta, and M. Friedman. 2008. Ever since Owen: Changing perspectives on the early evolution of tetrapods. *Annual Review of Ecology, Evolution, and Systematics* 39:571–592.

Coffin, M. F., and O. Eldholm. 1991. *Large igneous provinces: JOI/USSAC workshop report* (Technical Report 114). Austin: University of Texas at Austin Institute for Geophysics.

Coffin, M. F., and O. Eldholm. 1994. Large igneous provinces: Crustal structure, dimensions, and external consequences. *Review of Geophysics* 32:1–36.

Copper, P. 1977. Paleolatitudes in the Devonian of Brazil and the Frasnian-Famennian mass extinction. *Palaeogeography, Palaeoclimatology, Palaeoecology* 21:165–207.

Copper, P. 1986. Frasnian-Famennian mass extinctions and cold water oceans. *Geology* 14:835–838.

Copper, P. 1998. Evaluating the Frasnian-Famennian mass extinction: Comparing brachiopod faunas. *Acta Palaeontologia Polonica* 43:137–154.

Copper, P. 2002. Silurian and Devonian reefs: 80 million years of global greenhouse between two ice ages. *SEPM Special Publications* 72:181–238.

Courtillot, V. E. 1999. *Evolutionary catastrophes: The science of mass extinction.* Cambridge: Cambridge University Press.

Courtillot, V. E., V. A. Kravchinsky, X. Quidelleur, P. R. Renne, and D. P. Gladkochub. 2010. Preliminary dating of the Viluy traps (Eastern Siberia): Eruption at the time of Late Devonian extinction events? *Earth and Planetary Science Letters* 300:239–245.

Courtillot, V. E., and P. R. Renne. 2003. On the ages of flood basalt events. *Comptes Rendus Geoscience* 335:113–140.

Croswell, K. 2003. *Magnificent Mars.* New York: Free Press.

Daeschler, E. B., J. A. Clack, and N. H. Shubin. 2009. Late Devonian tetrapod remains from Red Hill, Pennsylvania, USA: How much diversity? *Acta Zoologica* 90(Supplement 1):306–317.

D'Arrigo, R., R. Seager, J. E. Smerdon, A. N. LeGrande, and E. R. Cook. 2011. The anomalous winter of 1783–1784: Was the Laki eruption or an analog of the 2009–2010 winter to blame? *Geophysical Research Letters* 38. doi:10.1029/2011GL046696.

DeConto, R. M., and D. Pollard. 2003. Rapid Cenozoic glaciation of Antarctica induced by declining atmospheric CO_2. *Nature* 421:245–249.

Demko, T. M., R. F. Dubiel, and J. T. Parrish. 1998. Plant taphonomy in incised valleys: Implications for interpreting paleoclimate from fossil plants. *Geology* 26:1119–1122.

Dennison, R. 1978. Placodermi. In *Handbook of paleoichthyology,* ed. H. P. Schultze, 2:1–128. Stuttgart: Gustav Fischer Verlag.

Dennison, R. 1979. Acanthodii. In *Handbook of paleoichthyology,* ed. H. P. Schultze, 5:1–62. Stuttgart: Gustav Fischer Verlag.

Dickens, G. R. 2001. Modeling the global carbon cycle with a gas hydrate capacitor across the latest Paleocene thermal maximum. In *Natural gas hydrates: Occurrence, distribution and detection,* ed. C. K. Paull and W. P. Dillon. American Geophysical Union Geophysical Monographs 124:19–38. New York: Wiley.

Dickens, G. R., J. R. O'Neil, D. K. Rea, and R. M. Owen. 1995. Dissociation of oceanic methane hydrate as a cause of the carbon isotope excursion at the end of the Paleocene. *Paleoceanography* 10:965–971.

DiMichele, W. A., and R. W. Hook. 1992. Paleozoic terrestrial ecosystems. In *Terrestrial ecosystems through time,* ed. A. K. Behrensmeyer, J. D. Damuth, W. A. DiMichele, R. Potts, H.-D. Sues, and S. L. Wing, pp. 205–325. Chicago: University of Chicago Press.

DiMichele, W. A., I. P. Montañez, C. J. Poulsen, and N. J. Tabor. 2009. Climate and vegetational regime shifts in the late Paleozoic ice age earth. *Geobiology* 7:200–226.

DiMichele, W. A., H. W. Pfefferkorn, and R. A. Gastaldo. 2001. Response of Late Carboniferous and Early Permian plant communities to climatic change. *Annual Review of Earth and Planetary Sciences* 29:461–487.

DiMichele, W. A., and T. L. Phillips. 1996. Climate change, plant extinctions and vegetational recovery during the middle-late Pennsylvanian transition: The case of tropical peat-forming environments in North America. In *Biotic recoveries from mass extinctions* (Special Publication 102), ed. M. B. Hart, 210–221. London: Geological Society of London.

DiMichele, W. A., T. L. Phillips, and R. A. Peppers. 1985. The influence of climate and depositional environment on the distribution and evolution of Pennsylvanian coal swamp plants. In *Geological factors in the evolution of plants*, ed. B. Tiffney, 223–256. New Haven: Yale University Press.

DiMichele, W. A., N. J. Tabor, D. S. Chaney, and W. J. Nelson. 2006. From wetlands to wet spots: Environmental tracking and the fate of Carboniferous elements in Early Permian tropical floras. In *Wetlands through time* (Special Paper 399), eds. S. F. Greb and W. A. DiMichele, 223–248. Boulder, Colo: Geological Society of America.

Dobretsov, N. L., and V. A. Vernikovsky. 2001. Mantle plumes and their geologic manifestations. *International Geological Review* 43:771–787.

Donoghue, M. J. 2005. Key innovations, convergence, and success: Macroevolutionary lessons from plant phylogeny. *Paleobiology* 31(Supplement):77–93.

Dorrington, G. E. 2015. Heavily loaded flight and limits to the maximum size of dragonflies (Anisoptera) and griffenflies (Meganisoptera). *Lethaia* 49:261–274.

Droser, M. L., D. J. Bottjer, P. M. Sheehan, and G. R. McGhee. 2000. Decoupling of taxonomic and ecologic severity of Phanerozoic marine mass extinctions. *Geology* 28:675–678.

Dudley, R. 1998. Atmospheric oxygen, giant Paleozoic insects and the evolution of aerial locomotor performance. *Journal of Experimental Biology* 201:1043–1050.

Ehrmann, W. U., and A. Mackensen. 1992. Sedimentological evidence for the formation of an East Antarctic ice sheet in Eocene/Oligocene time. *Palaeogeography, Palaeoclimatology, Palaeoecology* 93:85–112.

Elrick, M., and B. Witzke. 2016. Orbital-scale glacio-eustacy in the Middle Devonian detected using oxygen isotopes of condont apatite: Implications for long-term greenhouse-icehouse climatic transitions. *Palaeogeography, Palaeoclimatology, Palaeoecology* 445:50–59.

Enos, P. 1995. The Permian of China. In *The Permian of northern Pangea*: Vol. 2. *Sedimentary basins and economic resources*, ed. P. A. Scholle, T. M. Peryt, and D. S. Ulmer-Scholle, 225–256. Berlin: Springer Verlag.

Epshteyn, O. G. 1981a. Middle Carboniferous ice-marine deposits of northeastern U.S.S.R. In *Earth's pre-Pleistocene glacial record*, ed. M. J. Hambrey and W. B. Harland, 268–269. Cambridge: Cambridge University Press.

Epshteyn, O. G. 1981b. Late Permian ice-marine deposits of northeastern U.S.S.R. In *Earth's pre-Pleistocene glacial record*, ed. M. J. Hambrey and W. B. Harland, 270–273. Cambridge: Cambridge University Press.

Erwin, D. H. 1993. *The great Paleozoic crisis: Life and death in the Permian*. New York: Columbia University Press.

Erwin, D. H. 2015. *Extinction: How life on earth nearly ended 250 million years ago*. Princeton: Princeton University Press.

Erwin, D. H., M. Laflamme, S. M. Tweedt, E. A. Sperling, D. Pisani, and K. J. Peterson. 2011. The Cambrian conundrum: Early divergence and later ecological success in the early history of animals. *Science* 334:1091–1097.

Falcon-Lang, H. J. 2004. Pennsylvanian tropical rain forests responded to glacial-interglacial rhythms. *Geology* 32:689–692.

Falcon-Lang, H. J., and W. A. DiMichele. 2010. What happened to the coal forests during Pennsylvanian glacial periods? *Palaios* 25:611–617.

Falkowski, P., M. Katz, A. Milligan, K. Fennel, B. Cramer, M. P. Aubry, R. A. Berner, and W. M. Zapol. 2005. The rise of atmospheric oxygen levels over the past 205 million years and the evolution of large placental mammals. *Science* 309:2202–2204.

Fan, J. S., J. K. Rigby, and J. W. Qi. 1990. The Permian reefs of south China and comparisons with the Permian reef complex of the Guadalupe mountains, west Texas and New Mexico. *Brigham Young University Geology Studies* 36:15–55.

Farinella, P., L. Foschini, C. Froeschlé, R. Gonczi, T. J. Jopek, G. Longo, and P. Michel. 2001. Probable asteroidal origin of the Tunguska Cosmic Body. *Astronomy and Astrophysics* 377:1081–1097.

Fielding, C. R., T. D. Frank, L. P. Birgenheier, M. C. Rygel, A. T. Jones, and J. Roberts. 2008. Stratigraphic record and facies associations of the Late Paleozoic Ice Age in eastern Australia (New South Wales and Queensland). In *Resolving the Late Paleozoic Ice Age in time and space* (Special Paper 441), ed. C. R. Fielding, T. D. Frank, and J. L. Isbell, 41–57. Boulder, Colo.: Geological Society of America.

Fielding, C. R., T. D. Frank, and J. L. Isbell. 2008. The Late Paleozoic Ice Age—a review of current understanding and synthesis of global climate patterns. In *Resolving the Late Paleozoic Ice Age in time and space* (Special Paper 441), ed. C. R. Fielding, T. D. Frank, and J. L. Isbell, 343–354. Boulder, Colo.: Geological Society of America.

Filer, J. K. 2002. Late Frasnian sedimentation cycles in the Appalachian Basin: Possible evidence for high frequency eustatic sea-level changes. *Sedimentary Geology* 154:31–52.

Finnegan, S., K. Bergmann, J. Eiler, D. Jones, D. Fike, I. Eisenman, N. Hughes, A. Tripati, and W. Fischer. 2011. The magnitude and duration of Late Ordovician–Early Silurian glaciation. *Science* 331:903–906.

Flanagan, R. 1995. Mass murderers of the Devonian. *Earth* 4(August): 22.

Flower, B. P., and J. P. Kennett. 1995. Middle Miocene deepwater paleoceanography in the southwest Pacific: Relations with the East Antarctica ice sheet development. *Paleoceanography* 10:1095–1112.

Flügel, E., and W. Kiessling. 2002. Patterns of Phanerozoic reef crises. In *Phanerozoic reef patterns*, ed. W. Kiessling, E. Flügel, and J. Golonka. *SEPM Special Publications* 72:691–733.

Flügel, E., and G. D. Stanley. 1984. Re-organization, development and evolution of post-Permian reefs and reef-organisms. *Paleontolographica Americana* 54:177–186.

Fois, E., and M. Gaetani. 1984. The recovery of reef-building communities and the role of cnidarians in carbonate sequences of the Middle Triassic (Anisian) in the Italian Dolomites. *Paleontolographica Americana* 54:191–200.

Frakes, L. A., J. E. Francis, and J. I. Syktus. 1992. *Climate modes of the Phanerozoic.* Cambridge: Cambridge University Press.

Frank, T. D., L. P. Birgenheier, I. P. Montañez, C. R. Fielding, and M. C. Rygel. 2008. Late Paleozoic climate dynamics revealed by comparison of ice-proximal stratigraphic and ice-distal isotopic records. In *Resolving the Late Paleozoic Ice Age in time and space* (Special Paper 441), ed. C. R. Fielding, T. D. Frank, and J. L. Isbell, 331–342. Boulder, Colo.: Geological Society of America.

Ganino, C., and N. T. Arndt. 2009. Climate changes caused by degassing of sediments during the emplacement of large igneous provinces. *Geology* 37:323–326.

Garbelli, C., L. Angiolini, and S.-Z. Shen. 2017. Biomineralization and global change: A new perspective for understanding the end-Permian extinction. *Geology* 45:19–22.

Gastaldo, R. A., E. Purkyňová, and Z. Šimůnek. 2009. Megafloral perturbation across the Enna marine zone in the Upper Silesian basin attests to Late Mississippian (Serpukhovian) deglaciation and climate change. *Palaios* 24:351–366.

Gastaldo, R. A., E. Purkyňová, Z. Šimůnek, and M. D. Schmitz. 2009. Ecological persistence in the Late Mississippian (Serpukhovian, Namurian A) megafloral record of the Upper Silesian basin, Czech Republic. *Palaios* 24:336–350.

Gastaldo, R. A., I. M. Stevanović-Walls, W. N. Ware, and S. F. Greb. 2004. Community heterogeneity of Early Pennsylvanian peat mires. *Geology* 32:693–696.

Georgiev, S., H. J. Stein, J. L. Hannah, B. Bingen, H. M. Weiss, and S. Piasecki. 2011. Hot acidic Late Permian seas stifle life in record time. *Earth and Planetary Science Letters* 310: 389–400.

Giles, P. S. 2009. Orbital forcing and Mississippian sea level change: Time series analysis of marine flooding events in the Visean Windsor Group of eastern Canada and implications for Gondwana glaciation. *Bulletin of Canadian Petroleum Geology* 57:449–471.

González-Bonorino, G., and N. Eyles. 1995. Inverse relation between ice extent and the late Paleozoic glacial record of Gondwana. *Geology* 23:1015–1018.

Gradstein, F. M., J. G. Ogg, M. Schmitz, and G. Ogg. 2012. *The geologic time scale 2012*. Amsterdam: Elsevier B. V.

Gradstein, F. M., J. G. Ogg, and A. G. Smith. 2004. *A geologic time scale 2004*. Cambridge: Cambridge University Press.

Graham, J. B., N. M. Aguilar, R. Dudley, and C. Gans. 1997. The Late Paleozoic atmosphere and the ecological and evolutionary physiology of tetrapods. In *Amniote origins: Completing the transition to land*, ed. S. S. Sumida and K. L. M. Martin, 141–167. San Diego: Academic Press.

Graham, J. B., R. Dudley, N. M. Aguilar, and C. Gans. 1995. Implications of the late Palaeozoic oxygen pulse for physiology and evolution. *Nature* 375:117–120.

Grahn, Y., and F. Paris. 2011. Emergence, biodiversification and extinction of the chitinozoan group. *Geological Magazine* 148:226–236.

Grasby, S. E., H. Sanei, and B. Beauchamp. 2011. Catastrophic dispersion of coal fly ash into oceans during the latest Permian extinction. *Nature Geoscience* 4:104–107.

Grasby, S. E., W. Shen, R. Yin, J. D. Gleason, J. B. Blum, R. F. Lepak, J. P. Hurley, and B. Beauchamp. 2017. Isotope signatures of mercury contamination in latest Permian oceans. *Geology* 45:55–58.

Greb, S. F., W. M. Andrews, C. F. Eble, W. DiMichele, C. B. Cecil, and J. C. Hower. 2003. Desmoinesian coal beds of the Eastern Interior and surrounding basins: The largest tropical peat mires in Earth history. In *Extreme depositional environments: Mega end members in geologic time* (Special Paper 370), ed. M. A. Chan and A. W. Archer, 127–150. Boulder, Colo.: Geological Society of America .

Grimaldi, D., and M. S. Engel. 2005. *Evolution of the insects*. Cambridge: Cambridge University Press.

Gulbranson, E. L., I. P. Montañez, M. D. Schmitz, C. O. Limarino, J. L. Isbel, S.A. Marenssi, and J. L. Crowley. 2010. High-precision U-Pb calibration of Carboniferous glaciation and climate history, Paganzo Group, NW Argentina. *Geological Society of America Bulletin* 122:1480–1498.

Hallam, A., and P. G. Wignall. 1997. *Mass extinctions and their aftermath*. Oxford: Oxford University Press.

Harris, M. T., and P. M. Sheehan. 1998. Early Silurian stratigraphic sequences of the eastern Great Basin (Utah and Nevada). In *Silurian cycles: Linking dynamic stratigraphy with atmospheric and oceanic changes* (New York State Museum Bulletin 491), ed. E. Landing and M. E. Johnson, 51–61. Albany: New York State Museum.

Harrison, J. F., A. Kaiser, and J. M. VandenBrooks. 2010. Atmospheric oxygen level and the evolution of insect body size. *Proceedings of the Royal Society* B277:1937–1946.

Hayward, B. W. 2002. Late Pliocene to Middle Pleistocene extinctions in deep-sea benthic foraminifera (*Stilostomella* extinction) in the southwest Pacific. *Journal of Foraminiferal Research* 32:274–307.

Heckel, P. H. 1991. Lost Branch Formation and revision of upper Desmoinesian stratigraphy along midcontinent Pennsylvanian outcrop belt. *Kansas Geological Survey Geology Series* 4:1–68.

Heckel, P. H. 2008. Pennsylvanian cyclothems in Midcontinent North America as far-field effects of waxing and waning of Gondwana ice sheets. In *Resolving the Late Paleozoic Ice Age in time and space* (Special Paper 441), ed. C. R. Fielding, T. D. Frank, and J. L. Isbell, 275–289. Boulder, Colo.: Geological Society of America.

Hillenius, W. J. 1992. The evolution of nasal turbinates and mammalian endothermy. *Paleobiology* 18:17–29.

Hillenius, W. J. 1994. Turbinates in therapsids: Evidence for late Permian origins of mammalian endothermy. *Evolution* 48:207–229.

Horton, D. E., C. J. Poulsen, I. P. Montañez, and W. A DiMichele. 2012. Eccentricity-paced late Paleozoic climate change. *Palaeogeography, Palaeoclimatology, Palaeoecology* 331–332:150–161.

Hoskins, D. M., J. D. Inners, and J. A. Harper. 1983. *Fossil collecting in Pennsylvania* (General Geology Report G40). 3d ed. Harrisburg: Pennsylvania Geological Survey.

House, M. R. 1988. Extinction and survival in the Cephalopoda. *Systematics Association Special Volumes* 34:139–154.

Irmis, R. B., and J. H. Whiteside 2012. Delayed recovery of non-marine tetrapods after the end-Permian mass extinction tracks global carbon cycle. *Proceedings of the Royal Society* B279:1310–1318.

Irving, E. 2008. Why Earth became so hot 50 million years ago and why it then cooled. *Proceedings of the National Academy of Sciences USA* 105:16061–16062.

Isaacson, P. E., E. Diaz-Martinez, G. W. Grader, J. Kalvoda, O. Babek, and F. X. Devuyst. 2008. Late Devonian–earliest Mississippian glaciation in Gondwanaland and its biogeographic consequences. *Palaeogeography, Palaeoclimatology, Palaeoecology* 268:126–142.

Isbell, J. L., A. S. Biakov, I. L. Vedernikov, V. I. Davydov, E. L. Gulbranson, and N. D. Fedorchuk. 2016. Permian diamictites in northeastern Asia: Their significance concerning the bipolarity of the late Paleozoic ice age. *Earth-Science Reviews* 154:279–300.

Isbell, J. L., M. F. Miller, K. L. Wolfe, and P. A. Lenaker. 2003. Timing of late Paleozoic glaciation in Gondwana: Was glaciation responsible for the development of northern hemisphere cyclothems? In *Extreme depositional environments: Mega end members in geologic time* (Special Paper 370), ed. M. A. Chan and A. W. Archer, 5–24. Boulder, Colo.: Geological Society of America.

Isozaki, Y., and D. Aljinovic. 2009. End-Guadalupian extinction of the Permian gigantic bivalve Alatoconchidae: End of gigantism in tropical seas by cooling. *Palaeogeography, Palaeoclimatology, Palaeoecology* 284:11–21.

Ivany, L. D., S. Van Simaeys, E. W. Domack, and S. D. Samson. 2006. Evidence for an earliest Oligocene ice sheet on the Antarctic Peninsula. *Geology* 34:377–380.

Jablonski, D. 1986. Background and mass extinctions: The alternation of macroevolutionary regimes. *Science* 231:129–133.

Jackson, E. L. 1982. The Laki eruption of 1783: Impacts on population and settlement in Iceland. *Geography* 67:42–50.

Joachimski, M. M., and W. Buggisch. 2002. Conodont apatite $\delta^{18}O$ signatures indicate climate cooling as a trigger of the Late Devonian mass extinction. *Geology* 30:711–714.

Kaiser, S. I., M. Aretz, and R. T. Becker. 2016. The global Hangenberg Crisis (Devonian-Carboniferous transition): Review of a first-order mass extinction. In *Devonian climate, sea level and evolutionary events* (Special Publications 423), ed. R. T. Becker, P. Königshof, and C. E. Brett, 387–437. London: Geological Society.

Kaiser, S. I., T. Steuber, R. T. Becker, and M. M. Joachimski. 2006. Geochemical evidence for major environmental change at the Devonian-Carboniferous boundary in the Carnic Alps and the Rhenish Massif. *Palaeogeography, Palaeoclimatology, Palaeoecology* 240:146–160.

Kalvoda, J. 1990. Late Devonian-Early Carboniferous paleobiogeography of benthic foraminifera and climatic oscillations. In *Extinction events in Earth history*, ed. E. G. Kauffman and O. H. Walliser, 183–187. Berlin: Springer Verlag.

Kammer, T. W., and D. L. Matchen. 2008. Evidence for eustacy at the Kinderhookian–Osagean (Mississippian) boundary in the United States: Response to late Tournaisian glaciation? In *Resolving the Late Paleozoic Ice Age in time and space* (Special Paper 441), ed. C. R. Fielding, T. D. Frank, and J. L. Isbell, 261–274. Boulder, Colo.: Geological Society of America.

Katz, M. E., K. G. Miller, J. D. Wright, B. S. Wade, J. V. Browning, B. S. Cramer, and Y. Rosenthal. 2008. Stepwise transition from the Eocene greenhouse to the Oligocene icehouse. *Nature Geoscience* 1:329–334.

Kemp, D. B., A. L. Coe, A. S. Cohen, and L. Schwark. 2005. Astronomical pacing of methane release in the Early Jurassic period. *Nature* 437:396–399.

Kemp, T. S. 2006. The origin and early radiation of the therapsid mammal-like reptiles: A palaeobiological hypothesis. *Journal of Evolutionary Biology* 19:1231–1247.

Kenrick, P., and P. R. Crane. 1997a. *The Origin and Early Diversification of Land Plants: A Cladistic Study*. Washington, D.C.: Smithsonian Institution Press.

Kenrick, P., and P. R. Crane. 1997b. The origin and early evolution of plants on land. *Nature* 389:33–39.

Kent, D. V., and G. Muttoni. 2008. Equatorial convergence of India and early Cenozoic climate trends. *Proceedings of the National Academy of Sciences USA* 108:21146–21151.

Kiessling, W., and C. Simpson. 2011. On the potential for ocean acidification to be a general cause of ancient reef crises. *Global Change Biology* 17:56–67.

Knoll, A. H., R. K. Bambach, D. E. Canfield, and J. P. Grotzinger. 1996. Comparative Earth history and the Late Permian mass extinction. *Science* 273:452–457.

Knoll, A. H., R. K. Bambach, J. L. Payne, S. Pruss, and W. W. Fischer. 2007. Paleophysiology and end-Permian mass extinction. *Earth and Planetary Science Letters* 256:295–313.

Kopp, R. E., J. L. Kirschvink, I. A. Hilburn, and C. Z. Nash. 2005. The Paleoproterozoic snowball Earth: A climate disaster triggered by the evolution of oxygenic photosynthesis. *Proceedings of the National Academy of Sciences USA* 102:11131–11136.

Kraus, O., and C. Brauckmann. 2003. Fossil giants and surviving dwarfs: Arthropleurida and Pselaphognatha (Atelocerata, Diplopoda); characters, phylogenetic relationships and construction. *Verhandlungen des naturwissenschaftlichen Vereins in Hamburg* 40:5–50.

Krings, M., H. Kerp, T. N. Taylor, and E. L. Taylor. 2003. How Paleozoic vines and lianas got off the ground: On scrambling and climbing Carboniferous–Early Permian pteridosperms. *Botanical Review* 69:204–224.

Kukalová-Peck, J. 1985. Ephemeroid wing venation based upon new gigantic Carboniferous mayflies and basic morphology, phylogeny, and metamorphosis of pterygote insects (Insecta, Ephemerida). *Canadian Journal of Zoology* 63:933–955.

Kukalová-Peck, J. 1987. New Carboniferous Diplura, Monura, and Thysanura, the hexapod ground plan, and the role of thoracic side lobes in the origin of wings (Insecta). *Canadian Journal of Zoology* 65:2327–2345.

Lanciki, A., J. Cole-Dai, M. H. Thiemens, and J. Savarino. 2012. Sulfur isotope evidence of little or no stratospheric impact by the 1783 Laki volcanic eruption. *Geophysical Research Letters* 39, L01806, doi:10.1029/2011GL050075.

Lane, N. 2002. *Oxygen: The molecule that made the world*. Oxford: Oxford University Press.

Lecointre, G., and H. Le Guyader. 2006. *The tree of life: A phylogenetic classification*. Cambridge, Mass.: Belknap Press of Harvard University Press.

Lenton, T. M., M. Crouch, M. Johnson, N. Pires, and L. Dolan. 2012. First plants cooled the Ordovician. *Nature Geoscience* 5:86–89.

Leonova, T. 2009. Ammonoid evolution in marine ecosystems prior to the Permian-Triassic crisis. *Paleontological Journal* 43:858–865.

Levinton, J. S. 1982. *Marine ecology*. Englewood Cliffs, NJ: Prentice-Hall.

Lewis, A. R., D. R. Marchant, A. C. Ashworth, L. Hedenäs, S. R Hemming, J. V. Johnson, M. J. Leng, M. L. Machlus, A. E. Newton, J. I. Raine, J. K. Willenbring, M. Williams, and P. A. Wolfe. 2008. Mid-Miocene cooling and the extinction of tundra in continental Antarctica. *Proceedings of the National Academy of Sciences USA* 105:10676–10680.

Little, E. L. 1980. *The Audubon Society field guide to North American trees*. New York: Knopf.

Long, J. A. 1993. Early-Middle Palaeozoic vertebrate extinction events. In *Palaeozoic vertebrate biostratigraphy and biogeography*, ed. J. A. Long, 54–63. London: Belhaven Press.

Lucas, S. G. 2004. A global hiatus in the Middle Permian tetrapod fossil record. *Stratigraphy* 1:47–64.

Lucas, S. G., and A. B. Heckert. 2001. Olson's gap: A global hiatus in the record of Middle Permian tetrapods. *Journal of Vertebrate Paleontology* 21:75A.

Lutz, R. A., and D. C. Rhoads. 1977. Anaerobosis and theory of growth line formation. *Science* 198:1222–1227.

Marynowski, L., and P. Filipiak. 2007. Water column euxinia and wildfire evidence during the deposition of the Upper Famennian Hangenberg event horizon from the Holy Cross Mountains (central Poland). *Geological Magazine* 144:569–595.

Marynowski, L., P. Filipiak, and M. Zatoń. 2010. Geochemical and palynological study of the Upper Famennian Dasberg event horizon from the Holy Cross Mountains (central Poland). *Geological Magazine* 147:527–550.

McClung, W. S., C. A. Cuffey, K. A. Ericksson, and D. O. Terry Jr. 2016. An incised valley fill and lowland wedges in the Upper Devonian Foreknobs Formation, central Appalachian Basin: Implications for Famennian glacioeustacy. *Palaeogeography, Palaeoclimatology, Palaeoecology* 446: 125–143.

McClung, W. S., K. A. Eriksson, D. O. Terry Jr., and C. A. Cuffey. 2013. Sequence stratigraphic hierarchy of the Upper Devonian Foreknobs Formation, central Appalachian Basin, USA: Evidence for transitional greenhouse to icehouse conditions. *Palaeogeography, Palaeoclimatology, Palaeoecology* 387:104–125.

McGhee, G. R., Jr. 1988. The Late Devonian extinction event: Evidence for abrupt ecosystem collapse. *Paleobiology* 14:250–257.

McGhee, G. R., Jr. 1996. *The Late Devonian mass extinction.* New York: Columbia University Press.

McGhee, G. R., Jr. 2001. The 'multiple impacts hypothesis' for mass extinction: A comparison of the Late Devonian and the late Eocene. *Palaeogeography, Palaeoclimatology, Palaeoecology* 176:47–58.

McGhee, G. R., Jr. 2011. *Convergent evolution: Limited forms most beautiful.* Cambridge, Mass.: MIT Press.

McGhee, G. R., Jr. 2013. *When the invasion of land failed: The legacy of the Devonian extinctions.* New York: Columbia University Press.

McGhee, G. R., Jr. 2014a. The Late Devonian (Frasnian/Famennian) mass extinction: A proposed test of the glaciation hypothesis. *Geological Quarterly* 58:263–268.

McGhee, G. R., Jr. 2014b. The search for sedimentary evidence of glaciation during the Frasnian/Famennian (Late Devonian) biodiversity crisis. *SEPM The Sedimentary Record* 12(June):4–8.

McGhee, G. R., Jr., M. E. Clapham, P. M. Sheehan, D. J. Bottjer, and M. L. Droser. 2013. A new ecological-severity ranking of major Phanerozoic biodiversity crises. *Palaeogeography, Palaeoclimatology, Palaeoecology* 370:260–270.

McGhee, G. R., Jr., P. M. Sheehan, D. J. Bottjer, and M. L. Droser. 2004. Ecological ranking of Phanerozoic biodiversity crises: Ecological and taxonomic severities are decoupled. *Palaeogeography, Palaeoclimatology, Palaeoecology* 211:289–297.

McGhee, G. R., Jr., P. M. Sheehan, D. J. Bottjer, and M. L. Droser. 2012. Ecological ranking of Phanerozoic biodiversity crises: The Serpukhovian (Early Carboniferous) crisis had a greater ecological impact than the end-Ordovician. *Geology* 40:147–150.

McKinney, M. L. 1990. Trends in body-size evolution. In *Evolutionary trends*, ed. K. J. McNamara, 75–118. Tucson: University of Arizona Press.

Meave, J., and M. Kellman. 1994. Maintenance of rain forest diversity in riparian forests of tropical savannas: Implications for species conservation during Pleistocene drought. *Journal of Biogeography* 21:121–135.

Melezhik, V. A. 2006. Multiple causes of Earth's earliest global glaciation. *Terra Nova* 18:130–137.

Metcalfe, I., J. L. Crowley, R. S. Nicoll, and M. Schmitz. 2015. High-precision U-Pb CA-TIMS calibration of Middle Permian to Lower Triassic sequences, mass extinction and extreme climate-change in eastern Australian Gondwana. *Gondwana Research* 28:61–81.

Mii, H.-S., E. L. Grossman, T. E. Yancey, B. Chuvashov, and A. Egorov. 2001. Isotopic records of brachiopod shells from the Russian Platform—evidence for the onset of mid-Carboniferous glaciation. *Chemical Geology* 175:133–147.

Milne, L., and M. Milne. 1980. *The Audubon Society field guide to North American insects and spiders.* New York: Knopf.

Misof, B., L. Shanlin, K. Meusemann, R. S. Peters, A. Donath, C. Mayer, P. B. Frandsen, J. Ware, T. Flouri, R. G. Beutel, and 91 other authors. 2014. Phylogenomics resolves the timing and pattern of insect evolution. *Science* 346:763–767.

Montañez, I. P., and P. E. Isaacson. 2013. A 'sedimentary record' of opportunities. *SEPM The Sedimentary Record* 11(January):4–9.

Montañez, I. P., and C. J. Poulsen. 2013. The Late Paleozoic Ice Age: An evolving paradigm. *Annual Review of Earth and Planetary Sciences* 41:629–656.

Moore, R. C., C. G. Lalicker, and A. G. Fischer. 1952. *Invertebrate fossils.* New York: McGraw-Hill.

Morzadec, P. 1992. Evolution des Asteropyginae (Trilobita) et variations eustatiques au Dévonien. *Lethaia* 25:85–96.

Moy-Thomas, J. A., and R. S. Miles. 1971. *Palaeozoic fishes.* 2d ed. London: Chapman and Hall.

Moysiuk, J., M. R. Smith, and J.-B. Caron. 2017. Hyoliths are Palaeozoic lophophorates. *Nature.* doi:10.1038/nature20804.

Mundil, R., K. R. Ludwig, I. Metcalfe, and P. R. Renne. 2004. Age and timing of the Permian mass extinction: U/Pb dating of closed-system zircons. *Science* 305:1760–1763.

Neale, G. 2010. How an Icelandic volcano helped spark the French Revolution. *Guardian,* 15 April 2010.

Nel, A., G. Fleck, R. Garrouste, G. Gand, J. Lapeyrie, S. M. Bybee, and J. Prokop. 2009. Revision of Permo-Carboniferous griffenflies (Insecta: Odonatoptera: Meganisoptera) based upon new species and redescription of selected poorly known taxa from Eurasia. *Palaeontographica Abteilung A* 289:89–121.

Nelsen, M. P., W. A. DiMichele, S. E. Peters, and C. K. Boyce. 2016. Delayed fungal evolution did not cause the Paleozoic peak in coal production. *Proceedings of the National Academy of Sciences USA* 113:2442–2447.

Newell, N. D. 1949. Phyletic size increase, an important trend illustrated by fossil invertebrates. *Evolution* 3:103–124.

Niklas, K. J. 1997. *The evolutionary biology of plants.* Chicago: University of Chicago Press.

Nutman, A. P., V. C. Bennett, C. R. L. Friend, M. J. Van Kranendonk, and A. R. Chivas. 2016. Rapid emergence of life shown by discovery of 3,700-million-year-old microbial structures. *Nature* 537:535–538.

Oliver, W. A., Jr., and A. E. H. Pedder. 1994. Crises in the Devonian history of the rugose corals. *Paleobiology* 20:178–190.

Palmer, D., M. Brasier, D. Burnie, C. Cleal, P. Crane, B. A. Thomas, C. Buttler, J. W. C. Cope, R. M. Owens, J. Anderson, R. Benson, S. Brusatte, J. Clack, K. Bennis-Bryan, C. Duffin, D. Hone, Z. Johanson, A. Milner, D. Naish, K. Parsons, D. Prothero, and X. Xing. 2012. *Prehistoric Life.* New York: Dorling Kindersley.

Paris, F., C. Girard, C. Feist, and T. Winchester-Seeto. 1996. Chitinozoan bio-event in the Frasnian/Famennian boundary beds at La Serre (Montagne Noire, Southern France). *Palaeogeography, Palaeoclimatology, Palaeoecology* 121:131–145.

Payne, J. L., and M. E. Clapham. 2012. End-Permian mass extinction in the oceans: An ancient analog for the twenty-first century? *Annual Review of Earth and Planetary Sciences* 40:89–111.

Payne, J. L., A. V. Turchyn, A. Paytan, D. J. DePaola, D. J. Lehrmann, M. Yu, and J. Wei. 2010. Calcium isotope constraints on the end-Permian mass extinction. *Proceedings of the National Academy of Sciences USA* 107:8543–8548.

Pfefferkorn, H. W. 2004. The complexity of mass extinction. *Proceedings of the National Academy of Sciences USA* 101:12779–12780.

Pfefferkorn, H. W., R. A. Gastaldo, W. A. DiMichele, and T. L. Phillips. 2008. Pennsylvanian tropical floras from the United States as a record of changing climate. In *Resolving the Late Paleozoic Ice Age in Time and Space* (Special Paper 441), ed. C. R. Fielding, T. D. Frank, and J. L. Isbell, 305–316. Boulder, Colo.: Geological Society of America.

Phillips, T. L., and W. A. DiMichele. 1992. Comparative ecology and life-history biology of arborescent lycopsids in Late Carboniferous swamps of North America. *Annals of the Missouri Botanical Garden* 79:560–588.

Phillips, T. L., and R. A. Peppers. 1984. Changing patterns of Pennsylvanian coal-swamp vegetation and implications of climatic control on coal occurrence. *International Journal of Coal Geology* 3:205–255.

Polozov, A. G., H. H. Svensen, S. Planke, S. N. Grishina, K. E. Fristad, and D. A. Jerram. 2016. The basalt pipes of the Tunguska Basin (Siberia, Russia): High temperature processes and volatile degassing into the end-Permian atmosphere. *Palaeogeography, Palaeoclimatology, Palaeoecology* 441:51–64.

Powell, M. G. 2005. Climatic basis for sluggish macroevolution during the late Paleozoic ice age. *Geology* 33:381–384.

Powell, M. G. 2008. Timing and selectivity of the Late Mississippian mass extinction of brachiopod genera from the central Appalachian basin. *Palaios* 23:525–534.

Prokop, J., A. Nel, and I. Hoch. 2005. Discovery of the oldest known Pterygota in the Lower Carboniferous of the Upper Silesian Basin in the Czech Republic (Insecta: Archaeorthoptera). *Geobios* 38:383–387.

Prothero, D. R. 1994. *The Eocene-Oligocene Transition: Paradise Lost*. New York: Columbia University Press.

Prothero, D. R., L. C. Ivany, and E. A. Nesbitt. 2003. *From Greenhouse to Icehouse: The Marine Eocene-Oligocene Transition*. New York: Columbia University Press.

Pu, J. P., S. A. Bowring, J. Ramezani, P. Myrow, T. D. Raub, E. Landing, A. Mills, E. Hodgin, and F. A. Macdonald. 2016. Dodging snowballs: Geochronology of the Gaskiers glaciation and the first appearance of the Ediacaran biota. *Geology* 44:955–958.

Pusz, A. E., R. C. Thunell, and K. G. Miller. 2011. Deep water temperature, carbonate ion, and ice volume changes across the Eocene-Oligocene climatic transition. *Paleoceanography* 26, PA2205, doi:10.1029/2010PA001950.

Qiao, L., and S.-Z. Shen. 2015. A global review of the Late Mississippian (Carboniferous) *Gigantoproductus* (Brachiopoda) faunas and their paleogeographical, paleoecological, and paleoclimatic implications. *Palaeogeography, Palaeoclimatology, Palaeoecology* 420:128–137.

Racki, G. 1998. Frasnian-Famennian biotic crisis: Undervalued tectonic control? *Palaeogeography, Palaeoclimatology, Palaeoecology* 141:177–198.

Racki, G., and P. W. Wignall. 2005. Late Permian double-phased mass extinction and volcanism: An oceanographic perspective. In *Understanding Late Devonian and Permian-Triassic Biotic and Climatic Events*, ed. D. J. Over, J. R. Morrow, and P. B. Wignall, 263–297. Amsterdam: Elsevier B. V.

Raup, D. M., and J. J. Sepkoski Jr. 1982. Mass extinctions in the marine fossil record. *Science* 215:1501–1503.

Raymo, M. E., and W. F. Ruddiman. 1992. Tectonic forcing of late Cenozoic climate. *Nature* 359:117–122.

Raymond, A., and C. Metz. 1995. Laurussian land-plant diversity during the Silurian and Devonian: mass extinction, sampling bias, or both? *Paleobiology* 21:74–91.

Raymond, A., and C. Metz. 2004. Ice and its consequences: Glaciation in the Late Ordovician, Late Devonian, Pennsylvanian-Permian, and Cenozoic compared. *Journal of Geology* 112:655–670.

Reichow, M. K., M. S. Pringle, A. I. Al'Mukhamedov, M. B. Allen, V. L. Andreichev, M. M. Buslov, C. E. Davies, G. S. Fedoseev, J. G. Fitton, S. Inger, A. Ya. Medvedev, C. Mitchell, V. N. Puchkov, I. Yu. Safonova, R. A. Scott, and A. D. Saunders. 2009. The timing and extent of the eruption of the Siberian Traps large igneous province: Implications for the end-Permian environmental crisis. *Earth and Planetary Science Letters* 277:9–20.

Retallack, G. J. 1999. Postapocalyptic greenhouse paleoclimate revealed by earliest Triassic paleosols in the Sydney Basin, Australia. *Geological Society of America Bulletin* 111:52–70.

Retallack, G. J., and A. H. Jahren. 2008. Methane release from igneous intrusion of coal during Late Permian extinction events. *Journal of Geology* 116:1–20.

Retallack, G. J., and E. S. Krull. 2006. Carbon isotope evidence for terminal-Permian methane outbursts and their role in extinctions of animals, plants, coral reefs, and peat swamps. In *Wetlands through Time* (Special Paper 399), ed. S. F. Greb and W. A. DiMichele, 249–268. Boulder, Colo: Geological Society of America.

Retallack, G. J., C. A. Metzger, T. Greaver, A. H. Jahren, R. M. H. Smith, and N. D. Sheldon. 2006. Middle-Late Permian mass extinction on land. *Geological Society of America Bulletin* 118:1398–1411.

Retallack, G. J., J. J. Veevers, and R. Morante. 1996. Global coal gap between Premian-Triassic extinction and Middle Triassic recovery of peat-forming plants. *Geological Society of America Bulletin* 108:195–207.

Rey, K., R. Amiot, F. Fourel, T. Rigaudier, F. Abdala, M. O. Day, V. Fernandez, F. Fluteau, C. France-Lanord, B. S. Rubidge, R. M. Smith, P. A. Vigliette, B. Zipfel, and C. Lécuyer. 2016. Global climate perturbations during the Permo-Triassic mass extinctions recorded by continental tetrapods from South Africa. *Gondwana Research* 37:384–396.

Ridgwell, A. 2005. A Mid-Mesozoic Revolution in the regulation of ocean chemistry. *Marine Geology* 217:339–357.

Robinson, J. M. 1990. Lignin, land plants and fungi: Biological evolution affecting Phanerozoic oxygen balance. *Geology* 15:607–610.

Ross, C. A. 1972. Paleobiological analysis of fusulinacean (Foraminiferida) shell morphology. *Journal of Paleontology* 46:719–728.

Rößler, R., Z. Feng, and R. Noll. 2012. The largest calamite and its growth architecture—*Arthropitys bistriata* from the Early Permian petrified forest of Chemnitz. *Review of Palaeobotany and Palynology* 185:64–78.

Rubinstein, C. V., P. Gerriene, G. S. de la Puente, R. A. Astini, and P. Steemans. 2010. Early Middle Ordovician evidence for land plants in Argentina (eastern Gondwana). *New Phytologist* 188:365–369.

Ruse, M. 2013. *The Gaia Hypothesis: Science on a Pagan Planet*. Chicago: University of Chicago Press.

Sage, R. F. 1999. Why C_4 photosynthesis? In *C4 Plant Biology*. ed. R. F. Sage and R. K. Monson. San Diego, Cal.: Academic Press, pp. 3–16.

Sahney, S. and M. J. Benton. 2008. Recovery from the most profound mass extinction of all time. *Proceedings of the Royal Society* B275:759–765.

Sahney, S., M. J. Benton, and H. J. Falcon-Lang. 2010. Rainforest collapse triggered Carboniferous tetrapod diversification in Euramerica. *Geology* 38:1079–1082.

Sallan, L. C., and M. I. Coates. 2010. End-Devonian extinction and a bottleneck in the early evolution of modern jawed vertebrates. *Proceedings of the National Academy of Sciences USA* 107:10131–10135.

Saltzman, M. R., L. A. González, and K. C. Lohman. 2000. Earliest Carboniferous cooling triggered by the Antler orogeny? *Geology* 28:347–350.

Sandberg, C. A., F. G. Poole, and J. G. Johnson. 1988. Upper Devonian of Western United States. In *Devonian of the World*. eds. N. J. McMillan, A. F. Embry, and D. J. Glass. *Canadian Society of Petroleum Geology Memoir 14, volume* I:184–220.

Saunders, A. D. 2005. Large igneous provinces: Origin and environmental consequences. *Elements* 1:259–263.

Saunders, W. B., and W. H. C. Ramsbottom. 1986. The mid-Carboniferous eustatic event. *Geology* 14:208–212.

Schmidt, A., B. Ostro, K. S. Carslaw, M. Wilson, T. Thordarson, G. W. Mann, and A. J. Simmons. 2011. Excess mortality in Europe following a future Laki-style Icelandic eruption. *Proceedings of the National Academy of Sciences USA* 108:15710–15715.

Schmidt, A., T. Thordarson, L. D. Oman, A. Robock, and S. Self. 2012. Climatic impact of the long-lasting 1783 Laki eruption: Inapplicability of mass-independent sulfur isotopic composition measurements. *Journal of Geophysical Research* 117, D23116, doi:10.1029/2012JD018414.

Schwarzacker, W. 1989. Milankovitch type cycles in the Lower Carboniferous of NW Ireland. *Terra Nova* 1:468–473.

Scott, A. C. and I. J. Glasspool. 2006. The diversification of Paleozoic fire systems and fluctuations in atmospheric oxygen concentration. *Proceedings of the National Academy of Sciences USA* 103:10861–10865.

Self, S., T. Thordarson, and M. Widdowson. 2005. Gas fluxes from flood basalt eruptions. *Elements* 1:283–287.

Sepkoski, J. J., Jr. 1981. A factor analytic description of the Phanerozoic marine fossil record. *Paleobiology* 7:36–53.

Sepkoski, J. J., Jr. 1984. A kinematic model of Phanerozoic taxonomic diversity: III. Post-Paleozoic families and mass extinctions. *Paleobiology* 10:246–267.

Sepkoski, J. J., Jr. 1990. Evolutionary faunas. In *Palaeobiology: A Synthesis*. ed. D. E. G. Briggs and P. R. Crowther. Oxford, England: Blackwell Scientific, pp. 37–41.

Sepkoski, J. J., Jr. 1996. Patterns of Phanerozoic extinction: A perspective from global data bases. In *Global Events and Event Stratigraphy*. ed. O. H. Walliser. Berlin, Germany: Springer-Verlag, pp. 35–51.

Sepkoski, J. J., Jr. 1998. Rates of speciation in the fossil record. *Philosophical Transactions of the Royal Society of London* B353:315–326.

Sepkoski, J. J., Jr., and A. I. Miller. 1985. Evolutionary faunas and the distribution of Paleozoic marine communities in space and time. In *Phanerozoic Diversity Patterns*. ed. J. W. Valentine. Princeton, New Jersey: Princeton University Press, pp. 153–190.

Servais, T., A. W. Owen, D. A. T. Harper, B. Kröger, and A. Munnecke. 2010. The Great Ordovician Biodiversification Event (GOBE): The palaeoecological dimension. *Palaeogeography, Palaeoclimatology, Palaeoecology* 294:99–119.

Shaw, S. R. 2014. *Planet of the Bugs: Evolution and the Rise of Insects*. Chicago, Illinois: niversity of Chicago Press.

Shear, W. A., and J. Kukalová-Peck. 1990. The ecology of Paleozoic terrestrial arthropods: The fossil evidence. *Canadian Journal of Zoology* 68:1807–1834.

Sheehan, P. M. 1996. A new look at Ecologic Evolutionary Units (EEUs). *Palaeogeography, Palaeoclimatology, Palaeoecology* 127:21–32.

Sheehan, P. M. 2001. The Late Ordovician mass extinction. *Annual Review of Earth and Planetary Sciences* 29:331–364.

Shellnutt, J. G. 2013. The Emeishan large igneous province: A synthesis. *Geoscience Frontiers* (2013), http://dx.doi.org/10.1016/j.gsf.2013.07.003.

Shen, S.-Z., J. L. Crowley, Y. Wang, S. A. Bowring, and D. H. Erwin. 2011. Calibrating the end-Permian mass extinction. *Science* 334:1367–1372.

Shevenell, A. E., J. P. Kennett, and D. W. Lea. 2004. Middle Miocene Southern Ocean cooling and Antarctic cryosphere expansion. *Science* 305:1766–1770.

Smith, L. B. and J. F. Read. 2000. Rapid onset of late Paleozoic glaciation on Gondwana: Evidence from Upper Mississippian strata of the Midcontinent, United States. *Geology* 28:279–282.

Smithson, T. R., S. P. Wood, J. E. A. Marshall, and J. A. Clack. 2012. Earliest Carboniferous tetrapod and arthropod faunas from Scotland populate Romer's Gap. *Proceedings of the National Academy of Sciences USA* 109:4532–4537.

Sobolev, S. V., A. V. Sobolev, D. V. Kuzmin, N. A. Krivolutskaya, A. G. Petrunin, N. T. Arndt, V. A. Radko, and Y. R. Vasiliev. 2011. Linking mantle plumes, large igneous provinces and environmental catastrophes. *Nature* 477:312–316.

Stanley, S. M. 2007. An analysis of the history of marine animal diversity. *Paleobiology Memoirs* 4:1–55.

Stanley, S. M., and M. G. Powell. 2003. Depressed rates of origination and extinction during the late Paleozoic ice age: A new state for the global marine ecosystem. *Geology* 31:877–880.

Stearn, C. W. 1987. Effect of the Frasnian-Famennian extinction event on the stromatoporoids. *Geology* 15:677–679.

Stein, W. E., C. M. Berry, L. VanA. Hernick, and F. Mannolini. 2012. Surprisingly complex community discovered in the mid-Devonian fossil forest at Gilboa. *Nature* 483:78–81.

Stein, W. E., F. Mannolini, L. V. Hernick, E. Landing, and C. M. Berry. 2007. Giant cladoxylopsid trees resolve the enigma of the Earth's earliest forest stumps at Gilboa. *Nature* 446:904–907.

Stewart, W. N. 1983. *Paleobotany and the evolution of plants*. Cambridge: Cambridge University Press.

Steyer, S. 2012. *Earth Before the Dinosaurs*. Bloomington: Indiana University Press.

Stone, R. 2004. Iceland's doomsday scenario? *Science* 306:1278–1281.

Strand, K., S. Passchier, and J. Näsi. 2003. Implications of quartz grain microtextures for onset Eocene/Oligocene glaciation in Prydz Bay, ODP Site 1166, Antarctica. *Palaeogeography, Palaeoclimatology, Palaeoecology* 198:101–111.

Streel, M., M. V. Caputo, S. Loboziak, and J. H. G. Melo. 2000. Late Frasnian-Famennian climates based on palynomorph analyses and the question of the Late Devonian glaciations. *Earth-Science Reviews* 52:121–173.

Streel, M., M. Vanguestaine, A. Pardo-Trujillo, and E. Thomalla. 2000. The Frasnian-Famennian boundary sections at Hony and Sinsin (Ardenne, Belgium): New interpretation based on quantitative analysis of palynomorphs, sequence stratigraphy and climatic interpretations. *Geologica Belgica* 3:271–283.

Sun, Y., M. M. Joachimski, P. B. Wignall, C. Yan, Y. Chen, H. Jiang, L. Wang, and X. Lai. 2012. Lethally hot temperatures during the Early Triassic greenhouse. *Science* 338:366–370.

Svensen, H., S. Planke, A. G. Polozov, N. Schmidbauer, F. Corfu, Y. Y. Podladchikov, and B. Jamtveit. 2009. Siberian gas venting and the end-Permian environmental crisis. *Earth and Planetary Science Letters* 277:490–500.

Tasch, P. 1973. *Paleobiology of the Invertebrates*. New York: John Wiley and Sons, Inc.

Taylor, T. N., and E. L. Taylor. 1993. *The Biology and Evolution of Fossil Plants*. Englewood Cliffs, N.J.: Prentice Hall.

Thordarson, T., and G. Larsen. 2007. Volcanism in Iceland in historical time: Volcano types, eruption styles and eruptive history. *Journal of Geodynamics* 43:118–152.

Thordarson, T., and S. Self. 2003. Atmospheric and environmental effects of the 1783–1784 Laki eruption: A review and reassessment. *Journal of Geophysical Research* 108, D14011, doi:10.1029/2001JD002042.

Thordarson, T., S. Self, N. Óskarsson, and T. Hulsebosch. 1996. Sulfur, chlorine, and flourine degassing and atmospheric loading by the 1783–1784 AD Laki (Skaftár Fires) eruption in Iceland. *Bulletin of Volcanology* 58:205–225.

Twitchett, R. J. 2007. The Lilliput effect in the aftermath of the end-Permian extinction event. *Palaeogeography, Palaeoclimatology, Palaeoecology* 252:132–144.

Ukstins-Peate, I., and S. E. Bryan. 2008. Re-evaluating plume-induced uplift in the Emeishan large igneous province. *Nature Geoscience* 1:625–629.

Ustritsky, V. I. 1973. Permian climate. In *The Permian and Triassic Systems and Their Mutual Boundary* (Memoir 2), ed. A. Logan and L. V. Hill, 733–744. Calgary: Canadian Society of Petroleum Geologists.

Vachard, D., L. Pille, and J. Gaillot. 2010. Palaeozoic foraminifera: Systematics, Palaeoecology, and responses to global changes. *Revue de Micropaleontologie* 53:209–254.

Van Valen, L. M. 1984. A resetting of Phanerozoic community evolution. *Nature* 307:50–52.

Visscher, H., C. V. Looy, M. E. Collinson, H. Brinkhuis, J. H. A. van Konijnenburg-van Cittert, W. M. Kürschner, and M. A. Sephton. 2004. Environmental mutagenesis during the end-Permian ecological crisis. *Proceedings of the National Academy of Sciences USA* 101:12952–12956.

Voigt, S., and M. Ganzelewski. 2010. Toward the origin of amniotes: Diadectomorph and synapsid footprints from the early Late Carboniferous of Germany. *Acta Palaeontologia Polonica* 55:57–72.

Wade, B. S., A. J. P. Houben, W. Qualijtaal, S. Schouter, Y. Rosenthal, K. G. Miller, M. E. Katz, J. D. Wright, and H. Brinkhuis. 2012. Multiproxy record of abrupt sea-surface cooling across the Eocene-Oligocene transition in the Gulf of Mexico. *Geology* 40:159–162.

Wade, N. 2006. *Before the Dawn: Recovering the Lost History of Our Ancestors*. New York: Penguin Press.

Walker, J. D., and J. W. Geissman. 2009. Geologic Time Scale. *Geological Society of America*, doi: 10.1130/2009.CTS004R2C.

Wang, Z., and A. Chen. 2001. Traces of arborescent lycopsids and dieback of the forest vegetation in relation to the terminal Permian mass extinction in North China. *Review of Palaeobotany and Palynology* 117:217–243.

Wang, X.-D., X.-J. Wang, F. Zhang, and H. Zhang. 2006. Diversity patterns of Carboniferous and Permian rugose corals in South China. *Geological Journal* 41:329–343.

Wang, Y., P. M. Sadler, S.-Z. Shen, D. H. Erwin, Y.-C. Zhang, X.-D. Wang, W. Wang, J. L. Crowley, and C. M. Henderson. 2014. Quantifying the process and abruptness of the end-Permian mass extinction. *Paleobiology* 40:113–129.

Ward, P. D. 2006. *Out of Thin Air: Dinosaurs, Birds, and Earth's Ancient Atmosphere.* Washington, D.C.: Joseph Henry.

Ward, P. D., D. R. Montgomery, and R. Smith. 2000. Altered river morphology in South Africa related to the Permian–Triassic extinction. *Science* 289:226–229.

Webby, B. D., F. Paris, M. L. Droser, and I. G. Percival. 2004. *The Great Ordovician Biodiversification Event.* New York: Columbia University Press.

Weems, R. E. 1992. The 'terminal Triassic catastrophic event' in perspective: A review of Carboniferous through Early Jurassic vertebrate extinction patterns. *Palaeogeography, Palaeoclimatology, Palaeoecology* 94:1–29.

Wei, F., Y. Gong, and H. Yang. 2012. Biogeography, ecology and extinction of Silurian and Devonian tentaculitoids. *Palaeogeography, Palaeoclimatology, Palaeoecology* 358–360:40–50.

Weidlich, O., 2002. Permian reefs re-examined: Extrinsic control mechanisms of gradual and abrupt changes during 40 my of reef evolution. *Geobios* 35(Supplement 1):287–294.

Wellman, C. H. 2010. The invasion of the land by plants: When and where? *New Phytologist* 188:306–309.

Wellman, C. H., P. L. Osterhoff, and U. Mohiuddin. 2003. Fragments of the earliest land plants. *Nature* 425:282–285.

Westerhold, T., T. Bickert, and U. Röhl. 2005. Middle to late Miocene oxygen isotope stratigraphy of ODP site 1085 (SE Atlantic): New constraints on Miocene climate variability and sea-level fluctuations. *Palaeogeography, Palaeoclimatology, Palaeoecology* 217:205–222.

Wignall, P. B. 2001. Large igneous provinces and mass extinctions. *Earth-Science Reviews* 53:1–33.

Wignall, P., Y. Sun, D. Bond, G. Izon, R. Newton, S. Védrine, M. Widdowson, J. Ali, X. Lai, H. Jiang, H. Cope, and S. Bottrell. 2009. Volcanism, mass extinction, and carbon isotope fluctuations in the Middle Permian of China. *Science* 324:1179–1182.

Wignall, P. B., and R. J. Twitchett. 2002. Extent, duration, and nature of the Permian–Triassic superanoxic event. *Geological Society of America Special Paper* 356:395–413.

Wilson, D. S., and B. P. Luyendyk. 2009. West Antarctic paleotopography estimated at the Eocene–Oligocene climate transition. *Geophysical Research Letters* 36, L16302, doi:10.1029/2009GL039297.

Wootton, R. J., J. Kukalová-Peck, D. J. S. Newman, and J. Muzón. 1998. Smart engineering in the Mid-Carboniferous: How well could Palaeozoic dragonflies fly? *Science* 282:749–751.

Wright, V. P., and S. D. Vanstone. 2001. Onset of late Palaeozoic glacio-eustacy and the evolving climates of low latitude areas: A synthesis of current understanding. *Journal of the Geological Society, London* 158:579–582.

Young, G. C. 2010. Placoderms (armored fish): Dominant vertebrates of the Devonian Period. *Annual Review of Earth and Planetary Sciences* 38:523–550.

Zachos, J. C., J. R. Breza, and S. W. Wise. 1992. Early Oligocene ice-sheet expansion on Antarctica: Stable isotope and sedimentological evidence from Kerguelen Plateau, southern Indian Ocean. *Geology* 20:569–573.

Zachos, J. C., M. Pagani, L. Sloan, E. Thomas, and K. Billups. 2001. Trends, rhythms, and aberrations across the Eocene-Oligocene transition in central North America. *Nature* 445:639–642.

Zhou, M.-F., J. Malpas, X.-Y. Song, P. T. Robinson, M. Sun, A. K. Kennedy, C. M. Lesher, and R. R. Keays. 2002. A temporal link between the Emeishan large igneous province (SW China) and the end-Guadalupian mass extinction. *Earth and Planetary Science Letters* 196:113–122.

Ziegler, W., and H. R. Lane. 1987. Cycles in conodont evolution from Devonian to mid-Carboniferous. In *Palaeobiology of Conodonts*, ed. R. J. Aldridge, 147–163. Chichester: Horwood Press.

Index